C0-GHU-313

WILLIAM F. MAAG LIBRARY
YOUNGSTOWN STATE UNIVERSITY

Selected Titles in This Subseries

45 **Ya. Eliashberg, D. Fuchs, T. Ratiu, and A. Weinstein, Editors,** Northern California symplectic geometry seminar (TRANS2/196)

44 **Alexander Astashkevich and Serge Tabachnikov, Editors,** Differential Topology, Infinite-Dimensional Lie Algebras, and Applications (D. B. Fuchs 60th Anniversary Collection) (TRANS2/194)

43 **A. Yu. Morozov and M. A. Olshanetsky, Editors,** Moscow Seminar in Mathematical Physics (TRANS2/191)

42 **S. Tabachnikov, Editor,** Differential and Symplectic Topology of Knots and Curves (TRANS2/190)

41 **V. Buslaev, M. Solomyak, and D. Yafaev, Editors,** Differential Operators and Spectral Theory (M. Sh. Birman's 70th anniversary collection) (TRANS2/189)

40 **M. V. Karasev, Editor,** Coherent Transform, Quantization, and Poisson Geometry (TRANS2/187)

39 **A. Khovanskiĭ, A. Varchenko, and V. Vassiliev, Editors,** Geometry of Differential Equations (TRANS2/186)

38 **B. Feigin and V. Vassiliev, Editors,** Topics in Quantum Groups and Finite-Type Invariants (Mathematics at the Independent University of Moscow) (TRANS2/185)

37 **Peter Kuchment and Vladimir Lin, Editors,** Voronezh Winter Mathematical Schools (Dedicated to Selim Krein) (TRANS2/184)

36 **V. E. Zakharov, Editor,** Nonlinear Waves and Weak Turbulence (TRANS2/182)

35 **G. I. Olshanski, Editor,** Kirillov's Seminar on Representation Theory (TRANS2/181)

34 **A. Khovanskiĭ, A. Varchenko, and V. Vassiliev, Editors,** Topics in Singularity Theory (TRANS2/180)

33 **V. M. Buchstaber and S. P. Novikov, Editors,** Solitons, Geometry, and Topology: On the Crossroad (TRANS2/179)

32 **R. L. Dobrushin, R. A. Minlos, M. A. Shubin, and A. M. Vershik, Editors,** Topics in Statistical and Theoretical Physics (F. A. Berezin Memorial Volume) (TRANS2/177)

31 **R. L. Dobrushin, R. A. Minlos, M. A. Shubin, and A. M. Vershik, Editors,** Contemporary Mathematical Physics (F. A. Berezin Memorial Volume) (TRANS2/175)

30 **A. A. Bolibruch, A. S. Merkur′ev, and N. Yu. Netsvetaev, Editors,** Mathematics in St. Petersburg (TRANS2/174)

29 **V. Kharlamov, A. Korchagin, G. Polotovskiĭ, and O. Viro, Editors,** Topology of Real Algebraic Varieties and Related Topics (TRANS2/173)

28 **L. A. Bunimovich, B. M. Gurevich, and Ya. B. Pesin, Editors,** Sinai's Moscow Seminar on Dynamical Systems (TRANS2/171)

27 **S. P. Novikov, Editor,** Topics in Topology and Mathematical Physics (TRANS2/170)

26 **S. G. Gindikin and E. B. Vinberg, Editors,** Lie Groups and Lie Algebras: E. B. Dynkin's Seminar (TRANS2/169)

25 **V. V. Kozlov, Editor,** Dynamical Systems in Classical Mechanics (TRANS2/168)

24 **V. V. Lychagin, Editor,** The Interplay between Differential Geometry and Differential Equations (TRANS2/167)

23 **Yu. Ilyashenko and S. Yakovenko, Editors,** Concerning the Hilbert 16th Problem (TRANS2/165)

22 **N. N. Uraltseva, Editor,** Nonlinear Evolution Equations (TRANS2/164)

Published Earlier as Advances in Soviet Mathematics

21 **V. I. Arnold, Editor,** Singularities and bifurcations, 1994

20 **R. L. Dobrushin, Editor,** Probability contributions to statistical mechanics, 1994

(Continued in the back of this publication)

Northern California Symplectic Geometry Seminar

American Mathematical Society
TRANSLATIONS
Series 2 • Volume 196

Advances in the Mathematical Sciences — 45

(*Formerly Advances in Soviet Mathematics*)

Northern California Symplectic Geometry Seminar

Ya. Eliashberg
D. Fuchs
T. Ratiu
A. Weinstein
Editors

American Mathematical Society
Providence, Rhode Island

ADVANCES IN THE MATHEMATICAL SCIENCES
EDITORIAL COMMITTEE

V. I. ARNOLD
S. G. GINDIKIN
V. P. MASLOV

2000 *Mathematics Subject Classification.* Primary 53Dxx, 58Dxx, 58Jxx.

ABSTRACT. The Northern California Symplectic Geometry Seminar was established in the Fall of 1989 to encourage ongoing interaction between geometers at Stanford University and the University of California campuses at Berkeley, Davis, and Santa Cruz. The lectures over the years, by visitors, local faculty, and advanced graduate students, have provided a panorama of developments in symplectic and contact geometry and topology, Poisson geometry, quantization theory, and applications.

The present volume contains papers by some of the seminar participants. They reflect the diversity of topics of the seminar, ranging from symplectic topology to Hamiltonian dynamics, from quantum cohomology and mirror symmetry to infinite-dimensional symplectic geometry, the theory of Hamiltonian group actions, and quantization.

The book can be used by researchers and graduate students working in symplectic geometry and topology, and applications.

Library of Congress Card Number 91-640741
ISBN 0-8218-2075-3
ISSN 0065-9290

Copying and reprinting. Material in this book may be reproduced by any means for educational and scientific purposes without fee or permission with the exception of reproduction by services that collect fees for delivery of documents and provided that the customary acknowledgment of the source is given. This consent does not extend to other kinds of copying for general distribution, for advertising or promotional purposes, or for resale. Requests for permission for commercial use of material should be addressed to the Assistant to the Publisher, American Mathematical Society, P. O. Box 6248, Providence, Rhode Island 02940-6248. Requests can also be made by e-mail to reprint-permission@ams.org.

Excluded from these provisions is material in articles for which the author holds copyright. In such cases, requests for permission to use or reprint should be addressed directly to the author(s). (Copyright ownership is indicated in the notice in the lower right-hand corner of the first page of each article.)

© 1999 by the American Mathematical Society. All rights reserved.
The American Mathematical Society retains all rights
except those granted to the United States Government.
Printed in the United States of America.

∞ The paper used in this book is acid-free and falls within the guidelines
established to ensure permanence and durability.
Visit the AMS home page at URL: http://www.ams.org/

10 9 8 7 6 5 4 3 2 1 04 03 02 01 00 99

QA
3
A572
Ser. 2
V. 196

Contents

WILLIAM F. MAAG LIBRARY
YOUNGSTOWN STATE UNIVERSITY

Introduction

The Northern California Symplectic Geometry Seminar was established in the Fall of 1989 to encourage ongoing interaction between geometers at Stanford University and the University of California campuses at Berkeley, Davis, and Santa Cruz. Since then, the seminar has met on the first Monday of each month from October to May (except January), alternately at Stanford and Berkeley, with two lectures at each meeting. Beginning in the Fall of 1992, after the tragic death of Andreas Floer, who was among the original participants in the seminar, one of the lectures at the October meeting has been designated as the Andreas Floer Memorial Lecture. The Floer Memorial Lectures have been given by Helmut Hofer (1992), Dusa McDuff (1993), Yasha Eliashberg (1994), Jon Wolfson (1995), Vladimir Arnold (1996), Simon Donaldson (1997), and Maxim Kontsevich (1998).

The lectures over the years, by visitors, local faculty, and advanced graduate students, have provided a panorama of developments in symplectic and contact geometry and topology, Poisson geometry, quantization theory, and applications.

All the authors of the current volume, except Marcin Poźniak, were participants and/or speakers at the seminar.

The paper by Ana Cannas da Silva and Victor Guillemin concerns symplectic orbifolds with a Hamiltonian action by a group G. The authors define a quantization space V which is a representation of G, and describe a decomposition of V into irreducible representations of G.

Simon Donaldson in his paper describes and studies the infinite dimensional symmetric space structure which arises naturally on the space of Kähler metrics with a fixed cohomology class.

The paper by Viktor Ginzburg is a survey of counter-examples to the "Hamiltonian Seifert conjecture" and more generally of Hamiltonian systems without periodic orbit on a compact energy level.

Alexander Givental presents a "shortcut" to his celebrated original proof of the Mirror Theorem for the case of quintic threefolds, thus giving a proof of the formula of Candelas, de la Ossa, Green and Parkes for the the number of rational curves of degree d on a quintic threefold in $\mathbb{C}^4$.

François Lalonde and Charles Pestieau prove that the basic symplectic inequalities – the non-squeezing theorem and the energy-capacity inequality – are in a certain sense stable under product with any closed symplectic manifold.

Dusa McDuff's paper is devoted to a construction of the virtual moduli cycle. Virtual moduli cycles are essential for the theory of Gromov-Witten invariants, Floer homology theory, etc. McDuff's paper provides further clarification of this important theory.

The paper by Richard Montgomery may serve as a good introduction to the emerging theory of the geometry and topology of Engel structures on 4-manifolds. In it Montgomery establishes some foundational facts in this theory.

Yongbin Ruan contributed to the volume an expository article about some newly discovered relations between quantum cohomology and birational geometry, and the surgery techniques which play a fundamental role in these discoveries.

Dietmar Salamon's paper complements his earlier joint work with S. Dostoglou, which contained a proof of the Atiyah-Floer conjecture based on the construction of a natural isomorphism between instanton and symplectic Floer homology theories for the mapping tori of symplectomophisms of surfaces. In the current paper, Salamon proves that this isomorphism interwines the product structures in the two theories.

The work of Paul Seidel concerns the topology of the group of symplectomorphisms of the product of complex projective spaces; it also partially extends earlier results of Gromov, Abreu and McDuff about the group of symplectic automorphisms of $S^2 \times S^2$. In particular, Seidel shows that the inclusion of the group of symplectic diffeomorphisms into the group of all diffeomorphisms induces a non-injective map on some homotopy groups.

Susan Tolman and Jonathan Weitsman consider a symplectic manifold with a Hamiltonian torus action, and give a simple proof of the Goresky-Kottwitz-MacPherson formula for the image of the homomorphism induced between equivariant cohomology of the manifold and the fixed point set of the torus action.

Marcin Poźniak was a PhD student of Dietmar Salamon. He died on 29 November 1996 as a result of a tragic accident. As Salamon remembers, "Poźniak began his graduate studies at Warwick in October 1989. For his MSc degree he worked on triangulations and Morse theory, and his PhD thesis, completed in 1994, was devoted to the theory of so-called clean Lagrangian intersections. After his doctoral studies Marcin Poźniak returned to his position at the Jagellonian University of Krakow. Marcin Poźniak was a very thoughtful person and a talented young mathematician. He had many friends among the students and staff at Warwick. We shall miss him very much." Because of his tragic death, Poźniak's PhD thesis remained unpublished, although its results have been used and quoted by many authors working in the area. Thus we consider it appropriate to publish Poźniak's thesis in this volume. The manuscript was prepared for publication by Davide Castelvecchi.

The Editors

Amer. Math. Soc. Transl.
(2) Vol. **196**, 1999

Quantization of Symplectic Orbifolds and Group Actions

Ana Cannas da Silva and Victor Guillemin

ABSTRACT. To a compact symplectic orbifold with a hamiltonian action of a group G (and with other compatible data), we can associate a quantization space V which is a representation of G. This paper reports on the decomposition of V into irreducible representations of G in terms of data at the fixed points, and on the semiclassical limit of this decomposition, for the case where G is a compact connected semisimple Lie group.

1. Introduction

Let (M, ω) be a compact symplectic orbifold such that $[\omega]$ is the Chern class of a complex line bundle L over M. Choose an almost complex structure J on M compatible with ω (i.e. such that $(u, v) \mapsto \omega(u, Jv)$ is a positive metric), and choose an hermitian connection on L with curvature given by ω. To such data we can associate a first order elliptic differential operator, called the Dolbeault–Dirac operator [**8, 6**]:

$$\not{\partial} : C^\infty(M; \wedge^{0,\mathrm{even}} T^*M \otimes L) \longrightarrow C^\infty(M; \wedge^{0,\mathrm{odd}} T^*M \otimes L).$$

Inspired by Bott [**2**], we define **quantization** of (M, ω) to be the (K-theoretic) index of $\not{\partial}$, i.e.,

$$Q(M, \omega) := \mathrm{kernel}\,\not{\partial} - \mathrm{cokernel}\,\not{\partial}.$$

This index depends only on (M, ω) and not on the auxiliary data.

If a group G acts on M, and if the action lifts to L and preserves all data (in particular, the action must be hamiltonian [**12**]), then the quantization space becomes a linear representation of G. We will first address the following question: What is the decomposition of $Q(M, \omega)$ into irreducible representations of G?

As explained in this paper, not only does the answer depend exclusively on data at the fixed points, but also this answer can be given by a very explicit formula useful for computations. The prototype for this formula is a formula of Kostant [**15**] for multiplicities of group representations. The prototype for the proof is an observation of Cartier that we will now review.

From now on, let G be a compact connected semisimple Lie group with a maximal torus T. Let $\rho : G \to \mathrm{GL}(V)$ be an irreducible representation with

1991 *Mathematics Subject Classification.* Primary 58G10.

©1999 American Mathematical Society

maximal weight λ. As a T-representation, V breaks up into weight spaces,

$$V = \bigoplus_{\text{weights } \mu} V_\mu,$$

where $\mathrm{mult}(\mu) := \dim V_\mu$. Kostant [**15**] gave a formula for the multiplicity of a weight μ in ρ:

$$\mathrm{mult}(\mu) = \sum_{\omega \in W} (-1)^\omega \mathcal{P}(\omega(\delta + \lambda) - (\delta + \mu)), \tag{1.1}$$

where W is the Weyl group, $\delta = \frac{1}{2}\sum_{\alpha\in\Delta^+} \alpha$, Δ^+ is a chosen set of positive roots, and $\mathcal{P}(\nu)$ is the number of ways that ν may be partitioned into a sum of positive roots:

$$\mathcal{P}(\nu) := \left\{ (n_\alpha \in \mathbb{Z}_0^+) \mid \nu = \sum_{\alpha\in\Delta^+} n_\alpha \alpha \right\}.$$

Cartier [**7**] showed that the Kostant formula is equivalent to the Weyl character formula, which states that

$$\chi(\exp \xi) = \frac{\sum_{\omega\in W}(-1)^\omega e^{(\delta+\lambda)(\omega\xi)}}{e^{\delta(\xi)}\prod_{\alpha\in\Delta^+}(1 - e^{-\alpha(\xi)})}, \tag{1.2}$$

χ being the character of ρ and $\xi \in \mathrm{Lie}(T)$. Cartier's simple argument starts by expanding the Weyl denominator in a power series, to obtain

$$\frac{1}{\prod_{\alpha\in\Delta^+}(1 - e^{-\alpha})} = \sum_{\text{weights } \nu} \mathcal{P}(\nu) e^{-\nu},$$

where $\mathcal{P}$ is the Kostant partition function. Now the Weyl formula can be written

$$\sum_\mu \mathrm{mult}(\mu) e^{\mu(\xi)} = \sum_{\omega\in W}\sum_\nu (-1)^\omega \mathcal{P}(\nu) e^{(\delta+\lambda)(\omega\xi)} e^{-(\delta+\nu)(\xi)}.$$

Formula (1.1) follows from equating the coefficients of $e^{\mu(\xi)}$.

The upshot of this argument is that formulas (1.1) and (1.2) are dual to each other by a Fourier expansion. This is less trivial than it sounds: there are several ways of expanding the Weyl denominator into a Fourier series, and for some of these expansions the coefficient of e^ν will be given by a divergent infinite sum.

In §4 we obtain a formula (Theorem 1) resembling Kostant's for the multiplicity of a weight μ in $Q(M, \omega)$ for an action of a torus with the fixed points not necessarily isolated. We start from the equivariant index formula [**1**] modified for orbifolds [**19, 8, 17**], and apply a suitable Fourier expansion. Our proof uses the language of associated orbifolds explained in §A.3; Appendix A also reviews other relevant orbifold definitions. To ensure that each coefficient in the Fourier series is given by a convergent series, in §3 we define a polarization [**10**]. §2 describes how to handle an action of a compact connected semisimple Lie group by a combination of tricks.

In §5 we compute the *semiclassical limit* of the multiplicity formula (4.1). Namely, we replace the line bundle L by its nth tensor power, to get a T-invariant elliptic first order differential operator $\not{\partial}^{(n)}$. Setting $\mathrm{mult}^{(n)}(\mu)$ to be the multiplicity with which μ occurs as a weight of the representation of T on

$$\mathrm{kernel}\,\not{\partial}^{(n)} - \mathrm{cokernel}\,\not{\partial}^{(n)},$$

we show that, as the integer n tends to infinity, the function of μ given by

$$n^{-d}\mathrm{mult}^{(n)}(n\mu),$$

where $d = \frac{1}{2}\dim M - \dim T$, tends to the Radon–Nikodým derivative for the Duistermaat–Heckman measure relative to the Lebesgue measure on the dual of the Lie algebra of T: Theorem 2 of §5 and Theorem 3 of §6. The proof of Theorem 2 relies on complex analysis and on elementary results of invariant theory for finite groups; the proof of Theorem 3 relies on Fourier calculus and is analogous to the manifold proof in [**6**].

The case of torus actions on symplectic manifolds with only isolated fixed points was solved by the second author with Lerman and Sternberg [**10**]; their multiplicity formula was derived from the Atiyah–Bott–Lefschetz formula by an argument along the lines of Cartier's. Later he and Prato [**11**] tackled the case of nonabelian actions (the argument in this paper is different from theirs and is an expansion of the remarks in [**6**]). The present paper is a continuation of [**6**], which contained the corresponding results for the case of manifolds. The extension to orbifolds is important, since not only are the generic reduced spaces of symplectic manifolds orbifolds, but also the generic reduced spaces of symplectic orbifolds are still orbifolds. Number-theoretic applications of multiplicity formulas for orbifolds can be found in [**5**].

2. Quantization of nonabelian group actions

Let $\rho : G \to \mathrm{GL}(V)$ be a representation of a compact connected semisimple Lie group G with a maximal torus T. The dimension of the space V^G of G-invariants is given by

$$\begin{aligned}\dim V^G &= \int_G \chi(g)dg \\ &= \frac{1}{|W|}\int_T \underbrace{\prod_{\text{roots } \alpha} (1 - e^{\alpha(\xi)})}_{\sum_\mu c_\mu e^{\mu(\xi)}} \chi(t)\, dt \qquad \text{where } \exp\xi = t \\ &= \frac{1}{|W|}\sum_\mu c_\mu \dim V_\mu.\end{aligned}$$

The middle equality follows from Weyl's integral formula (see, for instance, [**3**]), the others follow from orthogonality relations for characters, and the coefficients c_μ are constants. Hence, the dimension of the space of G-invariants is a linear combination of the multiplicities for the maximal torus with coefficients independent of the G-representation.[1]

Now let $V = Q(M, \omega)$ be the quantization space of a compact symplectic orbifold as described in §1, and let $\mathcal{O}$ be a coadjoint orbit of G equipped with its canonical Lie–Poisson symplectic form, $\omega^{\mathcal{O}}$. The coadjoint action of G is hamiltonian with moment map given by inclusion. By the Borel-Weil theorem, any irreducible complex G-representation $\rho : G \to \mathrm{GL}(V^\rho)$ is the quantization space of

[1]We thank Michèle Vergne for showing us this trick.

a unique (integral) coadjoint orbit $\mathcal{O}$,

$$Q(\mathcal{O}, \omega^{\mathcal{O}}) \simeq V^{\rho}.$$

The "shifting trick" [**12**] is the observation that

$$\dim \left(Q(M \times \mathcal{O}, \omega - \omega^{\mathcal{O}})\right)^G = \text{ multiplicity of } \rho \text{ in } Q(M, \omega).$$

We conclude that multiplicities for a nonabelian group action are linear combinations of multiplicities for the action of a maximal torus. From now on we restrict our attention to actions of tori.

3. Polarization

Let (M, ω) be a compact symplectic orbifold such that $[\omega]$ is the Chern class of a complex line bundle L over M. Suppose that a torus T acts on (M, ω), and that this action is effective, lifts to L and has a moment map Φ. Choose an almost complex structure J on M compatible with ω, and choose an hermitian connection on L with curvature given by ω, both invariant under T.

The normal bundle NF to a connected component F of the fixed point set splits into a direct sum of vector subbundles

$$E_{F,1} \oplus \cdots \oplus E_{F,m}, \tag{3.1}$$

m depending on F, such that the isotropy representation of T on $E_{F,j}$ has a fixed rational weight $\alpha_{F,j} \in \text{Lie}(T)^*$, where $\alpha_{F,j} \neq \alpha_{F,k}$ for $j \neq k$. Since T is compact, each F is non-degenerate, in the sense that all $\alpha_{F,j} \neq 0$. Hence, we can polarize these weights as in [**10**] by choosing an element v of Lie(T) such that $\alpha_{F,j}(v) \neq 0$ for all i, j, and setting

$$\alpha_{F,j}^+ = \varepsilon_{F,j} \alpha_{F,j}, \qquad \text{where} \qquad \varepsilon_{F,j} = \text{sign } \alpha_{F,j}(v).$$

The polarized weights all lie in the half-space $\xi(v) > 0$.

Let $n_{F,j}$ be half of the real rank of the vector bundle $E_{F,j}$, and let

$$\alpha_F^\# = \sum_{\varepsilon_{F,j}=-1} n_{F,j} \alpha_{F,j}, \qquad n_F = \sum_{\varepsilon_{F,j}=-1} n_{F,j}.$$

For every m-tuple of non-negative integers, $k = (k_1, \ldots, k_m)$, let $E_F(k)$ be the tensor product

$$\left(\bigotimes_{j=1}^m \mathcal{S}^{k_j}(E_{F,j}^+)\right)^* \otimes \left(\bigotimes_{\varepsilon_{F,j}=-1} \wedge^{n_{F,j}}(E_{F,j}^+)\right)^*,$$

where $\mathcal{S}^{k_j}(E_{F,j}^+)$ is the k_j-th symmetric power of $E_{F,j}^+$ and

$$E_{F,j}^+ = \begin{cases} E_{F,j} & \text{if } \varepsilon_{F,j} = 1, \\ E_{F,j}^* & \text{if } \varepsilon_{F,j} = -1. \end{cases}$$

4. Multiplicity formula for orbifolds

Given a weight μ of T, let $\Delta_F(\mu)$ be the convex polytope in $\mathbb{R}^m$ consisting of all m-tuples, $(s_1, \ldots, s_m)$, $s_i \geq 0$, for which

$$\Phi_F - \sum_j s_j \alpha_{F,j}^+ = \mu,$$

where Φ_F is the value of the moment map Φ on the connected component F of the fixed point set. In this expression $\alpha^+_{F,j}$ and Φ_F are rational, whereas μ is integral. The fact that the $\alpha^+_{F,j}$'s are polarized (see §4) implies that $\Delta_F(\mu)$ is compact.

THEOREM 1. *The multiplicity with which μ occurs as a weight of the representation of T on $Q(M,\omega)$ is equal to the sum*

$$\mathrm{mult}(\mu) = \sum_F (-1)^{n_F} \mathcal{N}_F(\mu),$$

where

$$\mathcal{N}_F(\mu) = \sum_{k\in\Delta_F(\mu-\alpha_F^{\#})} \mathrm{KRR}(F, E_F(k)\otimes L_F), \tag{4.1}$$

$\mathrm{KRR}(F, E_F(k)\otimes L)$ being the Kawasaki-Riemann-Roch number [**14**] *of the orbifold vector bundle $E_F(k)\otimes L_F$ over F, with respect to the almost complex structure induced on F by J.*

PROOF. The equivariant index theorem for orbifolds [**19, 8**] states that, for a generic ξ in $\mathrm{Lie}(T)$ (generic in the sense that $\exp\xi$ generates a dense subgroup of T), the trace of $\exp\xi$ on the (virtual) vector space $Q(M,\omega)$ is equal to the sum over the fixed point components F of "local traces" $\chi_F(\xi)$, whose definition we will now undertake.

Let $\widehat{F}$ be the orbifold associated to F, as described in §A.3; $\widehat{F}$ will in general have various connected components of possibly different dimensions, depending on the canonical stratification of F (see §A.2). Let $L_{\widehat{F}}$, $E_{\widehat{F},j}$ and $E^+_{\widehat{F},j}$ be the pull-backs of $L|_F$, $E_{F,j}$ and $E^+_{F,j}$ by the natural map $\mu:\widehat{F}\to F$; according to (3.1), the pull-back of the normal bundle NF splits into $E_{\widehat{F},1}\oplus\cdots\oplus E_{\widehat{F},m}$. These bundles over $\widehat{F}$ have canonical automorphisms as explained in §A.4, which we denote by

$$A_F := A(L_{\widehat{F}}), \qquad A_{F,j} := A(E_{\widehat{F},j}), \qquad A^+_{F,j} := A(E^+_{\widehat{F},j}).$$

Finally, let $-2\pi i\Omega_{F,j}$ be the curvature form associated with an hermitian connection on $E_{F,j}$, and define $\Omega^+_{F,j}$, $\Omega_{\widehat{F},j}$ and $\Omega^+_{\widehat{F},j}$ similarly for the bundles $E^+_{F,j}$, $E_{\widehat{F},j}$ and $E^+_{\widehat{F},j}$, respectively.

The local traces are then

$$\chi_F(\exp\xi) = e^{\Phi_F(\xi)}\int_{\widehat{F}} \frac{1}{m_{\widehat{F}}}\cdot\frac{A_F\exp(\omega_{\widehat{F}})\mathrm{Todd}(\widehat{F})}{D_{\widehat{F}}\prod_j\det(I - A^{-1}_{F,j}\exp(-\alpha_{F,j}(\xi)I - \Omega_{\widehat{F},j}))}, \tag{4.2}$$

where $\omega_{\widehat{F}}$ is the pull-back of the symplectic form ω to $\widehat{F}$, $m_{\widehat{F}}$ is the order function of $\widehat{F}$ (see §A.2) and $D_{\widehat{F}}$ is the associated characteristic form defined in §A.4 – the innocent-looking $D_{\widehat{F}}$ is actually an elaborate characteristic form involving the normal bundles of the orbifold strata inside F.

If $\varepsilon_{F,j} = -1$, we rewrite the jth term in the denominator in polarized form as

$$\begin{aligned}(-1)^{n_{F,j}}\det A^{-1}_{F,j}\, e^{-n_{F,j}\alpha_{F,j}(\xi)}\det\exp(-\Omega_{\widehat{F},j})\\ \times\det(I - A_{F,j}\exp(\alpha_{F,j}(\xi)I + \Omega_{\widehat{F},j})).\end{aligned} \tag{4.3}$$

Let $E^{\#}_{\widehat{F}}$ be the line bundle

$$\Big(\bigotimes_{\varepsilon_{F,j}=-1}\wedge^{n_{F,j}}(E^+_{\widehat{F},j})\Big)^*$$

having curvature form $-2\pi i\Omega^{\#}_{\widehat{F}}$ and canonical automorphism

$$A^{\#}_{F} = \Big(\prod_{\varepsilon_{F,j}=-1} \det A^{+}_{F,j} \Big)^{-1}.$$

Since $\exp(\operatorname{trace} B) = \det(\exp B)$ for an endomorphism B, and the curvature form satisfies $\operatorname{trace}\Omega(E) = \Omega(\wedge^n E)$, where n is the rank of the vector bundle E, if we substitute (4.3) into (4.2), we can rewrite (4.2) in polarized form:

$$\begin{aligned} \chi_F(\exp\xi) &= (-1)^{n_F} e^{(\Phi_F - \alpha^{\#}_F)(\xi)} \\ &\times \int_{\widehat{F}} \frac{1}{m_{\widehat{F}}} \cdot \frac{A_F A^{\#}_F \exp\left(\omega_{\widehat{F}} + \Omega^{\#}_{\widehat{F}}\right) \operatorname{Todd}(\widehat{F})}{D_{\widehat{F}} \prod_j \det(I - (A^{+}_{F,j})^{-1} \exp(-\alpha^{+}_{F,j}(\xi) I - \Omega^{+}_{\widehat{F},j}))}. \end{aligned} \tag{4.4}$$

By Lemma (B.1) of [**6**], we can expand (4.4) into an infinite trigonometric series

$$\chi_F(\exp\xi) = (-1)^{n_F} \sum_k c_k\, e^{(\Phi_F + \alpha^{\#}_F - k_1\alpha^{+}_{F,1} - \cdots - k_m\alpha^{+}_{F,m})(\xi)}, \tag{4.5}$$

summed over all non-negative integer m-tuples k, where c_k equals

(4.6)

$$\int_{\widehat{F}} \frac{1}{m_{\widehat{F}}} \cdot \frac{(\prod_j \operatorname{trace} \mathcal{S}^{k_{F,j}}((A^{+}_{F,j})^{-1} \exp(-\Omega^{+}_{\widehat{F},j})))A_F A^{\#}_F \exp(\omega_{\widehat{F}} + \Omega^{\#}_{\widehat{F}})\operatorname{Todd}(\widehat{F})}{D_{\widehat{F}}}.$$

By the definition of the associated Chern character (see §A.4), formula (4.6) is simply

$$\int_{\widehat{F}} \frac{1}{m_{\widehat{F}}} \cdot \frac{\widehat{\operatorname{Ch}}(E_{\widehat{F}}(k) \otimes L_{\widehat{F}})\operatorname{Todd}(\widehat{F})}{D_{\widehat{F}}},$$

which is the Riemann–Roch number of the orbifold line bundle $E_F(k) \otimes L_F$ over F, as proved by Kawasaki [**14**].

On the other hand, for $\xi \in \operatorname{Lie}(T)$ the trace of $\exp\xi$ on the (virtual) vector space $Q(M,\omega)$ is equal to

$$\sum_{\text{weights } \mu} \operatorname{mult}(\mu) e^{\mu(\xi)}, \tag{4.7}$$

and by comparing (4.5) with (4.7) one obtains the identity (4.1). □

REMARK 1. In the manifold case, we recover the result in [**6**], where the Kawasaki–Riemann–Roch number is the usual Riemann–Roch number:

$$\operatorname{RR}(F, E_F(k) \otimes L_F) = \int_F \operatorname{Ch}(E_F(k) \otimes L_F)\operatorname{Todd}(F).$$

5. Semiclassical limit

Let n be a positive integer. Replacing the line bundle L by its nth tensor power, one gets, as in §1, a T-invariant elliptic first order differential operator $\bar{\partial}^{(n)}$. Denote by $\operatorname{mult}^{(n)}(\mu)$ the multiplicity with which μ occurs as a weight of the representation of T on $\operatorname{kernel}\bar{\partial}^{(n)} - \operatorname{cokernel}\bar{\partial}^{(n)}$, and let $d = \frac{1}{2}\dim M - \dim T$.

THEOREM 2. *As n tends to infinity, the quantity $n^{-d}\mathrm{mult}^{(n)}(n\mu)$ tends to*

$$\sum_{F\subseteq M^T}(-1)^{\sigma_F}\int_{\Delta_F(\mu)}\mathrm{Res}_F(s)ds, \tag{5.1}$$

where $\mathrm{Res}_F(s)$ is the sum of residues of

$$e^{\sum s_j z_j}\frac{1}{|\Gamma_F|}\int_F\frac{\exp\omega_F}{c_{F,1}(z_1)\dots c_{F,m}(z_m)}, \tag{5.2}$$

Γ_F being the structure group of F and $c_{F,j}(z)$ the Chern polynomial of $E^+_{F,j}$.

PROOF. By Theorem 1, $\mathrm{mult}^{(n)}(n\mu)$ is equal to the sum

$$\sum_F(-1)^{n_F}\mathcal{N}_F^{(n)}(n\mu),$$

where

$$\mathcal{N}_F^{(n)}(n\mu)=\sum\int_{\widehat{F}}\frac{1}{m_{\widehat{F}}}\cdot\frac{\widehat{\mathrm{Ch}}(E_{\widehat{F}}(k)\otimes L^n_{\widehat{F}})\mathrm{Todd}(\widehat{F})}{D_{\widehat{F}}}, \tag{5.3}$$

summed over all non-negative integral solutions k of the equation

$$n\Phi_F-k_1\alpha^+_{F,1}-\dots-k_m\alpha^+_{F,m}+\alpha^\#_F=n\mu.$$

(If we replace L by L^n, we must replace ω by $n\omega$, Φ by $n\Phi$ and A_F by A^n_F.) We set $\dim F=2p$ and $\dim\Delta_F(\mu)=q$. By (5.3), we have

$$n^{-d}\mathcal{N}^{(n)}(n\mu)=n^{-(d-p)}\sum_k\int_{\widehat{F}}n^{-p}\frac{1}{m_{\widehat{F}}}\cdot\frac{\widehat{\mathrm{Ch}}(E_{\widehat{F}}(k)\otimes L^n_{\widehat{F}})\mathrm{Todd}(\widehat{F})}{D_{\widehat{F}}}. \tag{5.4}$$

LEMMA 5.5. *Up to an error of order $O\left(\frac{1}{n}\right)$, the sum in (5.4) equals*

$$\sum_k\frac{1}{|\Gamma_F|}\sum_{g\in\Gamma_F}\rho^n_{F,0}(g)\rho^\#_F(g)\int_F\exp\omega_F\prod_j\mathrm{trace}\,\mathcal{S}^{k_j}\left(\rho^+_{F,j}(g)\exp\left(-\frac{\Omega^+_{F,j}}{n}\right)\right), \tag{5.6}$$

where $\rho_{F,0}$, $\rho^\#_F$ and $\rho^+_{F,j}$ are the representations of the structure group Γ_F of F on the orbifold charts of L, $E^\#$ and E^+_j over F.

PROOF. Let $\widehat{F}_l$ be a connected component of $\widehat{F}$ and let $2p_l=\dim\widehat{F}_l$. With ω and A_F replaced by $n\omega$ and A^n_F in (4.6), the integrand in this expression can be expanded into a sum of terms of the form

$$\pm\frac{1}{m_{\widehat{F}_l}}\cdot\frac{A^n_F A^\#_F\mathcal{A}(n\omega_{\widehat{F}})^r\wedge\Omega_{i_1}\wedge\dots\wedge\Omega_{i_s}\wedge(\Omega^\#_{\widehat{F}})^\nu\wedge T^\mu}{D_{\widehat{F}_l}},$$

where $\mathcal{A}$ is a factor involving the A^+_j's, Ω_{i_a} is a coefficient of the curvature form Ω^+_{F,i_a}, and T^μ is the component of degree 2μ of $\mathrm{Todd}(\widehat{F}_l)$. However, this term can only contribute to the integral if $r+s+\nu+\mu=p_l$, in which case it can be rewritten as

$$\pm\frac{1}{m_{\widehat{F}_l}}\cdot\frac{A^n_F A^\#_F\mathcal{A}\,n^{p_l}\,\omega^r_{\widehat{F}}\wedge(\Omega_{i_1}/n)\wedge\dots\wedge(\Omega_{i_s}/n)\wedge(\Omega^\#_{\widehat{F}}/n)^\nu\wedge T^\mu/n^\mu}{D_{\widehat{F}_l}}.$$

When multiplied by n^{-p} (as in (5.4)), the terms in this sum for which ν or μ is positive or $p>p_l$ can be discarded since they contribute errors of order $O(\frac{1}{n})$. We

WILLIAM F. MAAG LIBRARY
YOUNGSTOWN STATE UNIVERSITY

are left with the components of $\widehat{F}$ whose dimension is $2p$. Such components are indexed by the conjugacy classes of Γ_F (see §A.3). Hence the sum in (5.4) is equal to

$$\sum_k \sum_{\gamma\in\mathrm{Conj}(\Gamma_F)} \frac{1}{m_{\widehat{F}_\gamma}} \int_{\widehat{F}_\gamma} \frac{A_F^n A_F^{\#} \exp\omega_{F_\gamma} \prod_j \operatorname{trace} \mathcal{S}^{k_j}\left(\left(A_{F,j}^+\right)^{-1} \exp\left(-\frac{\Omega_{F,j}^+}{n}\right)\right)}{D_{\widehat{F}_\gamma}}.$$

On the component $\widehat{F}_\gamma$ corresponding to $\gamma \in \mathrm{Conj}(\Gamma_F)$, we have

$$A_F = \rho_{F,0}(\gamma), \quad A^{\#} = \rho_F^{\#}(\gamma), \quad \text{and} \quad A_j^+ \text{ is conjugate to } \rho_j^+(\gamma)$$

(we use the same symbol for a class function on a group and the induced function on conjugacy classes of that group). We also have $D_{\widehat{F}_\gamma} = 1$, since $N_{\widehat{F}_\gamma}$ is trivial, $m_{\widehat{F}_\gamma}$ is the common order of the centralizers of elements in the conjugacy class γ, and any class function χ satisfies

$$\sum_{\gamma\in\mathrm{Conj}(\Gamma_F)} \frac{\chi(\gamma)}{m_{\widehat{F}_\gamma}} = \frac{1}{|\Gamma_F|} \sum_{g\in\Gamma_F} \chi(g).$$

Finally, as the natural immersions $\mu : \widehat{F}_\gamma \to F$ are bijective, the integral over $\widehat{F}_\gamma$ coincides with the integral over F. □

Since the action of T is effective, we have $q = m - \dim T$, and hence $d - p = q + \sum_{j=1}^m (n_j - 1)$. By Lemma 5.5 and Lemma B.1, (5.4) is, up to an error of order $O(\frac{1}{n})$, equal to

$$\text{(5.7)} \qquad n^{-q} \sum_k \mathcal{W}_{F,k} \mathrm{Res}\left[e^{\frac{k_1}{n} z_1 + \cdots + \frac{k_m}{n} z_m} \int_F \frac{\exp\omega_F}{\prod_j \det(z_j I + \Omega_{F,j}^+)} \right],$$

where the operator Res takes the sum of residues and the number $\mathcal{W}_{F,k}$ equals

$$\frac{1}{|\Gamma_F|} \sum_{g\in\Sigma_F} \rho_{F,0}^n(g) \rho_F^{\#}(g) \prod_j (\rho_{F,j}^+(g))^{-k_j},$$

Σ_F being the abelian subgroup of Γ_F consisting of the elements g for which each $\rho_{F,j}^+(g)$ is a scalar multiple of the identity. By the orthogonality relations for characters we know that

$$\sum_{g\in\Sigma_F} \left(\rho_{F,0}^n \rho_F^{\#} \prod_j (\rho_{F,j}^+)^{-k_j} \right)(g) = 0$$

unless $(\rho_{F,0}^n \rho_F^{\#} \prod_j (\rho_{F,j}^+)^{-k_j})$ is the trivial character of Σ_F, in which case this sum is $|\Sigma_F|$. Hence, (5.7) is equal to

$$\text{(5.8)} \qquad n^{-q} \sum_k \frac{|\Sigma_F|}{|\Gamma_F|} \mathrm{Res}\left[e^{\sum_j \frac{k_j}{n} z_j} \int_F \frac{\exp\omega_F}{\prod_j \det(z_j I + \Omega_{F,j}^+)} \right],$$

summed over all k satisfying both

$$\Phi_F - \frac{k_1}{n}\alpha_{F,1}^+ - \cdots - \frac{k_m}{n}\alpha_{F,m}^+ + \frac{\alpha_F^{\#}}{n} = \mu$$

and

$$\rho^n_{F,0}\rho^{\#}_F \prod_j (\rho^+_{F,j})^{-k_j} = \mathrm{id} \tag{5.9}$$

in the character group of Σ_F. By representation theory for finite abelian groups, condition (5.9) picks out a sublattice of order $|\Sigma_F|$ inside $\mathbb{Z}^m$. Therefore, as n tends to infinity (5.8) tends to the integral

$$\int_{\Delta_F(\mu)} \operatorname{Res}\left[e^{sz}\int_F \frac{1}{|\Gamma_F|}\cdot\frac{\exp\omega_F}{c_1(z_1)\dots c_m(z_m)}\right]ds, \tag{5.10}$$

where $c_j(z_j) := \det(z_j I + \Omega^+_{F,j})$ is the Chern polynomial of $E^+_{F,j}$. □

REMARK 2. When the dimension of $\Delta_F(\mu)$ is $q > 0$, the expression (5.8) converges pointwise to (5.10) for all rational μ. However, when $q = 0$ it converges to (5.10) in a weak sense: for μ rational and any continuous test function f, the distribution (5.8) evaluated at f tends to the evaluation of the distribution (5.10) at f, as n tends to infinity.

6. Duistermaat–Heckman measure

In [**6**] it was proved that, in the case of manifolds, the function of μ defined by (5.1) is the Radon–Nikodým derivative

$$\frac{d\mu_{\mathrm{DH}}}{d\mu_{\mathrm{Lebesgue}}} \tag{6.1}$$

where μ_{DH} is the Duistermaat–Heckman measure and μ_{Lebesgue} is the standard Lebesgue measure on $\mathrm{Lie}(T)^*$ (suitably normalized). The same is true for orbifolds.

THEOREM 3. *The piecewise polynomial function defined by* (5.1) *is the Radon–Nikodým derivative* (6.1).

The proof is analogous to that in [**6**]. Instead of the usual abelian localization formula, the orbifold version should be used [**17**], where the contributions of the fixed point components are affected by a factor equal to the inverse of the order of the orbifold structure group of that component [**4**].

Appendix A. Orbifolds

A.1. An n-dimensional *orbifold* M is a Hausdorff topological space $|M|$ equipped with a complete atlas of orbifold charts, that is, a collection of triples $(\widetilde{\mathcal{U}}, \Gamma, \phi)$, where $\widetilde{\mathcal{U}}$ is a connected open subset of $\mathbb{R}^n$, Γ is a finite group acting on $\widetilde{\mathcal{U}}$ by linear transformations,[2] and $\phi : \widetilde{\mathcal{U}} \to |M|$ is a continuous Γ-invariant map inducing a homeomorphism from $\widetilde{\mathcal{U}}/\Gamma$ to $\mathcal{U} := \phi(\widetilde{\mathcal{U}})$, satisfying appropriate compatibility conditions and such that the $\mathcal{U}$'s cover $|M|$ and this collection of charts is maximal; more details and a review of orbifold differential geometry can be found in [**4, 8, 13, 16, 17, 18**]. For instance, a smooth function on an orbifold is a collection of smooth invariant functions on each orbifold chart which agree on overlaps, and similarly for many other concepts.

---

[2]We assume that the set of all fixed points of Γ has codimension at least two. We do *not* assume the action of Γ to be effective. The result of the action of $g \in \Gamma$ on $u \in \mathcal{U}$ is denoted by $g \cdot u$.

A.2. Let M and N be two orbifolds with a continuous inclusion of the underlying topological spaces. Suppose that M has an orbifold atlas where on each chart $(\widetilde{\mathcal{U}}, \Gamma, \phi)$ the pre-image of N, if not empty, is given by the intersection of $\widetilde{\mathcal{U}}$ with a linear subspace V of $\mathbb{R}^n$. Let Γ_V be the subgroup of those elements in Γ whose actions preserve V. Then N is a *suborbifold* of M if the collection $\{(\widetilde{\mathcal{U}} \cap V, \Gamma_V, \phi|_{\widetilde{\mathcal{U}} \cap V})\}$, together with the induced injections, forms an atlas of orbifold charts for N.

If $(\widetilde{\mathcal{U}}, \Gamma, \phi)$ is an orbifold chart for M, the *structure group* Γ_p of a point $p \in \mathcal{U}$ is the isotropy group of a pre-image of p under ϕ; Γ_p is defined up to isomorphism. The type of the structure group induces a *canonical stratification* of M into suborbifolds. On each connected component of M, there is an open dense set of points for which the order of the structure group is minimal, called the *principal stratum* of M. When M is connected, the *structure group* Γ_M of M is the (abstract) isotropy group of its principal stratum; the *order* of M is the order of Γ_M. When M is not connected the orders of its connected components define the locally constant *order function* $m_M : M \to \mathbb{N}$.

A.3. Given an orbifold M, the *associated orbifold* [**14, 9**] $\widehat{M}$ has charts $(\widetilde{\mathcal{V}}, \Gamma, \psi)$ built as follows: For each chart $(\widetilde{\mathcal{U}}, \Gamma, \phi)$ of M, let

$$\widetilde{\mathcal{V}} = \{(u, g) \in \widetilde{\mathcal{U}} \times \Gamma \mid g \cdot u = u\}.$$

Γ acts on $\widetilde{\mathcal{V}}$ via $h \cdot (u, g) = (h \cdot u, hgh^{-1})$. Let $\mathcal{V} = \widetilde{\mathcal{V}}/\Gamma$ be the space of orbits with projection $\psi : \widetilde{\mathcal{V}} \to \mathcal{V}$. The orbifold charts $(\widetilde{\mathcal{V}}, \Gamma, \psi)$ inherit the compatibility conditions from the $(\widetilde{\mathcal{U}}, \Gamma, \phi)$. As a set,

$$\widehat{M} = \bigsqcup_{p \in M} \mathrm{Conj}(\Gamma_p),$$

where $\mathrm{Conj}(\Gamma_p)$ is the set of conjugacy classes in Γ_p. For example, when M is a teardrop orbifold (i.e. M is homeomorphic to a sphere and has just one singular point p with structure group $\mathbb{Z}/n$ for some n), then $\widehat{M}$ has n components: one component diffeomorphic to M plus $n-1$ points each with structure group $\mathbb{Z}/n$, corresponding to p paired with the nonzero elements of $\mathbb{Z}/n$. Associated orbifolds may be viewed as a bookkeeping device which allows for concise orbifold index formulas.

A.4. A rank k *orbifold vector bundle* $\pi : E \to M$ over an orbifold M is a collection of Γ-equivariant rank k vector bundles $\pi_{\widetilde{\mathcal{U}}} : \widetilde{E_{\mathcal{U}}} \to \widetilde{\mathcal{U}}$ on each chart $(\widetilde{\mathcal{U}}, \Gamma, \phi)$, together with suitable compatibility conditions. The fibers $\pi^{-1}(p)$ are in general diffeomorphic to some quotient of a k-dimensional vector space by the action of the structure group Γ_p. Additional structures on an orbifold bundle are defined as collections of Γ-invariant objects on each $\pi_{\widetilde{\mathcal{U}}} : \widetilde{E_{\mathcal{U}}} \to \widetilde{\mathcal{U}}$ agreeing on overlaps.

An orbifold vector bundle $E \to \widehat{M}$ over an associated orbifold has a *canonical automorphism* $A(E)$, given on each chart $(\widetilde{\mathcal{V}}, \Gamma, \psi)$ by the natural action of $g \in \Gamma$ on the fiber of $\widetilde{E_{\mathcal{V}}}$ above $(u, g) \in \widetilde{\mathcal{V}}$ [**17**].

Suppose that M has an almost complex structure. Then the normal bundle $N_{\widehat{M}}$ of the natural immersion $\mu : \widehat{M} \to M$ can be endowed with an hermitian connection; let $-2\pi i\Omega(N_{\widehat{M}})$ be its curvature. The *associated characteristic form* (appearing in the index formulas) is

$$D_{\widehat{M}} = \det\left(I - A^{-1}(N_{\widehat{M}})\exp(-\Omega(N_{\widehat{M}}))\right).$$

The *associated Chern character* of a complex vector bundle $E \to \widehat{M}$ with curvature $-2\pi i\Omega$ with respect to an hermitian connection is

$$\widehat{\mathrm{Ch}}(E) = \mathrm{trace}\,(A(E)\exp\Omega).$$

Appendix B. Integral estimate

LEMMA B.1. *Let V be a d-dimensional complex vector space, and let A, B be commuting elements of* $\mathrm{GL}(V)$*, with A diagonalizable. Then*

$$\text{(B.2)}\quad n^{-(d-1)}\mathrm{trace}\,\mathcal{S}^k(A\exp\tfrac{B}{n}) = \begin{cases} \lambda^k \mathrm{Res}\left[\dfrac{e^{\frac{d+k-1}{n}z}}{\det(z-B)}\right]\left(1+O\left(\frac{1}{n}\right)\right) & \textit{if } A = \lambda I, \\ O\left(\frac{1}{n}\right) & \textit{otherwise}, \end{cases}$$

where the $O(\frac{1}{n})$'s are uniform in k, and the operator Res *takes the integral along a contour about the origin containing the zeroes of* $\det(z-B)$ *and divides it by $2\pi i$.*

PROOF. Without loss of generality, we can assume that A and B are simultaneously diagonalizable with eigenvalues $\lambda_1, \dots, \lambda_d$ (of A) and $\mu_1, \dots, \mu_d$ (of B), and that $e^{\mu_1}, \dots, e^{\mu_d}$ are distinct. Let C be a contour about the origin containing $\lambda_1, \dots, \lambda_d$. For n large enough, the left hand side of (B.2) is equal to the contour integral

$$n^{-(d-1)}\frac{1}{2\pi i}\int_C \frac{z^{d+k-1}}{\prod_j\left(z - \lambda_j e^{\mu_j/n}\right)}dz.$$

The pole at $\lambda_j e^{\mu_j/n}$ has residue

$$\text{(B.3)}\qquad n^{-(d-1)}\frac{\left(\lambda_j e^{\mu_j/n}\right)^{d+k-1}}{\prod_{i\neq j}\left(\lambda_j e^{\mu_j/n} - \lambda_i e^{\mu_i/n}\right)},$$

where

$$\lambda_j e^{\mu_j/n} - \lambda_i e^{\mu_i/n} = \begin{cases} \lambda_j \frac{\mu_j-\mu_i}{n}\left(1+O(\frac{1}{n})\right) & \text{if } \lambda_j = \lambda_i, \\ (\lambda_j - \lambda_i)\left(1+O(\frac{1}{n})\right) & \text{if } \lambda_j \neq \lambda_i. \end{cases}$$

There are $(d-1)$ factors in the denominator of (B.3). Therefore, if A has at least two different eigenvalues, all residues (B.3) are of order $O(\frac{1}{n})$. If $A = \lambda I$, then we express $n^{-(d-1)}\mathrm{trace}\,\mathcal{S}^k(\exp\frac{B}{n})$ as in Theorem (B2) of [**6**]. □

By analyticity, (B.2) is valid for any endomorphism $B : V \to V$, not necessarily in $\mathrm{GL}(V)$. When k is of order $O(n)$ (i.e. $k \to \infty$ as $n \to \infty$, but $\frac{k}{n} = O(1)$), (B.2) remains true if all eigenvalues of A have absolute value at most 1.

References

[1] M. Atiyah and G. Segal, *The index of elliptic operators*, II, Ann. of Math., **87** (1968), 531–545

[2] R. Bott, *On induced representations*, The Mathematical Heritage of Hermann Weyl (Durham, NC, 1987), Proc. Sympos. Pure Math., vol. 48, Amer. Math. Soc., Providence, RI, 1988, pp.1–13.

[3] T. Bröcker and T. tom Dieck, *Representations of compact Lie groups*, Graduate Texts in Mathematics, vol. 98, Springer-Verlag, New York, 1995.
[4] A. Canas da Silva, *Multiplicity formulas for orbifolds*, Ph.D. Thesis, M.I.T., 1996.
[5] ———, *Atiyah-Bott-Lefschetz for orbifolds and Dedekind sums*, Preprint, 1998.
[6] A. Canas da Silva and V. Guillemin, *On the Kostant multiplicity formula for group actions with non-isolated fixed points*, Adv. Math. **123** (1996), 1–15.
[7] J. Cartier, *On H. Weyl's character formula*, Bull. Amer. Math. Soc. **67** (1961), 228–230.
[8] J. J. Duistermaat, *The heat kernel Lefschetz fixed point formula for the Spin-c Dirac operator*, Birkhäuser, Boston, 1996.
[9] C. Farsi, *K-theoretical index theorems for orbifolds*, Quart. J. Math. Oxford Ser. (2) **43** (1992), 183–200.
[10] V. Guillemin, E. Lerman, and S. Sternberg, *On the Kostant multiplicity formula*, J. Geom. Phys. **5** (1988), 721–750.
[11] V. Guillemin and E. Prato, *Heckman, Kostant, and Steinberg formulas for symplectic manifolds*, Adv. Math. **82** (1990), 160–179.
[12] V. Guillemin and S. Sternberg, *Geometric quantization and multiplicities of group representations*, Invent. Math. **67** (1982), 515–538.
[13] A. Haefliger and E. Salem *Actions of tori on orbifolds*, Ann. Global Anal. Geom. **9** (1991), 37–59.
[14] T. Kawasaki, *The Riemann–Roch theorem for complex V-manifolds*, Osaka J. Math. **16** (1979), 151–159.
[15] B. Kostant, *A formula for the multiplicity of a weight*, Trans. Amer. Math. Soc. **93** (1959), 53–73.
[16] E. Lerman and S. Tolman, *Hamiltonian torus actions on symplectic orbifolds and toric varieties*, Trans. Amer. Math. Soc. **349** (1997), 4201–4230.
[17] E. Meinrenken, *Symplectic surgery and the* Spin^c*-Dirac operator*, Adv. Math. **134** (1998), 240–277.
[18] I. Satake, *On a generalization of the notion of manifold*, Proc. Nat. Acad. Sci. U.S.A. **42** (1956), 359–363.
[19] M. Vergne, *The equivariant index formula on orbifolds*, Duke Math. J. **82** (1996), 637–652.

Department of Mathematics, University of California, Berkeley, CA 94720
E-mail address: `acannas@math.berkeley.edu`

Department of Mathematics, Massachusetts Institute of Technology, Cambridge, MA 02139
E-mail address: `vwg@math.mit.edu`

Amer. Math. Soc. Transl.
(2) Vol. **196**, 1999

Symmetric Spaces, Kähler Geometry and Hamiltonian Dynamics

S. K. Donaldson

Preface

This article was written in the Spring of 1997. Some while after completing it I learnt of previous work by other authors which overlaps substantially. The "symmetric space" structure which is the main topic of the article was discovered by S. Semmes:

Complex Monge–Ampère and symplectic manifolds, Amer. J. Math. **114** (1992), 495–550.

I am grateful to J.-P. Bourguignon for informing me about this reference. Bourguignon tells me that he and Mabuchi were developing similar ideas at about the same time.

The "moment map" point of view which is mentioned in the last part of the paper was discovered by A. Fujiki:

The moduli spaces and Kähler metrics of polarised algebraic varieties, Sugaku **42** (1990), 231–243; English transl., Sugaku Expositions **5** (1992) 173–191.

I hope that it may still be worth making the present article available in its original form, although it would of course have been written rather differently if I had been aware of these earlier papers.

§1. Introduction

A Riemannian manifold is called a locally symmetric space if its curvature tensor is covariant constant; $\nabla R = 0$. Such symmetric spaces were classified, up to coverings, by E. Cartan, and the theory is tightly bound up with the classification of semi-simple Lie groups [**11**]. One of the memorable features of the theory is that the (irreducible) symmetric spaces occur in pairs of "compact" and "non-compact" type. The compact type have positive sectional curvature and the non-compact type have negative sectional curvature. In particular, let G be a compact Lie group with a bi-invariant Riemannian metric. This is an example of a symmetric space of compact type; the curvature tensor is given by the formula

$$R(X,Y)Z = \tfrac{1}{4}[[X,Y],Z], \tag{1}$$

and the sectional curvature by

$$K(X,Y) = \tfrac{1}{4}|[X,Y]|^2. \tag{2}$$

©1999 American Mathematical Society

(Here X, Y, Z are tangent vectors which, by left or right translation, can be taken to lie in the Lie algebra $\mathfrak{g}$ of G.) The non-compact dual H of G is obtained as follows. The Lie group G has a *complexification* G^c—a complex Lie group containing G as a real subgroup and with Lie algebra $\mathfrak{g}^c = \mathfrak{g} \otimes_{\mathbf{R}} \mathbf{C}$—and H is the homogeneous space $H = G^c/G$ with the metric induced from the invariant pseudo-Riemannian metric on G^c. The tangent space of H at the identity coset is $\mathfrak{g}^c/\mathfrak{g} \cong \mathfrak{g}$, and the curvature of H is given by the formulas

$$R(X,Y)Z = -\tfrac{1}{4}[[X,Y],Z], \tag{3}$$

$$K(X,Y) = -\tfrac{1}{4}|[X,Y]|^2. \tag{4}$$

Cartan's classification applies of course to *finite dimensional* symmetric spaces. The object of this paper is to point out that some of the same ideas appear naturally in an *infinite dimensional* situation. Let (M^{2n}, ω) be a compact symplectic manifold, so that the space $C^\infty(M)$ of functions on M is a Lie algebra under the Poisson bracket $\{\ ,\ \}$, defined by

$$\{f,g\}\omega^n = df \wedge dg \wedge \omega^{n-1}. \tag{5}$$

There is also an invariant L^2 inner product on $C^\infty(M)$:

$$\langle f,g\rangle = \frac{1}{n!}\int_M fg\ \omega^n. \tag{6}$$

We can write $C^\infty(M) = C_0^\infty(M) \oplus \mathbf{R}$, where $C_0^\infty(M)$ is the space of functions of integral zero, the L^2-orthogonal complement of the constants. The subalgebra $C_0^\infty(M)$ is the Lie algebra of an infinite dimensional Lie group $\mathcal{G}_0$. If $H^1(M;\mathbf{R}) = 0$ this is just the identity component of the *symplectomorphism group* SDiff of (M,ω). In general $\mathcal{G}_0$ is the group of "exact" symplectomorphisms, see [**16**] for example. It is sometimes convenient to suppose that $[\omega] \in H^2(M;\mathbf{R})$ is an integral class and to fix a complex Hermitian line bundle $L \to M$ with a unitary connection whose curvature is $-2\pi i\omega$. Then the full space $C^\infty(M)$ is the Lie algebra of a group $\mathcal{G}$: the group of connection-preserving Hermitian bundle maps from L to L (which necessarily cover exact symplectomorphisms of M).

Now the groups $\mathcal{G}, \mathcal{G}_0$ have bi-invariant metrics defined by the L^2 inner product on their Lie algebras, so it is no surprise that they furnish examples of infinite dimensional symmetric spaces. The two cases are scarcely different, since there is a local Riemannian (and Lie group) isomorphism

$$\mathcal{G} \approx \mathcal{G}_0 \times S^1. \tag{7}$$

It is clear that the formal part of the familiar theory goes over to these infinite dimensional spaces, so the curvature tensor of $\mathcal{G}$ is given by substituting the Poisson bracket into (1). Similarly, the geodesics in $\mathcal{G}$ are the translates of 1-parameter subgroups, the stuff of Hamiltonian dynamics. What is less obvious is that, if M is a Kähler manifold, there are "negatively curved duals" of these spaces, as we shall now describe.

Suppose that V is a compact complex n-manifold which admits a Kähler metric ω_0. The "$\overline{\partial}\partial$-lemma" asserts that any other Kähler metric cohomologous to ω_0 can be expressed via a Kähler potential; define $\mathcal{H}$ to be the space of Kähler potentials

$$\mathcal{H} = \{\phi \in C^\infty(V) : \omega_\phi = \omega_0 + i\overline{\partial}\partial\phi > 0\}. \tag{8}$$

This space has a better description in the case when $[\omega_0]$ is an integral class, so there is a corresponding holomorphic line bundle $L \to V$. Then we can identify $\mathcal{H}$ with the space of Hermitian metrics on L having positive curvature (or, more precisely, the compatible unitary connection has positive curvature). For if h_0 is a metric with curvature $-2\pi i\omega_0$, the curvature of $e^{2\pi\phi}h_0$ is $-2\pi i(\omega_0 + i\bar{\partial}\partial\phi)$ [**10,** §0.5]. The advantage of this second description is that it avoids the apparent dependence of (8) on the base point ω_0. The real numbers act on $\mathcal{H}$, by addition of constants, and we define $\mathcal{H}_0 = \mathcal{H}/\mathbf{R}$, which can be viewed as the *space of Kähler metrics on* V, in the given cohomology class.

Each Kähler potential $\phi \in \mathcal{H}$ gives a measure $d\mu_\phi = \frac{1}{n!}\omega_\phi^n$ on V. We define a Riemannian metric on the infinite dimensional manifold $\mathcal{H}$ using the L^2-norms furnished by these measures—a tangent vector $\delta\phi$ to $\mathcal{H}$ at a point $\phi \in \mathcal{H}$ is just a function on V, and we set

$$\|\delta\phi\|_\phi^2 = \int_V (\delta\phi)^2 d\mu_\phi. \tag{9}$$

Our main result is

THEOREM 1. *The Riemannian manifold $\mathcal{H}$ is an infinite dimensional symmetric space; it admits a Levi–Civita connection whose curvature is covariant constant. At a point $\phi \in \mathcal{H}$ the curvature is given by*

$$R_\phi(\delta_1\phi, \delta_2\phi)\delta_3\phi = -\tfrac{1}{4}\{\{\delta_1\phi, \delta_2\phi\}_\phi, \delta_3\phi\}_\phi,$$

where $\{\ ,\ \}_\phi$ is the Poisson bracket on $C^\infty(V)$ of the symplectic form ω_ϕ.

(Recall that in infinite dimensions the usual argument gives the uniqueness of a Levi–Civita [i.e. torsion-free, metric-compatible] connection, but not the existence in general.) The formula for the curvature of $\mathcal{H}$ entails that the sectional curvature is non-positive, given by

$$K_\phi(\delta_1\phi, \delta_2\phi) = -\tfrac{1}{4}\|\{\delta_1\phi, \delta_2\phi\}_\phi\|_\phi, \tag{10}$$

and, comparing with (1)–(4), these formulas certainly suggest that $\mathcal{H}$ should be regarded as the negatively curved dual of the group $\mathcal{G}$.

In the first half of this paper we dicuss the proof of this theorem. In the second half we explore the geometry of the space $\mathcal{H}$, in particular the geodesic equation. We state a number of natural conjectures, or questions, which we hope may be interesting to investigate further, and we outline the relevance of these ideas to well-established problems in Kahler geometry. In particular we will see that plausible results on the existence of geodesics would give a proof of the uniqueness of constant scalar curvature Kahler metrics.

§2. First proof

We will now give a direct proof of Theorem 1. The first step is to find the geodesic equation in $\mathcal{H}$. This is the Euler–Lagrange equation, $\delta E = 0$, for the energy functional

$$E = \int_0^1 \int_V \dot{\phi}^2 d\mu_\phi dt \tag{11}$$

on 1-parameter families $\phi(t)$ in $\mathcal{H}$, with fixed end points. The first variation of the measure is given by

$$d\mu_{\phi+\psi} = (1 + \tfrac{1}{2}\Delta\psi)\ d\mu_\phi + O(\psi^2),$$

where $\Delta = \Delta_\phi$ is the Laplacian on V of the metric ω_ϕ. So the first variation of E under a small variation $\psi(t)$ is

$$\delta E = \int_0^1 \int_V 2\dot\phi\dot\psi + \tfrac{1}{2}\dot\phi^2\Delta\psi\ d\mu_\phi = \textstyle\int_0^1 \int_V \psi(-2\frac{d}{dt}(\dot\phi d\mu_\phi) + \frac{1}{2}\Delta(\dot\phi^2)d\mu_\phi),$$

and the geodesic equation is

$$\frac{d}{dt}(\dot\phi d\mu_\phi) = \tfrac{1}{4}\Delta(\dot\phi^2)d\mu_\phi.$$

The t-derivative of $d\mu_\phi$ is $\frac{1}{2}\Delta\dot\phi\ d\mu_\phi$, and

$$\Delta(\dot\phi^2) = 2\dot\phi\Delta\dot\phi - 2|\nabla\dot\phi|^2_\phi,$$

so the geodesic equation reduces to

$$\ddot\phi = -\tfrac{1}{2}|\nabla\dot\phi|^2_\phi. \tag{12}$$

This geodesic equation shows us how to define a connection on the tangent bundle of $\mathcal{H}$. The notation is simplest if one thinks of such a connection as a way of differentiating vector fields along paths. Thus if $\phi(t)$ is any path in $\mathcal{H}$ and $\psi(t)$ is a field of tangent vectors along the path (that is, a function on $V \times [0,1]$), we define the covariant derivative along the path to be

$$D_t\psi = \frac{\partial\psi}{\partial t} + \tfrac{1}{2}(\nabla\psi, \nabla\dot\phi)_\phi. \tag{13}$$

This connection is torsion-free because in the canonical "coordinate chart", which represents $\mathcal{H}$ as an open subset of $C^\infty(V)$, the "Christoffel symbol"

$$\Gamma : C^\infty(V) \times C^\infty(V) \to C^\infty(V)$$

at ϕ is just

$$\Gamma(\psi_1, \psi_2) = \tfrac{1}{2}(\nabla\psi_1, \nabla\psi_2)_\phi,$$

which is symmetric in ψ_1, ψ_2. The connection is metric-compatible because

$$\begin{aligned}\frac{d}{dt}\|\psi\|^2_\phi = \frac{d}{dt}\int_V \psi^2 d\mu_\phi &= \int_V 2\frac{\partial\psi}{\partial t}\psi + \tfrac{1}{2}\psi^2\Delta(\dot\phi)d\mu_\phi \\ &= \int_V 2\frac{\partial\psi}{\partial t}\psi + \tfrac{1}{2}(\nabla(\psi^2), \nabla\dot\phi)\ d\mu_\phi \\ &= 2\int_V (\frac{\partial\psi}{\partial t} + \tfrac{1}{2}(\nabla\psi, \nabla\dot\phi)\ )\psi d\mu_\phi = 2\langle D_t\psi, \psi\rangle\end{aligned} \tag{14}$$

At this stage we could go on to compute the curvature of $\mathcal{H}$. But before doing this we will pause to explain the really significant property of the connection (13), which gets to the heart of the link betweeen Kähler and symplectic geometries (and which gives, by the way, more insight into the manipulations leading to (14)). By Moser's theorem [**17**], the symplectic manifolds (V, ω_ϕ) are all symplectically equivalent. The proof in this case is somewhat easier than the general situation considered by Moser.

Let $\phi(t)$ be a path starting at 0 in $\mathcal{H}$ and consider the t-dependent vector field

$$X_t = \tfrac{1}{2}\nabla_{\omega(t)}\dot{\phi},$$

where $\nabla_{\omega(t)}$ is the gradient operator defined by the metric $\omega(t) = \omega_0 + i\bar{\partial}\partial\phi(t)$. Then for fixed t the Lie derivative of $\omega(t)$ along X_t is

$$L_{X_t}(\omega(t)) = d(X_t \perp \omega(t))$$

But $X_t \perp \omega(t) = \frac{1}{2}Id\dot{\phi}(t)$, where I is the action of the complex structure on 1-forms, so

$$L_{X_t}\omega(t) = \tfrac{1}{2}(dId)\dot{\phi} = -i\bar{\partial}\partial\dot{\phi}.$$

On the other hand, the t-derivative of $\omega(t)$ is obviously $i\bar{\partial}\partial\dot{\phi}$. So if $f_t : V \to V$ is the 1-parameter family of diffeomorphisms obtained by integrating X_t (with $f_0 = 1_V$), we have

$$\frac{d}{dt}f_t^*\Big(\omega(t)\Big) = f_t^*(L_{X_t}\omega(t) + \frac{d\omega(t)}{dt}\Big) = 0,$$

i.e. the diffeomorphism f_t gives the desired symplectomorphism from (V, ω_0) to $(V, \omega(t))$. Now let $\mathcal{Y} \subset \mathcal{H} \times \mathrm{Diff}(V)$ be the set of pairs (ϕ, f) such that $f^*(\omega_\phi) = \omega_0$. This is a principle bundle over $\mathcal{H}$ with structure group the group SDiff(V) of symplectomorphisms of (V, ω_0). Then the discussion above shows that our connection on the tangent space of $\mathcal{H}$ is induced from an SDiff connection on $\mathcal{Y} \to \mathcal{H}$ via the action of SDiff on the vector space $C^\infty(V)$; that is, we have a connection-preserving bundle isomorphism

$$T\mathcal{H} = \mathcal{Y} \times_{\mathrm{SDiff}} C^\infty(V).$$

In this framework the 1-parameter family of diffeomorphisms f_t above appears as the horizontal lift of the path $\phi(t)$ to $\mathcal{Y}$. The metric-preserving property (14) is just the fact that the action of SDiff on $C^\infty(V)$ preserves the L^2 norm.

If $H^1(V) \neq 0$, the group SDiff is not quite the same as the group $\mathcal{G}_0$ of exact symplectomorphisms considered in Section 1 (with $(M, \omega) = (V, \omega_0)$), but one can easily adjust the definitions to get a principle $\mathcal{G}_0$ bundle over $\mathcal{H}$ which descends to a $\mathcal{G}_0$ bundle $\mathcal{X}_0$ over $\mathcal{H}_0$. As usual, the discussion is cleaner in the case when $[\omega_0]$ is integral, so we have a holomorphic line bundle $L \to V$ and we can regard $\mathcal{H}$ as the space of metrics of positive curvature on L. Then we let $\mathcal{X}$ be the set of pairs $(h, \tilde{f})$, where h is a metric on L and $\tilde{f}$ is a connection-preserving Hermitian bundle map from (L, h_0) to (L, h), which necessarily covers a diffeomorphism f of V with $f^*(\omega_\phi) = \omega_0$. (More precisely, we should take $\mathcal{X}$ to be the connected component of $(h_0, 1_L)$ in this space.) Then $\mathcal{X}$ is a principle $\mathcal{G}$ bundle over $\mathcal{H}$, and our Levi–Civita connection is induced from a $\mathcal{G}$-connection on $\mathcal{X}$.

We now go back to compute the curvature tensor of $\mathcal{H}$. To do this we consider a 2-parameter family $\phi(s,t)$ in $\mathcal{H}$, and a vector field $\psi(s,t)$ along $\phi(s,t)$. We denote s and t derivatives by suffixes ϕ_s, etc., where convenient. The curvature is given by the commutator

$$R(\phi_s, \phi_t)\psi = (D_sD_t - D_tD_s)\psi. \tag{15}$$

Expanding out, this is

$$\begin{aligned} R(\phi_s, \phi_t)\psi = {}& \tfrac{1}{2}\left((\nabla\phi_s, \nabla\psi_t) - (\nabla\phi_t, \nabla\psi_s) + \tfrac{\partial}{\partial s}(\nabla\phi_t, \nabla\psi) - \tfrac{\partial}{\partial t}(\nabla\phi_s, \nabla\psi)\right) \\ & + \tfrac{1}{4}\left((\nabla\phi_s, \nabla(\nabla\phi_t, \nabla\psi)) - (\nabla\phi_t, \nabla(\nabla\phi_s, \nabla\psi))\right). \end{aligned} \tag{16}$$

Here we have simply written $(\ ,\ )$ for the inner product on cotangent vectors in V defined by $\omega_{\phi(s,t)}$. Thus, if a, b are fixed cotangent vectors, then

$$\frac{\partial}{\partial s}(a,b) = ((i\bar{\partial}\partial\phi_s)^\sharp, a\otimes b),$$

where $(i\bar{\partial}\partial\phi_s)^\sharp$ denotes the symmetric 2-tensor on V defined by the real $(1,1)$-form $i\bar{\partial}\partial\phi_s$ in the usual way. So we have

$$\frac{\partial}{\partial s}(\nabla\phi_t, \nabla\psi) = (\nabla\phi_{st}, \nabla\psi) + (\nabla\phi_t, \nabla\psi_s) + (i\bar{\partial}\partial\phi_s)^\sharp(\nabla\phi_t\otimes\nabla\psi),$$

and similarly for $\frac{\partial}{\partial t}(\nabla\phi_s, \psi)$. Then (16) reduces to

$$\begin{aligned} R(\phi_s,\phi_t)\psi &= \tfrac{1}{2}\left((i\bar{\partial}\partial\phi_s)^\sharp(\nabla\phi_t\otimes\nabla\psi) - (i\bar{\partial}\partial\phi_t)^\sharp(\nabla\phi_s\otimes\nabla\psi)\right) \\ &\quad + \tfrac{1}{4}\left((\nabla\phi_s, \nabla(\nabla\phi_t,\nabla\psi)) - (\nabla\phi_t, \nabla(\nabla\phi_s,\nabla\psi))\right). \end{aligned} \tag{17}$$

Now the second expression on the right hand side of (17) can be written as

$$\tfrac{1}{4}(\nabla\nabla\phi_s)(\nabla\phi_s\otimes\nabla\psi) - (\nabla\nabla\phi_t)(\nabla\phi_s\otimes\phi_t),$$

so

$$4R(\phi_s,\phi_t)\psi = P(\phi_s)(\nabla\phi_t\otimes\nabla\psi) - P(\phi_t)(\nabla\phi_s\otimes\nabla\psi), \tag{18}$$

where P is the differential operator, from functions to symmetric 2-tensors,

$$P(f) = 2(i\bar{\partial}\partial f)^\sharp - \nabla\nabla f. \tag{19}$$

On the other hand, we can write the Poisson bracket $\{\ ,\ \}$ using the complex structure structure I:

$$\{f,g\} = (\nabla f, I\nabla g) = -(I\nabla f, \nabla g).$$

So

$$\{\{\phi_s,\phi_t\},\psi\} = (\nabla(\phi_s, I\nabla\phi_t), I\nabla\psi) = Q(\phi_s)(\nabla\phi_t\otimes\nabla\psi) - Q(\phi_t)(\nabla\phi_s\otimes\nabla\psi),$$

where

$$Q(f) = (I\otimes I)(\nabla\nabla f).$$

The calculation of the curvature tensor is therefore completed by showing that the operators P and Q are the same. First, as a matter of linear algebra, the map $\frac{1}{2}(1 + I\otimes I)$ from $s^2(T^*V)$ to itself is the standard projection to the $(1,1)$ part. So the assertion is that for a Kahler manifold the operator $i\bar{\partial}\partial$, which is defined by the complex structure, can be obtained as a projection of the second covariant derivative $\nabla\nabla$, which uses the Levi–Civita connection. This can be seen easily, for example by working in an osculating coordinate system as in ([**10,** p.108]). So we have shown that $R(\phi_s,\phi_t)\psi = -\frac{1}{4}\{\{\phi_s,\phi_t\},\psi\}$, as desired.

The expression for the curvature tensor in terms of Poisson brackets shows that R is invariant under the action of the symplectomorphism group. Since the connection on $T\mathcal{H}$ is induced from an SDiff-connection, it follows that R is covariant constant, and hence $\mathcal{H}$ is indeed an infinite-dimensional symmetric space.

§3. The decomposition of $\mathcal{H}$

Here we clarify the relationship between the spaces $\mathcal{H}$ and $\mathcal{H}_0 = \mathcal{H}/\mathbf{R}$. There is obviously a decomposition of the tangent space:

$$(T\mathcal{H})_\phi = \left\{ \psi : \int_V \psi d\mu_\phi = 0 \right\} \oplus \mathbf{R}. \tag{20}$$

We claim that this corresponds to a Riemannian decomposition

$$\mathcal{H} = \mathcal{H}_0 \times \mathbf{R}. \tag{21}$$

At one level, this follows from the picture developed above, since the tangent space decomposition (20) is invariant under the symplectomorphism group, and is therefore covariant constant with respect to our torsion-free connection. However it is interesting to see this more explicitly, partly because we see the appearance of a functional I on the space of Kähler potentials, which is well known in the literature, see [**1**], [**21**] for example. The decomposition (20) give a 1-form α on $\mathcal{H}$ with

$$\alpha_\phi(\psi) = \int_V \psi d\mu_\phi, \tag{22}$$

and the point is that this 1-form is *closed*. Indeed

$$\alpha_{\phi+\tilde{\psi}}(\psi) - \alpha_\phi(\psi) = \frac{1}{2}\int_V \psi\Delta\tilde{\psi} + O(\tilde{\psi}^2), \tag{23}$$

so

$$(d\alpha)_\phi(\psi, \tilde{\psi}) = \frac{1}{2}\int_V \psi\Delta\tilde{\psi} - \tilde{\psi}\Delta\psi = 0 \tag{24}$$

since the Laplacian is self-adjoint. This means that there is a function $I : \mathcal{H} \to \mathbf{R}$ with $I(0) = 0$ and $dI = \alpha$, and it is this function which gives rise to the Riemannian decomposition (21). We call a Kähler potential ϕ *normalised* if $I(\phi) = 0$. Then any Kähler metric has a unique normalised potential, and the restriction of our metric on $\mathcal{H}$ to $I^{-1}(0)$ endows the space $\mathcal{H}_0$ of Kähler metrics with a Riemannian structure; this is independent of the choice of base point ω_0 and clearly makes $\mathcal{H}_0$ into a symmetric space. The functional I can be written more exlpicitly by integrating α along lines in $\mathcal{H}$ to give the formula

$$I(\phi) = \sum_{p=0}^{n} \frac{1}{(p+1)!(n-p)!} \int_V \omega_0^{n-p} (i\bar{\partial}\partial\phi)^p \ \phi. \tag{25}$$

§4. Second proof

We will now outline another proof of Theorem 1 which avoids detailed calculations and shows more clearly the analogy with ordinary symmetric spaces G^c/G.

The tangent bundle of a Lie group, finite or infinite dimensional, is trivialised by left-invariant vector fields, which are, of course, closed under Lie bracket. Conversely, suppose we have a manifold Z and a trivialisation $TZ = Z \times U$, so for each element u of the vector space U we have a vector field X_u on Z. Suppose that this collection of vector fields is closed under Lie bracket: this means that U becomes a Lie algebra with a bracket $[\ ,\ ]_U$ such that $X_{[u,v]_U} = [X_u, X_v]$. If Z is finite dimensional, one of Lie's basic integration theorems tells us that this structure arises from a Lie group structure on Z, but this integration theorem fails in infinite dimensions, as we shall see. Let us call a space Z with this structure an "infinitesimal Lie group". Now suppose that there is a free action of a genuine Lie group S on Z which induces a Lie algebra injection $\mathrm{Lie}(S) \subset U$. Then we can form the orbit space Z/S, which—leaving aside global topological questions—will obviously share some of the familiar properties of homogeneous spaces. In particular, suppose that Z is a "complexification" of S, i.e. that Z has a formally integrable complex structure, the Lie algebra U is complex, the vector fields X_u are holomorphic and U is the complexification of $\mathrm{Lie}(S)$. One way this can occur is when the Lie group S acts freely on a complex manifold T, preserving the complex structure, and Z is a submanifold of T with the property that for each z in Z

$$TZ_z = TO(z)_z \oplus ITO(z)_z, \tag{26}$$

where $O(z)$ is the S-orbit of z in T. In this case we just define U to be $\mathrm{Lie}(S) \otimes \mathbf{C}$ and construct the vector fields X_u by taking complex-linear combinations of the vector fields defining the infinitesimal action of S. (We only need T to have a formally integrable almost complex structure—this implies that the X_u are closed under Lie bracket.) Then we have

PROPOSITION 2. *If S is a Lie group with a bi-invariant metric and Z is an infinitesimal Lie group which is a complexification of S, then Z/S has the structure of a Riemannian symmetric space, with holonomy group S (acting via the adjoint representation) and with curvature $R(u_1, u_2)u_3 = -\frac{1}{4}[[u_1, u_2], u_3]$.*

To see this one just needs to see that the standard theory in finite dimensions only involves infinitesimal calculations—with vector fields and Lie algebras—and so applies equally well under our hypotheses. Of course the part of the standard theory which does *not* go over is the existence of a transitive isometry group of the Riemannian manifold Z/S.

To bring the discussion above to bear on our problem we want to exhibit the principle $\mathcal{G}$ bundle $\mathcal{X} \to \mathcal{H}$ as an infinitesimal complexification, and this can be done in two ways, each involving the action of $\mathcal{G}$ on a complex manifold as described above. In fact it will be easier to work with the smaller space $\mathcal{H}_0$, which is essentially the same by the discussion in Section 3 above.

The first approach was described in [**7**]. For simplicity we suppose that the group of holomorphic automorphisms of V is trivial and that $H^1(V) = 0$. We consider the space $\mathcal{J}$ of almost complex structures on V compatible with the symplectic form ω_0. This is the space of sections of a fibre bundle over V with fibre the complex homogeneous space $Sp(2n, \mathbf{R})/U(n)$, and $\mathcal{J}$ inherits a complex structure from that of the fibre. The symplectomorphism group $\mathcal{G}_0$ acts naturally on $\mathcal{J}$, preserving the complex structure. If we have a diffeomorphism $f : V \to V$ and an $\omega_\phi \in \mathcal{H}_0$ with $f^*(\omega_\phi) = \omega_0$, then the pull-back $f^*(I)$ defines a point in $\mathcal{J}$ and this gives an embedding $\lambda : \mathcal{X}_0 \to \mathcal{J}$ which one readily sees, as in [**7**], has the properties envisaged in the abstract picture above.

For the second approach we can drop the assumption on the holomorphic automorphisms. We consider the space $\mathrm{Maps}(V, V)$, which has a complex structure induced from the complex structure on V (a tangent vector to $\mathrm{Maps}(V, V)$ at a map $g : V \to V$ is a section of the complex vector bundle $g^*(TV)$). Now we define a map $\mu : \mathcal{X}_0 \to \mathrm{Maps}(V, V)$, sending a pair (f, ϕ) to f. This is an embedding, and the image $\mu(\mathcal{X}_0)$ consists of the difffeomorphisms f such that $(f^{-1})^*(\omega_0)$ has type (1,1). Again one sees that this image is a complexification of $\mathcal{G}_0$, where now $\mathcal{G}_0$ acts on Maps(V, V) by composition on the right. Finally one shows that the structure on $\mathcal{X}_0$ defined by the two embeddings λ, μ is the same: this just comes down to the fact that the infinitesimal action of the vector fields on the integrable almost complex structures is given by the $\overline{\partial}$-operator of the tangent bundle of V, which is complex-linear.

In the case when the complex dimension n of V is 1 (and, under our current hypotheses, V must be the Riemann sphere) the condition that $(f^{-1})^*(\omega_0)$ has type $(1,1)$ is vacuous, so in this case $\mathcal{X}_0$ can be identified with the space Diff(V) of diffeomorphisms of V (or, more precisely, the identity component in this space). One needs to beware that the "infinitesimal group" structure on Diff(V) that we are considering here is not the same as the genuine group structure on this space. This corresponds to the fact that there is a non-standard Lie bracket $[\ ,\ ]_I$ on the space of vector fields on the sphere. We express any vector field ξ as $\xi_1 + I\xi_2$, where ξ_1, ξ_2 are Hamiltonian, and then set

$$[\xi_1 + I\xi_2, \eta_1 + I\eta_2]_I = ([\xi_1, \eta_1] - [\xi_2, \eta_2]) + I\left([\xi_1, \eta_2] + [\xi_2, \eta_1]\right).$$

§5. The geodesic, WZW, and Monge–Ampère equations

We will now study the geodesic equation (12) in $\mathcal{H}$ in more detail, and interpret the solutions geometrically. Suppose ϕ_t, $t \in [0,1]$, is a path in $\mathcal{H}$. We can view this as a function on $V \times [0,1]$ and in turn as a function on $V \times [0,1] \times S^1$, with trivial dependence on the S^1 factor; that is, we define

$$\Phi(v, t, e^{is}) = \phi_t(v).$$

We regard the cylinder $A = [0,1] \times S^1$ as a Riemann surface with boundary in the standard way—so $t + is$ is a local complex coordinate. Let Ω_0 be the pull-back of ω_0 to $V \times A$ under the projection map, and put $\Omega_\Phi = \Omega_0 + i\overline{\partial}\partial\Phi$, a $(1,1)$-form on $V \times A$. Then we have

PROPOSITION 3. *The path ϕ_t satisfies the geodesic equation (12) if and only if $\Omega_\Phi^{n+1} = 0$ on $V \times A$.*

The proof is left as an exercise for the reader (or use Proposition 4 below).

We now turn to variational problems involving 2-dimensional domains. If R is a compact Riemann surface with boundary, one may consider the *harmonic* maps from R to a finite-dimensional symmetric space $H = G^c/G$. These are the critical points of the energy functional $E(f) = \frac{1}{2}\int_R |\nabla f|^2$, with Euler–Lagrange equation $d^*(Df) = 0$. But there is also a deformation of the harmonic map equation, which makes use of the special geometry of the target space. The tangent space of H at each point is modelled on the Lie algebra of G, so there is a natural Lie bracket $TH \times TH \to TH$ which yields a covariant constant 3-form θ on H. Given a fixed boundary map $\sigma : \partial R \to H$, we can define a functional E_σ^{WZW} on the space of maps from R with boundary value σ as follows. Choose a reference map $f_0 : R \to H$ with boundary value σ, and set

$$E_\sigma^{WZW}(f) = \tfrac{1}{2}\int_R |\nabla f|^2 + \int_Z \theta, \tag{27}$$

where Z is any 3-chain in H with boundary $f(R) - f_0(R)$. This is independent of the choice of Z, since H is simply connected and θ is closed, and the functional depends on f_0 only up to a constant. The Euler–Lagrange equation is, in a local complex coordinate $t + is$ on R,

$$d^*(Df) + [f_t, f_s] = 0. \tag{28}$$

We call this equation the WZW equation because similar ideas appear in the Wess–Zumino–Witten theory in mathematical physics.

Clearly we can take this formal set-up over to the case of maps from R to $\mathcal{H}$, which we can interpret as functions Φ on $V \times R$, such that $\Omega_\Phi = \Omega_0 + i\bar{\partial}\partial\Phi$ is positive on the slices $V \times \{z\}$, $z \in R$.

PROPOSITION 4. *A map from R to $\mathcal{H}$ satisfies the WZW equation if and only if $\Omega_\Phi^{n+1} = 0$.*

To see this we first compute the harmonic map "tension field" $d^*(DF)$. We work in a local complex coordinate $\tau = t + is$ on R. The usual calculation of the Euler–Lagrange equation, just as for the geodesic problem in Section 2, shows that the tension field is

$$\Phi_{ss} + \Phi_{tt} + \tfrac{1}{2}(|\nabla\Phi_s|_\phi^2 + |\nabla\Phi_t|_\phi^2),$$

where ∇ denotes differentiation in the V variable, and $|\ |_\phi$ is the metric on V defined by Ω_Φ for fixed s, t. So the WZW equation is

$$\Phi_{ss} + \Phi_{tt} + \tfrac{1}{2}(|\nabla\Phi_s|_\phi^2 + |\nabla\Phi_t|_\phi^2) + (\nabla\Phi_s, I\nabla\Phi_t)_\phi = 0. \tag{29}$$

It is now just a matter of algebra to show that this equation is precisely $\Omega_\Phi^{n+1} = 0$. We can work at a given point in $V \times R$ and choose local complex coordinates $z_\alpha = x_\alpha + iy_\alpha$ on V such that the metric ω_ϕ is standard at that point, in these coordinates. Then, at this point,

$$\Omega_\Phi^{n+1} \propto 2^{n-1} \det \begin{pmatrix} \Phi_{\tau\bar{\tau}} & \cdot & \Phi_{\tau\bar{z}_\alpha} & \cdot & \cdot \\ \cdot & \frac{1}{2} & 0 & \cdot & 0 \\ \cdot & \cdot & \frac{1}{2} & \cdot & 0 \\ \Phi_{\bar{\tau}z_\alpha} & 0 & \cdot & \frac{1}{2} & \cdot \\ \cdot & 0 & \cdot & \cdot & \frac{1}{2} \end{pmatrix} = \Phi_{\tau\bar{\tau}} - 2\sum_\alpha |\Phi_{\bar{\tau}z_\alpha}|^2.$$

This reduces to the left hand side of (29) when we use the identity

$$\begin{aligned} 8\sum_\alpha |\Phi_{\bar{\tau}z_\alpha}|^2 &= \frac{1}{2}\sum_\alpha (\Phi_{sx_\alpha} + \Phi_{ty_\alpha})^2 + (\Phi_{tx_\alpha} - \Phi_{sy_\alpha})^2 \\ &= \tfrac{1}{2}(|\nabla\Phi_s|^2 + |\nabla\Phi_t|^2) + (\nabla\Phi_s, I\nabla\Phi_t), \end{aligned}$$

and write $\Phi_{\tau\bar{\tau}} = \frac{1}{4}(\Phi_{ss} + \Phi_{tt})$.

The analogue of the WZW functional E^{WZW} for maps to $\mathcal{H}$ is a variant of the functional I introduced in Section 3. Given boundary data $\rho : \partial R \to \mathcal{H}$, we consider the set of functions Φ on $V \times R$ which agree with ρ on the boundary. Then we define the variation of I on this set by

$$\delta I_\rho = \frac{1}{(n+1)!}\int_{V \times R} \delta\Phi\ \Omega_\Phi^{n+1}, \tag{30}$$

where the variation $\delta\Phi$ vanishes on the boundary by hypothesis. This boundary condition means that the same argument as in Section 3 works to show that the formula defines a functional I_ρ, which again has a more explicit expression like (25). This functional I_ρ reduces to the energy functional on paths, by an integration by parts that we leave to the reader, in the case when R is the cylinder and we restrict to S^1-invariant data.

The equation $\Omega_\Phi^{n+1} = 0$ is a degenerate Monge–Ampère equation. A solution gives rise to a rather concrete geometric structure on $V \times R$, as is well-known in the literature in this area [**13**]. If $\Omega_\Phi^{n+1} = 0$ and Ω_Φ is strictly positive on the V-slices, then the null space $N \subset T(V \times R)$ of Ω_Φ at each point has real dimension 2. On the one hand this field of subspaces forms a foliation, since Ω_Φ is closed; on the other hand the subspaces are complex lines, since Ω_Φ has type $(1,1)$. So a solution of the WZW equations gives rise to a foliation of $V \times R$ whose leaves are complex curves transverse to the V-slices. This conclusion fits in tidily with a corresponding result in the finite-dimensional case. Suppose first that R is simply connected. Then a map $f : R \to G^c/G$ is a solution of the WZW equation if and only if it has a lift to a *holomorphic* map from R to the complex Lie group G^c. In the Kähler geometry case a foliation of $V \times R$, transverse to the V-slices, can be viewed as a map from R to $\mathrm{Diff}(V)$, and the condition that the leaves are complex curves is precisely saying that this map is holomorphic, with respect to the complex structure on $\mathrm{Diff}(V) \subset \mathrm{Maps}(V, V)$ considered in Section 4. The condition that Ω_Φ has type (1,1) then tells us that the image of the map lies in the subspace $\mathcal{Y} \subset \mathrm{Diff}(V)$, which is our analogue of the complex group G^c. In the case when R is not simply connected we can carry out this analysis on the universal cover; a solution of the finite-dimensional WZW equation on R gives an equivariant holomorphic map from $\tilde{R}$ to G^c, with respect to a representation $\pi_1(R) \to G$, which in turn can be viewed as a flat G-connection over R. In the Kähler geometry case we get a corresponding statement, where the representation $\pi_1(R) \to \mathrm{SDiff}(V)$ is the monodromy of the foliation. (One can also obtain an interpretation of the WZW equations in terms of holomorphic maps into $Sp(2n, \mathbf{R})/U(n)$, using the other point of view in Section 4.)

Let us now spell out what this means in the case of a geodesic ϕ_t, $t \in [0,1]$, in $\mathcal{H}$, starting at $\phi = 0$, say. We regard the time derivative $\dot{\phi}_0$ as a Hamiltonian on the symplectic manifold (V, ω_0), so it defines a Hamiltonian flow $g_s : V \to V$ in the usual way. For each point $v \in V$ let $\gamma_v : \mathbf{R} \to V$ be its trajectory under the flow: $\gamma_v(s) = g_s(v)$. Then the projection of the appropriate leaf of the foliation gives a *holomorphic* extension $\Gamma_v : [0,1] \times \mathbf{R} \to V$, with $\Gamma_v(0, s) = \gamma_v(s)$, such that the restriction of the Γ_v to $\{t\} \times \mathbf{R}$ is a trajectory of the Hamiltonian flow of $\dot{\phi}_t$ on the symplectic manifold (V, ω_{ϕ_t}). Moreover the map $f_t : V \to V$ defined by $f_t(v) = \Gamma_v(0, t)$ takes the Hamiltonian system $(V, \omega_0, \dot{\phi}_0)$ to the system $(V, \omega_{\phi_t}, \dot{\phi}_t)$. So we can think of the study of geodesics in $\mathcal{H}$ as a kind of "analytic continuation" of Hamiltonian dynamics, in which the time parameter is made complex, and the complex curves Γ_v give holomorphic cobordisms between the trajectories of the family of Hamiltonian systems.

§6. Examples of geodesics

(i) Suppose that a compact, connected, Lie group G acts on V preserving the complex structure and a Kähler metric ω_0. Then the complexification G^c acts holomorphically on V, and we get a map from the finite-dimensional space G^c/G to $\mathcal{H}_0$, taking the coset of $\alpha \in G$ to $\alpha^*(\omega_0)$. There is no loss in supposing that the action of G is infinitesimally effective, so the L^2 norm on Hamiltonians defines an invariant metric on G, and hence a metric on G^c/G, making it a symmetric space.

(Of course for a simple group this structure is unique up to scale.) Then G^c/G is isometrically embedded as a totally geodesic submanifold in $\mathcal{H}_0$, and any geodesic in G^c/G gives a geodesic in $\mathcal{H}$. We leave the simple verification to the reader.

(ii) Let h be a function on the standard sphere $S^2 \subset \mathbf{R}^3$ with $h(x,y,z) = z$ near to the poles $p_\pm = (0,0,\pm 1)$, and with no critical points apart from these poles. Fix an identification of the tangent space to S^2 at p_- with $\mathbf{C}$. For each $r \in (-1,1)$ the set $U_r = h^{-1}(r,1]$ is, differentiably, a disc with smooth boundary in S^2. By the Riemann mapping theorem there is a unique conformal equivalence $\alpha_r : D \to U_r$ from the unit disc $D \subset \mathbf{C}$, normalised so that $\alpha_r(0) = p_-$ and so that the derivative of α_r at 0 is real and positive with respect to the identification above. The map α_r extends smoothly to $\overline{\alpha}_r : \overline{D} \to \overline{U}_r$. In particular, rotation of the disc defines an action of the circle on the curve $\partial U_r = h^{-1}(r) \subset S^2$. When r is close to ± 1 this is just the restriction of the usual rotation of the sphere about the z-axis, since $h = z$ near $p_\pm$. It is easy to see that there is an area form ω_0, unique up to scale, such that the circle action on each curve $h^{-1}(r)$ is the Hamiltonian flow of the function h with respect to ω_0.

Now let $\overline{\beta}_r$ be the holomorphic map from the cylinder $[0,\infty) \times S^1$ to S^2 defined by composing $\overline{\alpha}_r$ with the standard identification $[0,\infty) \times S^1 \to \overline{D} \setminus \{0\}$. For each point $v \in S^2$ there is, by construction, a unique $e^{is(v)} \in S^1$ such that $\overline{\beta}_{h(v)}(0, e^{is(v)}) = v$. For $t > 0$, define a smooth map $f_t : S^2 \to S^2$ by

$$f_t(v) = \overline{\beta}_{h(v)}(t, e^{is(v)}).$$

The Schwartz lemma implies that f_t is a bijection, and one may also show that the derivative of f is everywhere invertible, so there is a smooth inverse f_t^{-1}. Now we set

$$\omega_t = (f_t^{-1})^*(\omega_0)$$

and write $\omega_t = \omega_0 + i\bar{\partial}\partial\phi_t$ for normalised potentials ϕ_t. Then ϕ_t is an infinite geodesic ray in $\mathcal{H}_0(S^2)$, starting at ω_0 and with initial tangent vector $\dot{\phi} = h$ at $t = 0$. (It seems very likely that the same construction works for any function h on S^2 with two non-degenerate critical points.)

One can argue similarly if k is another function on S^2 with critical points only at $p_\pm$, and with $h > k$ on $S^2 \setminus \{p_+, p_-\}$. Suppose for simplicity that

$$k(x,y,z) = \begin{cases} -1 + \rho(z+1) & \text{near } p_-, \\ 1 + \sigma(z-1) & \text{near } p_+, \end{cases}$$

with $0 < \rho < 1 < \sigma$. Then one can use the conformal equivalences between the annular regions

$$A_r = \{v \in S^2 : k(v) < r < h(v)\},$$

and standard annuli $\{w \in \mathbf{C} : 1 < |w| < R\}$, for suitable $R(r)$, to construct symplectic forms ω, ω' on S^2 and a geodesic from ω to ω' in $\mathcal{H}_0$ with tangent vector h at ω and k at ω'. Here the Hamiltonian system (S^2, ω, h) (which is isomorphic to (S^2, ω', k)) will not, in general, be periodic, since the modulus R of the annulus varies with r.

(iii) Consider the complex n-torus $T = \mathbf{C}^n/\mathbf{Z}n + i\mathbf{Z}n$, and let ω_0 be the standard flat metric. The group $\Gamma = (S^1)^n$ acts on T^n via translations in the Lagrangian subspace $i\mathbf{R}^n \subset \mathbf{C}^n$, and this induces an isometric action of Γ on the space $\mathcal{H}_0$ of Kahler metics on T; so the set $\mathcal{H}_0^\Gamma$ of invariant metrics is totally geodesic in $\mathcal{H}_0$. The Poisson bracket of Γ-invariant functions vanishes, and it follows that the induced metric on $\mathcal{H}_0^\Gamma$ is *flat*, in the sense that its Riemann curvature vanishes. One might expect (though this is probably not automatically true in infinite dimensions) that $\mathcal{H}_0^\Gamma$ is isometric to an open subset in a pre-Hilbert space, and we will now verify that this is the case. For simplicity we will take $n = 1$ and write evrything out explicitly—in terms of our general picture the point is that the group of exact symplectomorphisms of T which commute with Γ is an abelian group—just the space of maps from $\mathbf{R}^n/\mathbf{Z}^n$ to $(S^1)^n$—and it has a *bona fide* complexification $\mathrm{Maps}(\mathbf{R}^n/\mathbf{Z}^n, (\mathbf{C}^*)^n)$. The analysis below arises from writing out the equivalence between $\mathcal{H}_0^\Gamma$ and

$$\frac{\mathrm{Maps}(\mathbf{R}^n/\mathbf{Z}^n, (\mathbf{C}^*)^n)}{\mathrm{Maps}(\mathbf{R}^n/\mathbf{Z}^n, (S^1)^n)}$$

which we obtain from the set-up in Section 4.

Elements of $\mathcal{H}_0^\Gamma$ (when $n = 1$) can be defined by circle-invariant Kähler potentials. Explicitly, $\mathcal{H}_0^\Gamma$ can be viewed as the set of functions $\phi : \mathbf{R}/\mathbf{Z} \to \mathbf{R}$ with $1 - \phi_{xx} > 0$, normalised so that

$$\int_{\mathbf{R}/\mathbf{Z}} \phi + \tfrac{1}{2}\phi_x^2 \, dx = 0,$$

and with metric

$$g_{\mathcal{H}_0^\Gamma}(\delta\phi) = \int_{\mathbf{R}/\mathbf{Z}} (\delta\phi)^2 (1 - \phi_{xx}) dx.$$

Let U be the set of functions ψ on $\mathbf{R}/\mathbf{Z}$ with $1 + \psi_{\xi\xi} > 0$, normalised so that

$$\int_{\mathbf{R}/bZ} \psi d\xi = 0$$

and endowed with the flat metric

$$g_U(\delta\psi) = \int_{\mathbf{R}/\mathbf{Z}} (\delta\psi)^2 d\xi.$$

(It is convenient to use a different symbol, ξ, for the coordinate on the circle in the definition of U.) We will exhibit an isometry

$$\alpha : (\mathcal{H}_0^\Gamma, g_{\mathcal{H}_0^\Gamma}) \to (U, g_U).$$

For $\phi \in \mathcal{H}_0^\Gamma$ define $\xi = \xi(x)$ by

$$\xi(x) = x - \phi_x(x) \in \mathbf{R}/\mathbf{Z}.$$

Then $\frac{d\xi}{dx} = 1 - \phi_{xx} > 0$, so the map $\xi(x)$ is a diffeomorphism of the circle (it is a covering map of degree 1), with inverse $x = x(\xi)$.

Now let

$$A(\xi) = \left(\frac{\phi_{xx}}{1 - \phi_{xx}}\right)_{x(\xi)}.$$

Then

$$\int_{\mathbf{R}/\mathbf{Z}} A(\xi) d\xi = \int_{\mathbf{R}/\mathbf{Z}} \left(\frac{\phi_{xx}}{1 - \phi_{xx}}\right)(1 - \phi_{xx}) dx = 0,$$

so there is a unique function $\psi(\xi)$ with $\int \psi d\xi = 0$ and $\psi_{\xi\xi} = A(\xi)$. Then

$$1 + \psi_{\xi\xi} = \frac{1}{1 - \phi_{xx}} > 0,$$

so ψ is in U. We define $\alpha(\phi) = \psi$. Now suppose we make a small variation $\tilde{\phi} = \phi + \eta$, so $\tilde{\xi}(x) = \xi(x) - \eta_x$. It follows that

$$\tilde{x}(\xi) = x(\xi) + \left(\frac{\eta_x}{1-\phi_{xx}}\right)_{x(\xi)} + O(\eta^2)$$

and

$$\tilde{A}(\xi) = A(\xi) + \frac{\eta_x \phi_{xxx}}{(1-\phi_{xx})^3} + \frac{\eta_{xx}}{(1-\phi_{xx})^2} + O(\eta^2).$$

Now set $\tau(\xi) = \eta(x(\xi))$. The chain rule gives

$$\tau_{\xi\xi} = \frac{1}{1-\phi_{xx}} \frac{d}{dx}\left(\frac{\eta_x}{1-\phi_{xx}}\right) = \frac{\eta_x \phi_{xxx}}{(1-\phi_{xx})^3} + \frac{\eta_{xx}}{(1-\phi_{xx})^2}.$$

So $\tilde{A}(\xi) = A(\xi) + \tau_{\xi\xi} + O(\eta^2)$. Furthermore,

$$\int_{\mathbf{RZ}} \tau(\xi) d\xi = \int_{\mathbf{R}/\mathbf{Z}} \eta \, (1-\phi_{xx}) dx = O(\eta^2),$$

since ϕ and $\tilde{\phi}$ are normalised. In sum, the derivative of α maps η to τ and

$$\int \tau^2 d\xi = \int \eta^2 \frac{d\xi}{dx} dx = \int \eta^2 (1-\phi_{xx}) dx,$$

so α is an isometry. Finally we can write down the inverse map $\alpha^{-1} : U \to \mathcal{H}_0^\Gamma$. Given $\psi \in U$, we set $x(\xi) = \xi + \psi_\xi$ and $B(x) = \left(\frac{\psi_{xi\xi}}{1+\psi_{\xi\xi}}\right)_{x(\xi)}$. Then $\int B(x) dx = \int \psi_{\xi\xi} d\xi = 0$, so $B(x) = \phi_{xx}$ for a unique normalised ϕ, and

$$1 - \phi_{xx} = \frac{1}{1+\psi_{\xi\xi}} > 0.$$

Thus ψ lies in $\mathcal{H}_0^\Gamma$. It is easy to check that this construction is inverse to the previous one, hence completing the proof.

Notice that this transformation takes the geodesic equation in the invariant case, namely

$$\phi_{tt} = -\tfrac{1}{2}\phi_{xt}^2/(1-\phi_{xx}), \tag{31}$$

to the trivial equation $\psi_{tt} = 0$ in U; that is, the equation (31) (and more generally the *real* homogeneous Monge–Ampère equation in any dimension) can be solved explicitly, a fact which is perhaps not obvious.

§7. Existence and uniqueness questions

A Riemannian symmetric space G^c/G, like any complete simply connected Riemannian manifold of non-positive curvature, has the property that any two points can be joined by a unique geodesic. Likewise, if R is a compact Riemann surface-with-boundary and ρ is a smooth map from ∂R to G^c/G, then there is a unique solution of the WZW equation on R with boundary value ρ. (This is proved in [**8**] for the case when G is a unitary group, and the argument extends easily to the general case. When R is the disc the result is equivalent to a factorisation theorem for loop groups, and the solution can be given explicitly by following the argument in [**19**].) Moreover one can show that this solution gives the absolute minimum of the WZW functional E^{WZW}.

The geodesic result can be seen as a particular case of the 2-dimensional theory, by taking circle-invariant maps of the cylinder. Following through the analogy developed in this paper, and expressing things in terms of the Monge–Ampère equation, we are led to the following:

CONJECTURE/QUESTION 5. *Let R be a compact Riemann surface with boundary and $\rho : V \times \partial R \to \mathbf{R}$ a function such that $\omega_0 + i\bar{\partial}\partial\rho$ is a strictly positive $(1,1)$ form on each slice $V \times \{z\}$ for each fixed $z \in \partial R$. Let $\mathcal{S}_\rho$ be the set of functions Φ on $V \times R$ equal to ρ over the boundary and such that $\omega_0 + i\bar{\partial}\partial\Phi$ is strictly positive on every slice $V \times \{w\}, w \in R$. Then there is a unique solution of the Monge–Ampère equation $(\Omega_0 + i\bar{\partial}\partial\Phi)^{n+1} = 0$ in $\mathcal{S}_\rho$, and this solution realises the absolute minimum of the functional I_ρ.*

This question is a version of the Dirichlet problem for the degenerate Monge–Ampère equation, a topic around which there is a substantial literature; see [**1**], [**13**] for example. Note that regularity questions are very important in this theory, since the equation is not elliptic: it may be that it is better to work in some larger class of functions. But the statement above should indicate the general direction of the problem.

An affirmative answer to Question 5 would of course imply the corresponding statements for geodesics. In finite dimensions the other familiar existence result for geodesics is the initial value problem, that is, the definition of the exponential map. It is easy to see that the analogue of these results *fails* for $\mathcal{H}$. This is clear from the point of view of Section 5, since we can construct a Hamiltonian trajectory γ_v which does not extend to a holomorphic strip, even for a short "time" interval. In the case of the S^1 invariant metrics considered in Section 6 (iii) we can solve the initial value proble for a short time, since U is an open set, but not for all time, since U is a proper subset of a pre-Hilbert space. The lack of solutions to the initial value problem in $\mathcal{H}$ can be contrasted with the "dual" space $\mathcal{G}$, where the exponential map *can* be defined (*viz.* Hamiltonian dynamics), but points in $\mathcal{G}$ *cannot* in general be joined by a geodesic (there are exact symplectomorphisms arbitrarily close to the identity which are not obtained by Hamiltonian flows). The existence conjecture (5), if true, would thus lead to a pleasing symmetry between the two cases.

The only part of (5) which seems to be easily dealt with is the uniqueness. This can be proved by a maximum principle argument which we give now, although it is a variant of a standard argument in the literature on Monge–Ampère equations.

LEMMA 6. *Suppose $\tilde{\Omega}$ is a $(1,1)$ form on $V \times R$ such that $\tilde{\Omega} \geq 0$ on $V \times R$ and $\tilde{\Omega} > 0$ on each V-slice in $V \times R$. Suppose that $f : V \times R \to \mathbf{R}$ is a smooth function such that $(\tilde{\Omega} + i\bar{\partial}\partial f)^{n+1} = 0$ in $V \times R$. Then the maximum value of f is attained on the boundary.*

To prove this lemma, choose a function $\gamma : R \to \mathbf{R}$ with $i\bar{\partial}\partial\gamma > 0$ on R. Then for all $\epsilon > 0$ the form $\tilde{\Omega}_\epsilon = \tilde{\Omega} + \epsilon i\bar{\partial}\partial\gamma$ (where γ is regarded as a function on $V \times R$) is strictly positive on $V \times R$. Let $f_\epsilon = f - \epsilon\gamma$, so $(\tilde{\Omega}_\epsilon + i\bar{\partial}\partial f_\epsilon)^{n+1} = 0$. At an interior maximum of f_ϵ we have $i\bar{\partial}\partial f_\epsilon \geq 0$, and this would give a contradiction, since we would have $\tilde{\Omega}_\epsilon + i\bar{\partial}\partial f_\epsilon > 0$; so the maximum value of f_ϵ is attained on the boundary. Now take the limit as ϵ tends to 0 to deduce, just as in the proof of the ordinary maximum principle, that the maximum value of f is attained on the boundary.

COROLLARY 7. *If one solution Φ to the boundary value problem in Conjecture/Question 5 exists, it is unique.*

For, if Φ' were another solution, we could take $\tilde{\Omega} = \Omega_0 + i\overline{\partial}\partial\Phi$ and $f = \Phi' - \Phi$ in Lemma 6 and deduce that $f \leq 0$. Changing the roles of Φ and Φ', we deduce that $f \geq 0$, so we must have $f = 0$.

In a similar way we get another extremal characterisation of solutions, again following standard lines, cf. [**13**]. Given boundary data ρ, let $\mathcal{S}^+(\rho)$ be the set of functions Ψ on $V \times R$ such that $\Psi = \rho$ on the boundary and $\Omega_0 + i\overline{\partial}\partial\Psi > 0$ in $V \times R$. This set is not empty, for we can first produce a function Ψ_0 equal to ρ on the boundary with $\Omega_0 + i\overline{\partial}\partial\Psi_0$ strictly positive on V-slices, using the fact that $\mathcal{H}$ is contractible, and then find Ψ in $\mathcal{S}^+(\rho)$ in the form $\Psi = \Psi_0 + K\gamma$, where γ is a function on R as in the proof of Lemma 6, but chosen so that in adddition $\gamma = 0$ on ∂R, and the parameter K is made large.

PROPOSITION 8. (i) *Given ρ as in Conjecture/Question 5 and any $\Psi \in \mathcal{S}^+(\rho)$, the minimum value of Ψ is attained on the boundary. In particular, for any $p \in V \times R$,*

$$\inf_{\Psi \in \mathcal{S}^+(\rho)} \Psi(p) > -\infty.$$

(ii) *If a solution $\Phi \in \mathcal{S}(\rho)$ of the boundary value problem in Conjecture/Question 5 exists, then, for all $p \in V \times R$,*

$$\Phi(p) = \inf_{\Psi \in \mathcal{S}^+(\rho)} \Psi(p).$$

To prove (i) we just use the fact that $i\overline{\partial}\partial\Psi$ is strictly positive in the R-slices to see that Ψ has no interior minimum. Part (ii) follows immediately from Lemma 6, together with fact that, if Φ is a solution and we choose a function γ as above, vanishing on the boundary, then $\Phi + \epsilon\gamma$ lies in $\mathcal{S}^+_\rho$, for any $\epsilon > 0$.

Of course we could try to view the infimum appearing in Proposition 8 as *defining* a "generalised solution" of the Dirichlet problem.

The discussion above focuses on the solutions to the boundary value problem. On the other hand, leaving aside the questions of the existence of these extrema, it is interesting to study the infimum of the functional I_ρ. This is particularly so in the case of the geodesic problem, when the functional can be rewritten as the "energy" of a path. If these infima are strictly positive, for all choices of fixed, distinct, end points, they make $\mathcal{H}$ into a metric space, in the usual fashion. In this connection we will now observe that *assuming* the existence of minimising geodesics we can write down an explicit lower bound for the functional. To do this we observe that for any $\phi \in \mathcal{H}$

$$\int_V \phi d\mu_0 \leq I(\phi) \leq \int_V \phi d\mu_\phi.$$

These inequalities just express the *convexity* of the functional I; to derive them, differentiate along the path $t(\phi)$. It follows that if ϕ is normalised (i.e. $I(\phi) = 0$) and not identically zero, then it must take both strictly positive and negative values. Now suppose that $\phi_t, t \in [0,1]$, is a geodesic from 0 to ϕ, where ϕ is normalised. Then the length (or energy) of the geodesic is given by

$$L = \int_V \dot{\phi}^2 d\mu_{\phi_t},$$

for any $t \in [0,1]$. In particular, taking $t = 0$,

$$\sqrt{L} \geq M^{-1/2} \int_V |\dot\phi_0| d\mu_0 > M^{-1/2} \int_{\dot\phi_0 > 0} \dot\phi_0 d\mu_0,$$

where M is the volume of V (which is of course the same for all metrics in $\mathcal{H}$). Now the geodesic equation obviously gives $\ddot\phi \leq 0$, so $\phi \leq \dot\phi_0$. It follows that

$$\int_{\dot\phi_0 > 0} \dot\phi d\mu_0 \geq \int_{\phi > 0} \phi d\mu_0,$$

where the last term is strictly positive by the remarks above, and depends only on ϕ and not on the geodesic. A similar argument gives

$$\sqrt{L} > -M^{-1/2} \int_{\phi < 0} \phi d\mu_\phi.$$

So we are lead to a conjecture whose statement does not involve the existence of geodesics:

CONJECTURE/QUESTION 9. *If $\phi \in \mathcal{H}_0$ is normalised and $\tilde\phi_t$, $t \in [0,1]$, is any path from 0 to ϕ in $\mathcal{H}$, then*

$$\int_0^1 \int_V \left(\frac{d\tilde\phi}{dt}\right)^2 d\mu_{\tilde\phi_t} dt \geq M^{-1} \left(\max\left(\int_{\phi>0} \phi d\mu_0, -\int_{\phi<0} \phi d\mu_\phi\right)\right)^2.$$

The restriction to normalised potentials ϕ is not important, since we know that $\mathcal{H}$ splits as a product, and we could immediately write down a corresponding inequality, involving $I(\phi)$, for any $\phi \in \mathcal{H}$.

§8. Extremal Kähler metrics

In this section we will explain how the material we have developed in this paper is related to certain well-known topics in Kähler geometry. Consider first the question, posed by Calabi [**4**] and solved by Yau [**22**], of finding a Kähler metric in $\mathcal{H}_0$ with a prescribed volume form. A volume form gives a linear functional $\lambda : C^\infty(V) \to \mathbf{R}$, and the space $\mathcal{H}$ of Kähler potentials is an open subset of $C^\infty(V)$. By the definition of the functional I, finding a solution to the problem is the same as finding a $\phi \in \mathcal{H}_0$ such that the derivative of I at ϕ is λ; that is, we have to minimise the the linear functional λ over the convex set $\{\phi \in \mathcal{H} : I(\phi) \leq 0\}$. This variational formulation is well known; the point we want to bring out is that the prescribed-volume problem essentially involves the *affine* geometry of $\mathcal{H}$. (The problem discussed in the previous section is a variant of this prescribed-volume problem, in that we have fixed boundary values and the prescribed "volume form" is 0.)

There is another problem, again going back to Calabi [**5**], which is tightly bound up with the geometry of $\mathcal{H}$ as a symmetric space which we have discussed in this paper. This is the problem of finding an "extremal Kähler metric" in $\mathcal{H}$, which, by definition, means a metric of *constant scalar curvature*. This problem includes the renowned question of the existence of Kähler–Einstein metrics, in the case when $c_1(V)$ is a multiple of $[\omega_0]$. The link with the symmetric-space geometry of $\mathcal{H}$ arises from the fact that the scalar curvature can be viewed as a *moment map* for the action of $\mathcal{G}$ on the symplectic manifold $\mathcal{J}$. This means that, save for the absence of a genuine complexified group $\mathcal{G}^c$, the search for critical Kahler metrics can be fitted into a general pattern of problems involving "Kahler quotients", see [**9,** Chapter 6] or [**12**], for example. This is explained in [**7**], so we will not reproduce the discussion here. The main point is that one could hope that this approach may shed light on Yau's conjecture [**23**] relating the existence of critical Kähler metrics to stability in the sense of Hilbert schemes and geometric invariant theory. (In [**7**] this was referred to a Tian's conjecture, but Tian has kindly pointed out to the author that it is due to Yau. The original conjecture is stated only for the Kähler–Einstein case, but it seems natural to extend it to general critical metrics. Such an extension has been suggested by LeBrun, particularly in connection with ruled surfaces, where Burns and de Bartolomeis [**3**] and LeBrun [**14**] have shown that the existence of critical metrics is tied up with the stability of the corresponding vector bundles.)

To bring this down to earth, consider the question of the uniqueness of critical metrics. For any metric ϕ in $\mathcal{H}$ let $\mathcal{D}_\phi$ be the second-order differential operator from functions on V to sections of $s^2_{\mathbf{C}}(TV)$, given by

$$\mathcal{D}_\phi f = \overline{\partial}_{TV}((df)^\sharp),$$

where $(df)^\sharp$ is the vector field dual to the 1-form df under the metric ω_ϕ and $\overline{\partial}_{TV}$ is the $\overline{\partial}$-operator on the tangent bundle of V. (This operator is the "complex Hessian": in flat space it is given in local coordinates by $\mathcal{D}(f) = \frac{\partial^2 f}{\partial z_\alpha \partial z_\beta}$; in general $\mathcal{D}_\phi f$ is the component of the Riemannian Hessian $\nabla\nabla f$ complementary to $\overline{\partial}\partial f$.) Write $S(\phi)$ for the scalar curvature of a metric ω_ϕ.

PROPOSITION 10. *If ϕ_t is a geodesic in $\mathcal{H}$, then*

$$\frac{d}{dt}\int_V \dot{\phi} S(\phi_t) d\mu_{\phi_t} = \int_V |\mathcal{D}_{\phi_t}\dot{\phi}|^2 d\mu_{\phi_t}.$$

This equation can been seen most easily in the $\mathcal{J}$-description, with a fixed symplectic form and varying complex structure. Then we write $\mathcal{D}_J$ for the operator above formed using a complex structure J. The reason these operators are important is that they define the infinitesimal action of the symplectomorphism group on $\mathcal{J}$ (or, more precisely, on the subset of integrable structures, see the discussion in [**7**]). A geodesic in $\mathcal{H}$ goes over to a path $J(t)$ such that

$$\frac{dJ(t)}{dt} = I\mathcal{D}_{J(t)}H,$$

for some *fixed* function H, independent of t. So

$$\frac{d}{dt}\int_V HS d\mu = \int_V H\frac{dS}{dt} d\mu.$$

Now the moment map identity established in [**7**] states that for any variation δJ in $\mathcal{J}$, and any function h,

$$\int_V (\delta J, I\mathcal{D}_J h) d\mu = \int_V h\delta S \, d\mu,$$

so

$$\int_V H\frac{dS}{dt} d\mu = \int_V (I\mathcal{D}_J H, I\mathcal{D}_J H) d\mu = \|\mathcal{D}_J H\|^2,$$

and transforming back to the other description, with a fixed complex structure, this gives the equation stated in Proposition 10.

COROLLARY 11. *If ϕ_0, ϕ_1 are two critical Kähler metrics in $\mathcal{H}_0$ which can be joined by a geodesic, then there is a holomorphic automorphism α of V such that $\alpha^*(\omega_{\phi_0}) = \omega_{\phi_1}$.*

This follows from Proposition 10 because if ϕ_t is the geodesic the derivative of $\int_V S\dot{\phi} d\mu$ vanishes at the endpoints by hypothesis, and so we must have $\mathcal{D}_{\phi_t}\dot{\phi} = 0$ for all t. But this means that $\phi_t = \alpha_t^* \phi_0$, where $\alpha_t = \exp(t\xi)$ for a holomorphic vector field ξ on V, i.e. we are in the situation considered in example (i) of Section 6. Another way of expressing this corollary is that if the geodesic existence conjecture (5) is true then we can deduce the essential uniqueness of critical Kähler metrics. In the Kähler–Einstein case this uniqueness has been proved by Bando and Mabuchi [**2**], using a different method, but the question for general critical metrics seems to be open.

The significance of Proposition 10 becomes clearer if we recall that Mabuchi [**15**] has shown that one can define a functional, the "K-energy", on $\mathcal{H}$ by specifying its first variation to be

$$\delta K = \int_V \delta\phi \; S(\phi) \;\; d\mu_\phi.$$

Thus the critical Kähler metrics are the critical points of K on $\mathcal{H}_0$. (This definition is in a similar vein to that of the functional I: one has to check that the second variation is symmetric in its two arguments to see that K is well-defined. It fits into the general pattern of Kähler quotient theory, in that there is a functional defined in a similar way, using the moment map, for any Kähler quotient.) Then Corollary 11 is the statement that K is *convex* along geodesics in $\mathcal{H}$, a fact which again fits into the general pattern, compare [**9**], Chapter 6, for example.

We give two other illustrations of the use of this convexity of the Mabuchi fuctional.

First, we consider certain modifications of the geodesic equation given by the motion of a "particle" in $\mathcal{H}$ moving in the potential $-\lambda K$, where $\lambda \geq 0$ is a real parameter. The equation of motion is

$$\ddot{\phi} = -\tfrac{1}{2}|\dot{\phi}|_\phi^2 + \lambda S(\phi),$$

where S is the scalar curvature. Along such a path we have

$$\ddot{J} = \|\mathcal{D}_\phi \dot{\phi}\|^2 + \lambda \|S\|^2 \geq 0,$$

so these paths would do equally well for proving the uniqueness (Corollary 11). It seems reasonable to hope that any two points can be joined by a path of this kind, and the equation for $\lambda > 0$ has the virtue of being *elliptic* on $V \times [0,1]$.

Second, this point of view is fruitful in understanding the Calabi equation [**5**]

$$\dot{\phi} = -S(\phi) + S_0,$$

where S_0 is the average of the scalar curvature, which is a topological invariant of $(V, [\omega])$. This is the gradient flow equation of the functional K on $\mathcal{H}_0$, and on the other hand it generates the gradient flow of the functional $\|S\|^2$ on $\mathcal{J}$: the formal structure is the same as for the Hermitian Yang–Mills gradient flow discussed in [**9**], for example. Calabi's inequality

$$\frac{d}{dt}\|S - S_0\|^2 = -\|\mathcal{D}S\|^2$$

follows immediately from the moment map identity. (Chruściel [**6**] has proved that when V is a Riemann surface the initial value problem for this Calabi equation has a solution for all positive time, which converges to the constant curvature metric as $t \to \infty$.)

Turning to the existence of extremal Kähler metrics: one of the cornerstones of the finite dimensional Kähler quotient theory, in its algebro-geometric formulation as geometric invariant theory, is the "Hilbert criterion". This states that the stability of a complex orbit can be detected by looking at complex one-parameter subgroups $\mathbf{C}^* \subset G^c$. If one takes these ideas over to the case of $\mathcal{J}$, we do not have a genuine complex group $\mathcal{G}^c$, but the geodesics in $\mathcal{H}$ furnish a substitute for the 1-parameter subgroups. In this way one can formulate a statement which is the analogue of the Hilbert criterion for stability (where stability is interpreted in terms of zeros of the moment map) in our infinite dimensional situation:

CONJECTURE/QUESTION 12. *The following are equivalent:*
(1) *There is no critical Kähler metric in* $\mathcal{H}_0$.
(2) *There is an infinite geodesic ray* $\phi_t, t \in [0, \infty)$, *in* $\mathcal{H}$, *such that*

$$\int_V S\dot{\phi} d\mu_\phi < 0$$

for all $t \in [0, \infty)$.
(3) *For any point* $\phi \in \mathcal{H}_0$ *there is a geodesic ray as in* (2) *starting at* ϕ.

Notice that, by Proposition 10, it is sufficient that the inequality in part (2) of this conjecture should hold for *large* t; that is, the condition has to do with the asymptotics of the geodesic ray. A further step in this programme would be to find conditions for the existence of these rays in terms of the complex geometry of V, perhaps making contact with other work, by Tian [**20**], [**21**], Nadel [**18**], and others, on obstructions to the existence of Kähler–Einstein metrics.

References

1. T. Aubin, *Nonlinear analysis on manifolds: Monge–Ampère equations*, Springer-Verlag, Berlin, 1984.
2. S. Bando and T. Mabuchi, *Uniqueness of Einstein Kähler metrics modulo connected group actions*, Algebraic Geometry, Sendai, 1985 (T. Oda, ed.), Adv. Studies Pure Math., vol. 10, North-Holland, Amsterdam, 1987, pp. 11–40.
3. D. Burns and P. de Bartolomeis, *Stability of vector bundles and extremal metrics*, Invent. Math. **92** (1988), 403–407.
4. E. Calabi, *The space of Kähler metrics*, Proc. Internat. Congr. Math. (Amsterdam, 1954), Vol 2, Noordhoff, Groningen, and North-Holland, Amsterdam, 1956, pp. 206–207.
5. ———, *Extremal Kähler meterics*, Seminar on Differential Geometry (S.-T. Yau, ed.), Annals of Math. Stud., vol. 102, Princeton Univ. Press, Princeton, NJ, 1982, pp. 259–290.
6. P. Chruściel, *Semi-global existence and convergence of solutions of the Robinson-Trautman (2-dimensional Calabi) equation*, Comm. Math. Phys. **137** (1991), 289-313.
7. S. K. Donaldson, *Remarks on gauge theory, complex geometry and 4-manifold topology*, Fields Medalists' Lectures (M. F. Atiyah and D. Iagolnitzer, eds.), World Sci. Publ., Singapore, 1997, pp. 384–403.
8. ———, *Boundary value problems for Yang–Mills fields*, J. Geom. Phys. **8** (1992), 89–192.
9. S. K. Donaldson and P. B. Kronheimer, *The geometry of four-manifolds*, Oxford Univ. Press, Oxford, 1990.
10. P. Griffiths and J. Harris, *Principles of algebraic geometry*, Wiley, New York, 1978.
11. S. Helgason, *Differential geometry, Lie groups and symmetric spaces*, Academic Press, New York, 1978.

12. F. C. Kirwan, *The cohomology of quotients in symplectic and algebraic geometry*, Mathematical Notes, vol. 31, Princeton Univ. Press, Princeton, NJ, 1984.
13. M. Klimek, *Pluripotential theory*, Oxford Univ. Press, Oxford, 1991.
14. C. LeBrun, *Polarised 4-maniflds, extremal Kähler metrics and Seiberg–Witten theory*, Math. Res. Lett. **2**, 653–662.
15. T. Mabuchi, *K-energy maps integrating Futaki invariants*, Tôhoku Math. J. (2) **38** (1986), 245–257.
16. D. McDuff and D. Salamon, *Symplectic topology*, Oxford Univ. Press, Oxford, 1995.
17. J. Moser, *On the volume elements on a manifold*, Trans. Amer. Math. Soc. **120** (1965), 286–294.
18. A. Nadel, *Multiplier ideal sheaves and Kähler–Einstein metrics of positive scalar curvature*, Ann. of Math. (2) **132** (1990), 549–596.
19. A. Pressley and G. B. Segal, *Loop groups*, Oxford Univ. Press, Oxford, 1986.
20. G. Tian, *Kähler–Einstein metrics on algebraic manifolds*, Proc. Internat. Congr. Math. (Kyoto, 1990), Vol. 1, Springer-Verlag, Berlin, 1991, pp. 587–598.
21. ———, *Kähler–Einstein metrics with positive scalar curvature*, Invent. Math. **130** (1997), 1–37.
22. S.-T. Yau, *On the Ricci curvature of a compact Kähler manifold nd the complex Monge–Ampere equation*, Comm. Pure Appl. Math. **31** (1978), 339–441.
23. ———, *Open problems in differential geometry*, Chern—A Great Geometer of the Twentieth Century (S.-T. Yau, ed.), International Press, Hong Kong, 1992, pp. 275–319.

THE MATHEMATICAL INSTITUTE, OXFORD

Amer. Math. Soc. Transl.
(2) Vol. **196**, 1999

Hamiltonian Dynamical Systems Without Periodic Orbits

Viktor L. Ginzburg

ABSTRACT. The present paper is a review of counterexamples to the "Hamiltonian Seifert conjecture" or, more generally, of examples of Hamiltonian systems having no periodic orbits on a compact energy level. We begin with the discussion of the "classical" and volume preserving Seifert conjectures. Then a construction of counterexamples to the Hamiltonian Seifert conjecture in dimensions greater than or equal to six is outlined, mainly following the method introduced by the author of the paper. The Hamiltonian version of Seifert's theorem is also stated. A list of all currently known examples and constructions of Hamiltonian flows without periodic flows concludes the review.

1. Introduction

The existence problem for periodic orbits of vector fields or diffeomorphisms occupies one of the central places in the theory of dynamical systems and adjacent areas such as mechanics and symplectic geometry. This question is often well motivated for a particular system, but in many cases the answer gives very little information about the dynamics of the system in general. However, the methods developed to solve the existence problem may have an impact extending well beyond the scope of the original problem.

The search for dynamical systems without periodic orbits has been inspired by a few questions. One of them is to determine the limits of the existence theorems. For instance, Seifert's theorem on periodic orbits of vector fields on S^3 led to the famous Seifert conjecture, recently disproved by K. Kuperberg [**KuK**]. However, in addition to this, the systems found as a result of this search sometimes exhibit a new type of dynamics and extend our understanding of the qualitative behavior that can occur for a given class of flows.

In this review we focus on examples of Hamiltonian systems without periodic orbits on a compact energy level in the context of the Seifert conjecture. The paper is organized as follows.

In Section 2 we recall Seifert's theorem and the Seifert conjecture. We discuss the history of counterexamples to the Seifert conjecture, from Wilson's theorem

1991 *Mathematics Subject Classification.* Primary 57R30; Secondary 58F05.
The work is partially supported by the NSF and by the faculty research funds of the University of California, Santa Cruz.

©1999 American Mathematical Society

(with its proof) to the ultimate solution due to K. Kuperberg. We also briefly touch upon related results on volume preserving flows.

Section 3 is devoted to Hamiltonian vector fields. We outline the constructions of counterexamples to the Hamiltonian Seifert conjecture in dimensions greater than or equal to six, mainly following the method introduced by the author of the present review. We also state the Hamiltonian version of Seifert's theorem.

Finally, a list of all presently known constructions of Hamiltonian flows without periodic flows is presented in Section 4.

Acknowledgments. The author is deeply grateful to Ana Cannas da Silva, Carlos Gutierrez, Ely Kerman, Greg and Krystyna Kuperberg, Debra Lewis, Richard Montgomery, Marina Ratner, Andrey Reznikov, Claude Viterbo, and the referee for their advice, remarks, and useful discussions. He would also like to thank the Université Paris-Sud for its hospitality during the period when the work on this manuscript was started.

2. The Seifert Conjecture

2.1. Seifert's Theorem. The history of examples of dynamical systems without periodic orbits as we understand it today begins with a result of Seifert that a C^1-smooth vector field on S^3 which is C^0-close to the Hopf field has at least one periodic orbit, [**Sei**]. Later this theorem was generalized by Fuller, [**Fu**], as follows.

THEOREM 2.1. *Let $E \to B$ be a principal circle bundle over a compact manifold B with $\chi(B) \neq 0$. Let X be a C^1-smooth vector field C^0-close to the field X_0 generating the S^1-action on E. Then X has at least one periodic orbit.*

Today we know two approaches to proving existence theorems for periodic orbits such as Theorem 2.1, neither of which has been trivialized and made into a part of mathematical pop culture. The first approach, not counting the original Seifert proof, relies on the notion of the Fuller index, [**Fu**], an analogue of the Euler characteristic for periodic orbits. The second one, analytical, is due to Moser [**Mo2**]. Moser's method uses a version of an infinite-dimensional inverse function theorem or, more precisely, its proof. In fact, as has been recently noticed by Kerman, [**Ke1**], under the additional assumption that X is C^1-close to X_0 (and not only C^0-close), Moser's argument can be significantly simplified by applying the inverse function theorem in Banach spaces. This stronger closeness hypothesis holds in virtually all applications of the theorem known to the author. Note also that in results such as Theorem 2.1, bounding $X - X_0$ with respect to a higher order norm often considerably simplifies the proof.

A representative corollary of Theorem 2.1 (for which the C^1-closeness assumption is sufficient) is as follows: *Let $f\colon \mathbb{R}^{2n} \to \mathbb{R}$ be a smooth function having a non-degenerate minimum at the origin with, say, $f(0) = 0$. Assume that all eigenvalues of $d^2f(0)$ are equal. Then for every sufficiently small $\epsilon > 0$ the level $\{f = \epsilon\}$ carries a periodic orbit of the Hamiltonian flow of f.* In effect, the assumption that all eigenvalues of $d^2f(0)$ are equal is purely technical and immaterial. Moreover, there are at least n periodic orbits of the Hamiltonian flow on $\{f = \epsilon\}$ (see [**We1**], [**Mo2**], [**Bot**]). For example, when all eigenvalues are distinct, the existence of n periodic orbits readily follows from the inverse function theorem.

REMARK 2.2. It is interesting to point out that Seifert's theorem on $(4n+1)$-dimensional spheres (i.e., Theorem 2.1 for the Hopf fibration $S^{4n+1} \to \mathbb{CP}^{2n}$) can be proved by the standard methods of algebraic topology. Namely, let $D_x \subset S^{4n+1}$ with $x \in S^{4n+1}$ be a small embedded $4n$-dimensional disc transversal to the fibers and centered at x. We can choose D_x to depend smoothly on x. Let $P(x)$ be the first intersection with D_x of the integral curve of X through x. Clearly $P(x) = x$ if and only if the integral curve closes up after one revolution along the fiber. There exists a unique vector v_x tangent to D_x at x such that $P(x)$ lies on the geodesic in D_x beginning at x in the direction v_x and the distance from x to $P(x)$ is $\|v_x\|$. Hence, $v_x = 0$ if and only if $P(x) = x$. Thus it suffices to prove that the vector field v vanishes at least at one point on S^{4n+1}. Note also that v is normal to X_0 with respect to a suitably chosen metric, and so v and X_0 would be linearly independent if v did not vanish.

The sphere S^{4n+1} does not admit two linearly independent vector fields. (This is a very particular case of Adams' theorem that can be proved by using the Steenrod squares; see [**SW**].) Therefore, on S^{4n+1} the field v vanishes somewhere and X has a periodic orbit.

2.2. The Seifert Conjecture. In the same paper [**Sei**], where he proved his theorem discussed above, Seifert asked whether every non-singular vector field on S^3 has a periodic orbit. The hypothetically affirmative answer to this question has become known as the Seifert conjecture. The three dimensional sphere plays, of course, a purely historical role in the conjecture, and a similar question can be asked for other manifolds and also for more restricted classes of vector fields (e.g., real analytic, divergence free, or Hamiltonian) or for non-vanishing vector fields in a fixed homotopy class (see Remark 2.7).

The first counterexample to the generalized Seifert conjecture is due to Wilson, [**Wi**], who showed that the smooth Seifert conjecture fails in dimensions greater than three by proving the following fact.

THEOREM 2.3. *Let M be a compact connected manifold with $\chi(M) = 0$ and $m = \dim M \geq 4$. Then there exists a smooth vector field on M without periodic orbits and singular points.*

PROOF. Let v_0 be a vector field on M without zeros. The idea is to modify v_0 so as to eliminate periodic orbits. First assume that v_0 has a finite number of periodic orbits. (Note that this is not a generic property.) Then each of the periodic orbits can be eliminated by the following procedure, which is common to many constructions of vector fields without closed integral curves.

A plug is a manifold $P = B \times I$, where $I = [-1, 1]$ and B is a compact manifold of dimension $m - 1$ with boundary, and a non-vanishing vector field w on P with the following properties:[1]

1. *The boundary condition*: $w = \partial/\partial t$ near ∂P, where t is the coordinate on I. As a consequence, an integral curve of w can only leave the plug through $B \times \{1\}$.
2. *Existence of trapped trajectories*: There is a trajectory of w beginning on $B \times \{-1\}$ that never exits the plug. Such a trajectory is said to be trapped in P.

[1] This definition, more restrictive than that given in [**KuG, KuGK**], is only one of several existing definitions of plugs. See also [**KuK**], [**Gi2**], and references therein. However, the differences between these definitions seem to be of a technical rather than conceptual nature.

3. *Aperiodicity*: w has no periodic orbits in P. In other words, the "flow" of w is aperiodic.
4. *Matched ends or the entrance–exit condition*: If two points, $(x, -1)$, the "entrance", and $(y, 1)$, the "exit", are on the same integral curve of w, then $x = y$. Hence, every trajectory of w which enters and exists P has its exit point right above the entrance point (with I being regarded as the vertical direction).
5. *The embedding condition*: There exists an embedding $i\colon P \hookrightarrow \mathbb{R}^m$ such that $i_*(w) = \partial/\partial x_m$ near ∂P. In other words, near the boundary of P, the embedding i preserves the vertical direction.

Assuming that the plug has been constructed, we can use it to modify v_0. Namely, consider a small flow box near a point on a periodic orbit of v_0. We choose coordinates on the box so that $v_0 = \partial/\partial x_m$. Using i, we embed P into the box so that the trapped trajectory matches the periodic orbit of v_0. Let v be the vector field obtained by replacing v_0 by w inside of $i(P)$. By the first property of the plug (the boundary condition), v is smooth. By the third condition (aperiodicity) and the fourth (matched ends), v has one periodic orbit less than v_0. By applying this procedure to every periodic orbit of v_0, we eliminate all of them.

Let us now turn to the construction of a plug. First observe that when a plug P' satisfying all of the above conditions except the entrance–exit condition is constructed, it is easy to find a genuine plug P. Namely, P is the union of two copies of P' with the second copy put upside-down on the top of the first copy. Hence, from now on we may forget about the entrance–exit condition. This trick, with minor modifications, is present in all constructions of plugs.

The plug from [**Wi**], the so-called *Wilson's plug*, is the cylinder over the unit ball $D^{m-1} \subset \mathbb{R}^{m-1}$. Thus $B = D^{m-1}$ and $t = x_m$, and the plug is automatically embedded into $\mathbb{R}^m$. To define w, fix an embedding $\mathbb{T}^2 \hookrightarrow D^{m-1} \times 0$ whose normal bundle is trivial. (Hence the requirement that $m \geq 4$.) Then a tubular neighborhood U of $\mathbb{T}^2$ in P is diffeomorphic to $\mathbb{T}^2 \times D^{m-2}(\epsilon)$, where $D^{m-2}(\epsilon)$ is the $(m-2)$-dimensional ball of a small radius $\epsilon > 0$. Let w_1 be an irrational vector field on $\mathbb{T}^2$ and let f be a bump function on $D^{m-2}(\epsilon)$ equal to 1 at the center of the ball and vanishing near $\partial D^{m-2}(\epsilon)$. Since $\mathbb{T}^2$ is embedded into $D^{m-1} \times 0$, we may assume that each $y \times D^{m-2}(\epsilon)$, $y \in \mathbb{T}^2$, is parallel to the vertical direction, i.e., $\partial/\partial x_m$ is tangent to the fibers $y \times D^{m-2}(\epsilon)$ of the tubular neighborhood U. Now we set

$$w = f w_1 + (1 - f)\frac{\partial}{\partial x_m} \tag{1}$$

on U and $w = \partial/\partial x_m$ on $P \setminus U$. It is easy to check that (P, w) satisfies the plug conditions (except the entrance–exit condition). In particular, all trapped trajectories are asymptotic to $\mathbb{T}^2$. This completes the construction of the plug. In what follows we will refer to $\mathbb{T}^2$ with an irrational flow embedded into the plug as the *core* of the plug.

REMARK 2.4. Wilson's plug defined in this section is clearly only C^∞-smooth. However, its construction can be modified to apply in the real analytic category; see, e.g., [**Gh, KuGK**] and references therein.

When the periodic orbits of the flow of v_0 are not isolated, the plug P needs to be slightly altered. Namely, one can construct P so that the beginnings in $B \times \{-1\}$

of trapped trajectories form a set with non-empty interior. Then a finite number of plugs are inserted into M so as to interrupt every trajectory of v_0 (regardless of whether it is closed or not). As before, no new periodic orbit is created, but the original ones are eliminated. □

REMARK 2.5. In the plug used in the case where there are infinitely many periodic orbits, we clearly have $\operatorname{div} w \neq 0$. This amounts to the fact that in the volume preserving version of Wilson's argument, one has to start with a vector field having only a finite number of periodic orbits.

The next step after Wilson's result was the construction, due to Schweitzer [**Sc**], of a C^1-smooth non-vanishing vector field on S^3 (or on any compact three-dimensional manifold) without periodic orbits. The proof again is by inserting plugs. Schweitzer's plug uses as the core the closure of a trajectory of the Denjoy flow on $\mathbb{T}^2$ instead of an irrational flow. (Hence, only C^1-smoothness.) Since the trajectory is neither dense nor closed, for B one can take the torus $\mathbb{T}^2$ with a small open disc deleted in order to have the embedding condition satisfied. The vector field on P is then given by (1), where w_1 is the Schweitzer field on B and f is an appropriately chosen cut-off function. Schweitzer's construction was later improved by Harrison, [**Ha**], to obtain a $C^{2+\epsilon}$-smooth vector field.

Finally, a major breakthrough came when K. Kuperberg, [**KuK**], constructed a C^∞-smooth (and even real analytic) three-dimensional plug, thus rendering a non-singular real analytic flow without periodic orbits on every compact three-manifold (see also [**Gh**] and [**KuGK**]). Kuperberg's plug is entirely different from those described above. One begins with a "plug" P' satisfying all of the conditions of the plug but aperiodicity – there are exactly two periodic trajectories inside of P'. Then one builds P by inserting parts of P' into P' again (self-insertion) so that the vector field on the inserted parts matches the original vector field. The self-insertion is performed so as to guarantee that the resulting flow on P is free of periodic orbits.

REMARK 2.6. The common feature of the constructions described above (with the exception of Kuperberg's plug) is that a non-singular vector field without periodic orbits is used as the core of the plug in order to trap a trajectory. For example, in Wilson's plug the core flow is an irrational flow on the torus and in Schweitzer's plug the core is the Denjoy flow. Hence, one non-singular vector field without periodic orbits gives rise to a multitude of such vector fields with various higher dimensional phase spaces (proliferation of aperiodic flows). This idea is also applied to find counterexamples to the Seifert conjecture in other categories.

REMARK 2.7. The vector field v constructed in the proof of Theorem 2.3 is homotopic to the original vector field v_0 in the class of non-singular vector fields. This follows because the field w on Wilson's plug is homotopic to the vertical vector field. The same holds for many other constructions of plugs, including Kuperberg's plug. Hence, the homotopy type of a non-singular vector field is preserved while the vector field is altered by inserting the plugs to eliminate the periodic orbits.

2.3. Volume Preserving Flows. The Seifert conjecture for this class of flows seems to be rather similar to the Seifert conjecture in the smooth and real analytic categories. To be more precise (as was pointed out by A. Katok, [**Ka**]), *a divergence free smooth non-vanishing vector field v_0 on a compact manifold of dimension $m \geq$*

4 can be changed into one without periodic orbits, provided v_0 has only a finite number of periodic orbits. In fact, the field w given by (1) on Wilson's plug (for isolated periodic orbits) can be chosen divergence free. This yields a smooth volume preserving flow on S^{2n+1}, $2n+1 \geq 5$, without periodic orbits.

Much less is known in dimension three. Let us state two important results due to G. Kuperberg, [**KuG**].

The first one is that *every compact three-manifold M possesses a volume preserving C^∞-smooth flow with a finite number of periodic orbits and no fixed points*, [**KuG**]. This follows from the fact that M can be obtained from $\mathbb{T}^3$ by a series of Dehn surgeries (provided that M is orientable). Let us equip $\mathbb{T}^3$ with an irrational flow. A Dehn surgery can be interpreted as the insertion of a version of a smooth volume preserving plug P into $\mathbb{T}^3$. These plugs differ from those described in Section 2.2 in some essential ways. The plugs P are not aperiodic – each P carries exactly two periodic orbits. Moreover, the flow on P is not standard on the boundary to account for the Dehn twist in the surgery. Note that this method also yields a flow on M with exactly two periodic orbits.

The second result is that *every compact three-manifold possesses a non-vanishing divergence free C^1 vector field without periodic orbits*, [**KuG**]. In particular, when applied to S^3, this theorem gives a volume preserving C^1 counterexample to the Seifert conjecture. The proof of the theorem uses the previously mentioned result and a volume preserving version of Schweitzer's plug to eliminate periodic orbits. To construct such a plug, G. Kuperberg applies, in a non-trivial way, stretching in the vertical direction to compensate for the area change resulting from the Denjoy flow (see [**KuG**] for more details).

REMARK 2.8. There is a broad class of volume preserving flows without periodic orbits on homogeneous spaces of Lie groups. These flows arise as actions of unipotent subgroups. For example, the horocycle flow (Section 3.3) and its Hamiltonian analogues from Example 4.2 are among such flows. Deep results on the closures of orbits for these flows and on their invariant measures are obtained by Ratner (see [**Rat**] and the references therein).

3. Hamiltonian vector fields without periodic orbits

The question of (non-)existence of periodic orbits for Hamiltonian dynamical systems differs from that for general smooth dynamical systems in at least one important way – Hamiltonian systems tend to have a lot of periodic orbits. For example, it is reasonable to expect that a given Hamiltonian system with a proper smooth Hamiltonian has a closed orbit on every regular energy level. Taken literally, this statement is not correct in general. (For example, Zehnder, [**Ze**], found a Hamiltonian flow on $\mathbb{T}^{2n}$, $2n \geq 4$, with an irrational symplectic structure such that there are no periodic orbits for a whole interval of energy values; see Example 4.1 below.) However, periodic orbits are known to exist for almost all energy values for a broad class of symplectic manifolds. For instance, as shown by Hofer, Zehnder, and Struwe, [**HZ1, St**], almost all levels of a smooth proper function on $\mathbb{R}^{2n}$ carry at least one periodic orbit. A similar theorem holds for cotangent bundles, [**HV**]. (The reader interested in a detailed discussion and further references should consult [**HZ2**].)

Furthermore, according to the C^1-closing lemma of Pugh and Robinson, [**PR**], a periodic orbit can be created from a recurrent trajectory by a C^1-small smooth perturbation of the original dynamical system. (As Carlos Gutierrez pointed out to the author, the perturbation can in fact be made C^∞-smooth, [**Gu2**].) As a consequence of the closing lemma, a C^1-generic system has the union of its periodic trajectories dense in the set of its recurrent points. Both of these results hold for Hamiltonian systems, [**PR**].

The situation becomes more subtle when the C^1-topology is replaced by the C^k-topology with $k > 1$. The C^2-closing lemma has not been yet proved or disproved (see, e.g., [**Gu1**], [**Ca**], and [**AZ**] for partial results, examples, and references). In the Hamiltonian setting, the systems whose periodic trajectories are dense may no longer be generic if k is roughly speaking greater than the dimension of the manifold. For example, according to M. Herman, [**Her1, Her2**], Hamiltonian vector fields with Hamiltonians C^k-close to Zehnder's Hamiltonian on $\mathbb{T}^{2n}$ do not have periodic orbits for an interval of energy values, provided that the symplectic form satisfies a certain Diophantine condition and $k > 2n$. These examples are, however, in some sense exceptional. In fact, the theorems on the density of energy values for periodic orbits can be interpreted as saying that the Hamiltonian C^k-closing lemma holds in a very strong form for many symplectic manifolds. It is still not known if the C^k-closing lemma with $k \geq 2$ fails in general for Hamiltonian flows when the symplectic form is exact near the energy level.

Examples of Hamiltonian flows without periodic orbits on a compact energy level are scarce. In Section 4 we attempt to list all known constructions of such flows. In this section, we focus on the Hamiltonian Seifert conjecture and on one particular method to construct Hamiltonian vector fields without periodic orbits.

REMARK 3.1. There are a number of results concerning existence of periodic orbits on a fixed energy level. Recall that, according to Weinstein's conjecture, there is a periodic orbit on an energy level of contact type, [**We2**]. This conjecture was proved for $\mathbb{R}^{2n}$ by Viterbo in [**Vi1**], and then for many other symplectic manifolds (e.g., for cotangent bundles in [**HV**]). The reader interested in details and more up-to-date references should consult [**Ho**], [**HZ2**], or [**Vi2**]. Also note that in the context of contact topology and hydrodynamics the Seifert conjecture is discussed in [**EG**].

3.1. The Seifert conjecture for Hamiltonian vector fields. The Seifert conjecture can be extended to Hamiltonian flows in a number of ways. For example, one may ask if there is a proper smooth function on a given symplectic manifold (e.g., $\mathbb{R}^{2n}$) having a regular level without periodic orbits. Recall that a *characteristic* of a two-form η of rank $(2n-2)$ on a $(2n-1)$-dimensional manifold is an integral curve of the field of directions formed by the null-spaces $\ker \eta$. Thus, the question can be reformulated as whether or not in a given symplectic manifold there exists a regular compact hypersurface without closed characteristics. One can even ask whether the manifold admits a function with a sufficiently big set of energy levels without periodic orbits.

Let $i\colon M \hookrightarrow W$ be an embedded smooth compact hypersurface without boundary in a $2n$-dimensional symplectic manifold (W, σ).

THEOREM 3.2. *Assume that $2n \geq 6$ and that $i^*\sigma$ has only a finite number of closed characteristics. Then there exists a C^∞-smooth embedding $i'\colon M \hookrightarrow W$, C^0-close and isotopic to i, such that ${i'}^*\sigma$ has no closed characteristics.*

An irrational ellipsoid M in the standard symplectic vector space $\mathbb{R}^{2n} = W$ corresponds to a collection of n uncoupled harmonic oscillators whose frequencies are linearly independent over $\mathbb{Q}$, i.e., M is the unit level of a quadratic Hamiltonian with rationally independent frequencies. Thus, M carries exactly n periodic orbits. Applying Theorem 3.2, we obtain the following fact.

COROLLARY 3.3. *For $2n \geq 6$, there exists a C^∞-embedding $S^{2n-1} \hookrightarrow \mathbb{R}^{2n}$ such that the restriction of the standard symplectic form to S^{2n-1} has no closed characteristics.*

COROLLARY 3.4. *For $2n \geq 6$, there exists a C^∞-function $h\colon \mathbb{R}^{2n} \to \mathbb{R}$, C^0-close and isotopic (with a compact support) to a positive definite quadratic form, such that the Hamiltonian flow of h has no closed trajectories on the level set $\{h = 1\}$.*

REMARK 3.5. These results, proved in [**Gi2**] and, independently, by M. Herman [**Her3**], first required the ambient space to be at least eight-dimensional, i.e., $2n \geq 8$, in the C^∞-case. A $C^{2+\epsilon}$-hypersurface in $\mathbb{R}^6$ was found by M. Herman [**Her3**]. In [**Gi5**], Theorem 3.2 and its corollaries were extended to $2n = 6$.

REMARK 3.6. Almost nothing is known on how big the set of energy values $\mathcal{E}$ of the levels without periodic orbits can be. It is clear that, as in Corollary 3.4, there exists a function on $\mathbb{R}^{2n}$ for which $\mathcal{E}$ is infinite (but discrete). It also seems plausible that $\mathcal{E}$ can have limit points at critical values of the Hamiltonian. However, it is unknown whether or not $\mathcal{E}$ can have a limit point that is a regular value.

As is clear from the argument of [**Gi2, Gi5**], there exists a C^0-foliation of an open neighborhood of an ellipsoid in $\mathbb{R}^{2n}$ such that every leaf is C^∞-smooth and isotopic to the sphere and no leaf carries closed characteristics.

REMARK 3.7. Theorem 3.2 can be applied to any compact hypersurface with a finite number of closed characteristics. The only known examples of such hypersurfaces in $\mathbb{R}^{2n}$ that are not diffeomorphic to S^{2n-1} are the non-simply connected hypersurfaces constructed by Laudenbach, [**La**].

An alternative way to formulate the Hamiltonian Seifert conjecture is to consider an odd dimensional manifold with a maximally non-degenerate closed two-form (rather than the restriction of the symplectic form to a hypersurface). A result similar to Theorem 3.2, involving only two-forms on M but no symplectic embedding, is also correct.

Let $\dim M = 2n - 1$ and let η be a closed maximally non-degenerate (i.e., of rank $(2n-2)$) two-form on M. Recall that two such forms η and η' are said to be *homotopic* if there exists a family η_τ, $\tau \in [0, 1]$, of closed maximally non-degenerate forms connecting $\eta = \eta_0$ with $\eta' = \eta_1$ and such that all η_τ have the same cohomology class. The following theorem is proved in [**Gi2**] and [**Gi5**].

THEOREM 3.8. *Assume that $2n-1 \geq 5$ and that η has a finite number of closed characteristics. Then there exists a closed maximally non-degenerate 2-form η' on M which is homotopic to η and has no closed characteristics.*

REMARK 3.9. In fact, Theorem 3.8 is a corollary of Theorem 3.2. To derive Theorem 3.8 from Theorem 3.2, consider a pair (M^{2n-1}, η), where η is closed and maximally non-degenerate. Then, as follows from Gotay's coisotropic embedding theorem, [**Go**], there exist a symplectic manifold (W^{2n}, σ) and a proper embedding $M \hookrightarrow W$ such that $\sigma|_M = \eta$.

More explicitly, set $W = M \times (-1, 1)$ and let t be the coordinate on $(-1, 1)$. To construct σ, fix a 1-form α on M such that $\ker \alpha$ is everywhere transversal to the characteristics of η. Then $\sigma = \eta + d(\epsilon t \alpha)$ is symplectic on W, provided that $\epsilon > 0$ is small enough. (The author is grateful to Ana Cannas da Silva for this remark.)

Technically, however, it is more convenient to prove Theorem 3.8 first and then to modify its proof to obtain Theorem 3.2.

REMARK 3.10. Theorem 3.8 extends to the real analytic case: one can make the form η' real analytic, provided that Q and η are real analytic. The argument is the same as that used in the construction of a real analytic version of Wilson's, [**Wi**], or Kuperberg's, [**KuK**], flow (see [**Gh**] and [**KuGK**]).

3.2. The Hamiltonian Seifert Theorem. Before we turn to the outline of the proofs of Theorems 3.2 and 3.8, let us formulate a result which can be viewed as a Hamiltonian version of Theorem 2.1.

Recall that a periodic orbit of a vector field or an integral curve of a one-dimensional foliation is called non-degenerate if the linearization of its Poincaré return map does not have unity as an eigenvalue. Let (B, ω) be a compact symplectic manifold and let $\pi\colon M \to B$ be a principle S^1-bundle. For a one-form λ such that $d\lambda$ is C^0 sufficiently small, the closed two-form $\eta = \pi^*\omega + d\lambda$ on M is maximally non-degenerate (and homotopic to $\pi^*\omega$).

THEOREM 3.11 ([**Gi4**]). *If $d\lambda$ is C^0-small enough, the number of closed characteristics of η is strictly greater than the cup-length of B. If, in addition, all closed characteristics are non-degenerate, then the number of closed characteristics is greater than or equal to the sum of the Betti numbers of B.*

Theorem 3.11 generalizes some of the results (for equal eigenvalues) of [**We1, Mo2, Bot**]. A simple geometrical proof for the case where B is a surface and λ is C^2-small can be found in [**Gi1**]. Theorem 3.11 is closely related to the Arnold conjecture and to the problem of existence of periodic orbits of a charge in a magnetic field; [**Gi3**]. Note also that if $d\lambda$ is not C^0-small, but η is still non-degenerate and homotopic to $\pi^*\omega$, the form η may have no closed characteristics at all. The example is the horocycle flow described in the next section (see [**Gi3**] and [**Gi4**] for more details).

3.3. Proofs of Theorems 3.2 and 3.8 and symplectic embeddings. There are two essentially different methods to prove these theorems. Both approaches follow the same general line as Wilson's argument (i.e., the proof of Theorem 2.3), modified to make it work in the symplectic category, and vary only in the dynamics of the plug.

The plug introduced by M. Herman, [**Her3**], is the symplectization of Wilson's plug when $2n \geq 8$ or of the Schweitzer–Harrison plug when $2n = 6$ (hence the $C^{2+\epsilon}$-smoothness constraint) modified at infinity. In other words, these plugs are obtained by taking the induced flows on the cotangent bundles to the non-Hamiltonian plugs and altering these flows away from a neighborhood of the zero section while keeping all the properties of the plug.

The plugs from [**Gi2, Gi5**] are built similarly to Wilson's plug, but the entire construction is Hamiltonian. Let us give some more details on this method, focusing specifically on the proof of Theorem 3.8. (The passage to Theorem 3.2 is simply by showing that the same argument can be carried out for hypersurfaces. See [**Gi2**].)

In the construction of the plug P below we identify a neighborhood in M containing P with a small ball in $\mathbb{R}^{2n-1}$ equipped with the two-form σ induced by the canonical inclusion $\mathbb{R}^{2n-1} \subset \mathbb{R}^{2n}$.

The key element that makes Wilson's construction work in the Hamiltonian category is that the core of the plug (i.e., the flow replacing the irrational flow on the torus) is also chosen to be Hamiltonian. In other words, to have the aperiodicity condition satisfied, we need to find a Hamiltonian flow on a symplectic manifold having no periodic orbits on some compact energy level. Furthermore, to have the embedding condition satisfied, this energy level with the induced two-form has to be embeddable into $\mathbb{R}^{2n-1}$ to be made a part of the plug. (In particular, the symplectic form must be exact on a neighborhood of the energy level.) The Hamiltonian flow used [**Gi2, Gi5**] is the horocycle flow, which we will now describe.

Let Σ be a closed surface with a hyperbolic metric, i.e., a metric with constant negative curvature $K = -1$. Denote by λ the canonical Liouville one-form "$p\,dq$" on $T^*\Sigma$ and by Ω the pull-back to $T^*\Sigma$ of the area form on Σ. The form $\omega = d\lambda + \Omega$ is symplectic on $T^*\Sigma$ and exact on the complement of the zero section. The Hamiltonian flow of the standard metric Hamiltonian on $T^*\Sigma$ with respect to the twisted symplectic form ω is known to have no periodic orbits on the unit energy level $ST^*\Sigma$. Indeed, the restriction φ^t of this flow to $ST^*\Sigma$ is the standard horocycle flow. The fact that φ^t has no periodic orbits is a consequence of the classical result of Hedlund, [**Hed**], that the horocycle flow is minimal, i.e., all orbits are dense. We will return to the horocycle flow in Section 4.

To construct the plug, we then need to prove that there is a "symplectic" embedding of a neighborhood U of $ST^*\Sigma$ into $\mathbb{R}^{2n-1}$, i.e., an embedding $j\colon U \hookrightarrow \mathbb{R}^{2n-1}$ such that $j^*\sigma = \omega$. When $2n-1 \geq 7$, this readily follows from a general symplectic embedding theorem due to Gromov, [**Gr1, Gr2**].

When $2n-1 = 5$ an additional argument is required. Consider the forms $\omega_t = d\lambda + t\Omega$ on $T^*\Sigma$. These forms are symplectic for all t. The form ω_0 is the standard symplectic form on $T^*\Sigma$, and ω_1 is the twisted form ω. Note that if U is small enough, there exists a "symplectic" embedding $j_0\colon (U, \omega_0) \hookrightarrow (\mathbb{R}^5, \sigma)$. This is an easy consequence of the fact that there exists a Lagrangian immersion $\Sigma \to \mathbb{R}^4$, [**Gr1, Gr2**]. Note also that U can be replaced by a closed neighborhood of $ST^*\Sigma$. Then the existence of a symplectic embedding for $2n-1 = 5$ is a particular case of the following result improving the dimensional constraints from [**Gr1, Gr2**] by one.

Let U and V be manifolds of equal dimensions. The manifold U is assumed to be compact, perhaps with boundary, while V may be open but must be a manifold without boundary. Let σ be a symplectic form on V. Abusing notation, also denote by σ the pull-back of σ to $V \times \mathbb{R}$ under the natural projection $V \times \mathbb{R} \to V$.

THEOREM 3.12 ([**Gi5**]). *Let ω_t, $t \in [0,1]$, be a family of symplectic forms on U in a fixed cohomology class: $[\omega_t] = \text{const}$. Assume that there is an embedding $j_0\colon U \hookrightarrow V \times \mathbb{R}$ such that $j_0^*\sigma = \omega_0$. Then there exists an embedding $j\colon U \hookrightarrow V \times \mathbb{R}$, isotopic to j_0, with $j^*\sigma = \omega_1$.*

REMARK 3.13. Since $\sigma|_V$ is symplectic, the composition of j with the projection to V is necessarily an immersion. When $\partial U = \emptyset$, Theorem 3.12 follows immediately from Moser's theorem [**Mo1**].

The rest of the proof (i.e., the construction of the form η' on the complement of the core) proceeds according to the same scheme as the proof of Theorem 2.3 with more or less straightforward modifications; see [**Gi2**] for details. This completes the outline of the proofs of Theorems 3.2 and 3.8.

4. The list of Hamiltonian flows without periodic orbits.

In this section we list all examples known to the author of smooth Hamiltonian systems on symplectic manifolds (V, ω) having no periodic orbits on a compact regular level M. This list is similar to the one given in [**Gi5**] with the exception of Example 4.2.

The list is divided into two parts according to whether or not ω is exact near M. The reason for this division is that the qualitative behavior of the flows for which $[\omega|_M] \neq 0$ can be expected to differ in an essential way from the behavior of the flows with $[\omega|_M] = 0$. For instance, Zehnder's flow (Example 4.1) appears to be more robust than seems possible for "exact" flows.

Case 1: *The form ω is not exact.*

EXAMPLE 4.1 (Zehnder's torus, [**Ze**]). Let $2n \geq 4$. Consider the torus $V = \mathbb{T}^{2n}$ with an irrational translation-invariant symplectic structure ω. Choose a Hamiltonian H on V so that every level $\{H = c\}$ with $c \in (0.5, 1.5)$ is the union of two standard embedded tori $\mathbb{T}^{2n-1} \subset \mathbb{T}^{2n}$. Since ω is irrational, the characteristics of $\omega|_{\mathbb{T}^{2n-1}}$ form a quasiperiodic flow on $\mathbb{T}^{2n-1}$. Thus, none of the levels $\{H = c\}$ with $c \in (0.5, 1.5)$ carries a periodic orbit. Note that ω is not exact on any of these energy levels. As we have already pointed out, the flow in question exhibits remarkable stability properties according to a result of M. Herman, [**Her1, Her2**].

EXAMPLE 4.2. (The Hamiltonian horocycle flow in dimension $2n$). Let $\mathbb{C}H^n$ be the complex hyperbolic space equipped with its standard Kähler metric (see, e.g., Section XI.10 of [**KN**]). Pick a discrete subgroup Γ in the group $\mathrm{SU}(1,n)$ of Hermitian isometries of $\mathbb{C}H^n$ such that $B = \Gamma\backslash\mathbb{C}H^n$ is smooth and compact. (To see that Γ exists, recall that according to a theorem of Borel, [**Bor**], there is a discrete subgroup Γ such that $\Gamma\backslash\mathbb{C}H^n$ is compact. Then by Selberg's lemma, [**Sel**], Γ can be chosen so that B is smooth.)

Let H be the standard metric Hamiltonian and let $d\lambda$ be the standard symplectic structure on T^*B. Denote by Ω the pull-back to T^*B of the Kähler symplectic form on B. The Hamiltonian flow of H with respect to the twisted symplectic structure $\omega = \Omega + d\lambda$ has no periodic orbits on the level $M = \{H = 1\}$. This follows from the facts that this flow is generated by the (right) action of a unipotent one-parameter subgroup of $\mathrm{SU}(1,n)$ and that Γ contains no unipotent elements, [**KM**]; see also [**Rag**], [**Rat**], and the references therein.

When $n = 1$ this construction gives exactly the horocycle flow on the unit (co)tangent bundle to a surface $B = \Sigma$. If $n > 1$, the form ω is not exact in any neighborhood of M.

Note also that the Hamiltonian horocycle flows arise in the description of the motion of a charge in a magnetic field on the configuration space B; see [**Gi3, Gi4, Ke2**]. These are the only known "magnetic" flows without periodic orbits on an energy level.

Case 2: *The form ω is exact near the energy level.*

In this case we should distinguish whether $\dim V = 4$ cr $\dim V \geq 6$.

When $\dim V = 4$, the only known smooth example is the horocycle flow described as a Hamiltonian system in Section 3.3 and in Example 4.2 with $n = 1$ and $B = \Sigma$. To slightly generalize this construction, note that a neighborhood of $ST^*\Sigma$ can be identified with $V = \Gamma\backslash\mathrm{SU}(1,1) \times (1-\epsilon, 1+\epsilon)$, where $\Gamma = \pi_1(\Sigma)$, so that H becomes the projection to the second component. Then, instead of $\Gamma = \pi_1(\Sigma)$ we can take any discrete subgroup such that the quotient $\Gamma\backslash\mathrm{SU}(1,1)$ is compact and smooth. As follows from Remark 3.9, the flow on $\Gamma\backslash\mathrm{SU}(1,1)$ generated by the action of a unipotent subgroup of $\mathrm{SU}(1,1)$ is the Hamiltonian flow of H on $\{H = 1\}$ with respect to some symplectic form.

Finally, note that by Remark 3.9 the flows on three-manifolds constructed by G. Kuperberg and described above in Section 2.3 can also be thought of as Hamiltonian flows on symplectic manifolds. This gives a class of smooth Hamiltonian flows on symplectic manifolds with a finite number of closed orbits on a given energy level or C^1-flows without periodic orbits. It is not known if such a flow on S^3 can be obtained by a C^2-embedding of S^3 into $\mathbb{R}^4$.

When $\dim V \geq 6$, the only known examples are essentially those described in Section 3.1 or those obtained by iterations of their constructions. In other words, beginning with a flow with a finite number of periodic orbits, one can eliminate these orbits by using the plugs from [**Gi2, Gi5**] or [**Her3**]. The resulting manifold can also be used as the core of a plug in the same way as the horocycle flow. As in Remark 2.6, these new plugs can in turn be employed to construct Hamiltonian flows without periodic orbits, etc.

Since the starting point of this method is a flow with a finite number of periodic orbits, it makes sense to ask how many such flows are known in addition to those on irrational ellipsoids. For example, we have already mentioned the non-simply connected hypersurfaces in $\mathbb{R}^{2n}$ found by Laudenbach [**La**]. To produce more examples, we can apply the same method as described in the previous paragraph. Namely, if a Hamiltonian flow with a finite number of periodic orbits can be used as the core of the plug, the resulting "plug" will also have only a finite number of periodic orbits inside of it. Thus by inserting it into another flow with a finite number of periodic orbits, we can create yet one more flow with the same property. For example, by taking $S^1 \subset \mathbb{R}^2$ as the core, Cieliebak, [**Ci**], constructed embeddings $S^{2n-1} \subset \mathbb{R}^{2n}$, $2n \geq 4$, such that the pairs of closed characteristics are linked and knotted in an essentially arbitrary way. (Each plug results in two "parallel" closed orbits, but there are no constraints on knotting of these pairs or their linking.)

References

[AZ] S. Aranson and Zhuzhoma, *On the C^r-closing lemma on surfaces*, Russian Math. Surveys **43** (1988), no. 5, 209–210.

[Bor] A. Borel, *Compact Clifford–Klein forms of symmetric spaces*, Topology **2** (1963), 111–122.

[Bot] M. Bottkol, *Bifurcation of periodic orbits on manifolds and Hamiltonian systems*, J. Differential Equations **37** (1980), 12–22.

[Ca] C. Carroll, *Rokhlin towers and C^r closing for flows on T^2*, Ergodic Theory Dynamical Systems **12** (1992), 683–706.

[Ci] K. Cieliebak, *Symplectic boundaries: Creating and destroying closed characteristics*, Geom. Funct. Anal. **7** (1997), 269–321.

[EG] J. Etnyre and R. Ghrist, *Contact topology and hydrodynamics*, Preprint, 1997.

[Fu] F. B. Fuller, *An index of fixed point type for periodic orbits*, Amer. J. Math. **89** (1967), 133–148.

[Gh] E. Ghys, *Construction de champs de vecteurs sans orbite périodique (d'après Krystyna Kuperberg)*, Sém. Bourbaki, 1993–1994, Astérisque, no. 227, Soc. Math. France, Paris, 1995, pp. 283–307.

[Gi1] V. L. Ginzburg, *New generalizations of Poincaré's geometric theorem*, Funct. Anal. Appl. **21** (1987), 100–106.

[Gi2] ———, *An embedding $S^{2n-1} \to \mathbb{R}^{2n}$, $2n - 1 \geq 7$, whose Hamiltonian flow has no periodic trajectories*, Internat. Math. Res. Notices **1995**, 83–98.

[Gi3] ———, *On closed trajectories of a charge in a magnetic field. An application of symplectic geometry*, Contact and Symplectic Geometry, C.B. Thomas (ed.), Publ. Newton Inst., Cambridge Univ. Press, Cambridge, 1996, pp. 131–148.

[Gi4] ———, *On the existence and non-existence of closed trajectories for some Hamiltonian flows*, Math. Z. **223** (1996), 397–409.

[Gi5] ———, *A smooth counterexample to the Hamiltonian Seifert conjecture in $\mathbb{R}^6$*, Internat. Math. Res. Notices **1997**, 641–650.

[Go] M. Gotay, *On coisotropic imbeddings of presymplectic manifolds*, Proc. Amer. Math. Soc. **84** (1982), 111–114.

[Gr1] M. Gromov, *A topological technique for the construction of solution of differential equations and inequalities*, Proc. Internat. Congr. Math. (Nice, 1970), Vol. 2, Gauthier-Villars, Paris, 1971, pp. 221–225.

[Gr2] ———, *Partial differential relations*, Springer-Verlag, New York, 1986.

[Gu1] C. Gutierrez, *A counter-example to a C^2-closing lemma*, Ergodic Theory Dynamical Systems **7** (1987), 509–530.

[Gu2] ———, *Letter to the author*, 1998.

[Ha] J. Harrison, *A C^2 counterexample to the Seifert conjecture*, Topology **27** (1988), 249–278.

[Hed] G. A. Hedlund, *Fuchsian groups and transitive horocycles*, Duke Math. J. **2** (1936), 530–542.

[Her1] M.-R. Herman, *Exèmples de flots hamiltoniens dont aucune perturbation en topologie C^∞ n'a d'orbites périodiques sur un ouvert de surfaces d'énergies*, C. R. Acad. Sci. Paris Sér. I Math. **312** (1991), 989–994.

[Her2] ———, *Différentiabilité optimale et contre-exemples à la fermeture en topologie C^∞ des orbites récurrentes de flots hamiltoniens*, C. R. Acad. Sci. Paris Sér. I Math. **313** (1991), 49–51.

[Her3] ———, *Fax message to Eliashberg.*

[Ho] H. Hofer, *Pseudoholomorphic curves in symplectizations with applications to the Weinstein conjecture in dimension three*, Invent. Math. **114** (1993), 515–563.

[HV] H. Hofer and C. Viterbo, *The Weinstein conjecture for cotangent bundles and related results*, Ann. Scuola Norm. Sup. Pisa Cl. Sci. (4) **15** (1988), 411–445 (1989).

[HZ1] H. Hofer and E. Zehnder, *Periodic solutions on hypersurfaces and a result by C. Viterbo*, Invent. Math. **90** (1987), 1–9.

[HZ2] ———, *Symplectic invariants and Hamiltonian dynamics*, Birkhäuser, Basel, 1994.

[Ka] A. Katok, *Private communication*, 1990.

[KM] D. A. Kazhdan and G. A. Margulis, *A proof of Selberg's conjecture*, Mat. Sbornik (N.S.) **75** (1968), 162–168; English transl. in Math. USSR-Sb. **4** (1968).

[Ke1] E. Kerman, *Private communication*, 1998.

[Ke2] ———, *Periodic orbits of Hamiltonian flows near symplectic critical submanifolds*, Preprint, 1999.

[KN] S. Kobayashi and K. Nomizu, *Foundations of differential geometry.* Vols. 1, 2, Interscience, New York, 1963, 1969.

[KuG] G. Kuperberg, *A volume preserving counterexample to the Seifert conjecture*, Comment. Math. Helv. **71** (1996), 70–97.

[KuGK] G. Kuperberg and K. Kuperberg, *Generalized counterexamples to the Seifert conjecture*, Ann. of Math. (2) **144** (1996), 239–268.

[KuK] K. Kuperberg, *A smooth counterexample to the Seifert conjecture in dimension three*, Ann. of Math. (2) **140** (1994), 723–732.

[La] F. Laudenbach, *Trois constructions en topologie symplectique*, Ann. Fac. Sci. Toulouse Math. (6) **6** (1997), 697–709.

[Mo1] J. Moser, *On the volume elements on a manifold*, Trans. Amer. Math. Soc. **120** (1965), 286–294.

[Mo2] ———, *Periodic orbits near an equilibrium and a theorem by Alan Weinstein*, Comm. Pure Appl. Math. **29** (1976), 727–747.

[PR] C. Pugh and C. Robinson, *The C^1 closing lemma, including Hamiltonians*, Ergodic Theory Dynamical Systems **3** (1983), 261–313.

[Rag] M. S. Raghunathan, *Discrete subgroups of Lie groups*, Springer-Verlag, New York, 1972.

[Rat] M. Ratner, *On Raghunathan's measure conjecture*, Ann. of Math. (2) **134** (1991), 545–607.

[Sc] P. A. Schweitzer, *Counterexamples to the Seifert conjecture and opening closed leaves of foliations*, Ann. of Math. (2) **100** (1970), 229–234.

[Sei] H. Seifert, *Closed integral curves in 3-space and isotopic 2-dimensional deformations*, Proc. Amer. Math. Soc. **1** (1950), 287–302.

[Sel] A. Selberg, *On discontinuous groups in higher-dimensional symmetric spaces*, Contributions to Function Theory (Internat. Colloq. Function Theory, Bombay, 1960), Tata Inst. Fundamental Res., Bombay, 1960, pp.147-164.

[SW] N. E. Steenrod and J. H. C. Whitehead, *Vector fields on the n-sphere*, Proc. Nat. Acad. Sci. USA **37** (1951), 58–63.

[St] M. Struwe, *Existence of periodic solutions of Hamiltonian systems on almost every energy surfaces*, Bol. Soc. Brasil. Mat. **20** (1990), 49–58.

[Vi1] C. Viterbo, *A proof of Weinstein's conjecture in $\mathbb{R}^{2n}$*, Ann. Inst. H. Poincaré, Anal. Non Linéaire **4** (1987), 337–356.

[Vi2] ———, *Functors and computations in Floer cohomology*. Part I, Preprint, Université Paris-Sud, 1997.

[We1] A. Weinstein, *Normal modes for non-linear Hamiltonian systems*, Invent. Math. **20** (1973), 377–410.

[We2] ———, *On the hypothesis of Rabinowitz' periodic orbit theorem*, J. Differential Equations **33** (1979), 353–358.

[Wi] F. Wilson, *On the minimal sets of non-singular vector fields*, Ann. of Math. (2) **84** (1966), 529–536.

[Ze] E. Zehnder, *Remarks on periodic solutions on hypersurfaces*, Periodic Solutions of Hamiltonian Systems and Related Topics, (P. H. Rabinowitz et al., eds.), Reidel, Amsterdam, 1987, pp. 267–279.

Department of Mathematics, University of California at Santa Cruz, Santa Cruz, CA 95064

E-mail address: `ginzburg@math.ucsc.edu`

Amer. Math. Soc. Transl.
(2) Vol. **196**, 1999

The Mirror Formula for Quintic Threefolds

Alexander Givental

To D. B. Fuchs on his 60th birthday

The formula of Candelas, de la Ossa, Green and Parkes [**6**] expressing the virtual numbers n_d, $d = 1, 2, 3, \ldots$, of degree d holomorphic spheres on quintic threefolds in $\mathbb{C}P^4$ in terms of series solutions to the linear differential equation

$$(1) \qquad \left(q\frac{d}{dq}\right)^4 I = 5q\left(5q\frac{d}{dq}+1\right)\left(5q\frac{d}{dq}+2\right)\left(5q\frac{d}{dq}+3\right)\left(5q\frac{d}{dq}+4\right)I$$

has been intriguing algebraic and symplectic geometers since the beginning of the decade. The first proof of this formula was given two years ago in the extensive paper [**9**] among a number of other theorems on equivariant Gromov–Witten theory. Several authors managed to adjust the approach of that paper to complete intersections in homogeneous Kähler spaces [**13, 2, 14**], in toric manifolds [**11, 12**] and to symmetric products of Riemann surfaces [**4**].

We present here a shortcut to our original proof in the case of quintic threefolds. Several variants of the proof can be found in [**9, 11, 12, 17, 5, 20**] as particular cases of more general theorems. Yet it seems useful to illustrate all ingredients of the proof in the simplest nontrivial example.

We assume that the reader is familiar with generalities on orbifolds and orbibundles, equivariant cohomology and localization formulas, Kontsevich's moduli spaces of stable maps [**15, 3**], and with the formulation of the conjecture. We will concentrate therefore only on the issues relevant for the proof of the mirror formulas.

In the last section, called *Updates*, we outline several modifications of the proof available at the moment, and provide the corresponding references.

The linear sigma-model. A quintic threefold in $X = \mathbb{C}P^4$ is given by a generic degree 5 homogeneous equation $Q(x_1, \ldots, x_5) = 0$. A degree d parametrized curve $\mathbb{C}P^1 \to X$ is described by 5 relatively prime binary forms of degree d, which we will represent by the polynomials $x_1(\zeta), \ldots, x_5(\zeta)$ of degree $\leq d$ in the affine coordinate ζ on $\mathbb{C}P^1$. The curve sits on the quintic if and only if the degree $\leq 5d$ polynomial $Q(x_1(\zeta), \ldots, x_5(\zeta))$ vanishes identically. This identity yields $5d+1$ equations of degree 5 in the projective space $LX_d = \mathbb{C}P^{5d+4}$ of all (not necessarily

---

1991 *Mathematics Subject Classification.* Promary 14Jxx.

Research supported by NSF grant DMS-9704774.

©1999 American Mathematical Society

relatively prime) 5-tuples of degree $\leq d$ polynomials. Attempting to count degree d spheres on the quintics by means of intersection theory in LX_d, we will arrive at the answer $n_d = 5^{5d+1}$, which is meaningful, as has been explained by Morrison and Plesser in [**19**], but wrong.

Our approach to the quintic formula begins with the following observation [**10**]. The variety given in LX_d by the equations $Q(x(\zeta)) \equiv 0$ is invariant with respect to the Möbius transformation group of $\mathbb{C}P^1$, and one can employ *equivariant* intersection theory for curve counting. The maximal torus S^1 in the Möbius group acts via $\zeta \mapsto \zeta \exp i\phi$ on the space LX_d. The cohomology algebra $H^*_{S^1}(LX_d)$ is generated by the equivariant Chern class $-p$ of the Hopf bundle over LX_d. The polynomial Q defines an invariant section $Q(x(\zeta))$ of an equivariant $5d+1$-dimensional bundle LV_d over LX_d. The following formal series encodes complete equivariant topological information about equivariant Euler classes of these bundles:

$$L(q,z) := \sum_{d=0}^{\infty} q^d \int_{LX_d} e^{pz} Euler(LV_d).$$

The integral in this formula means evaluation of an equivariant cohomology class on the invariant fundamental class of the manifold, and it takes values in the coefficient algebra $H^*(BS^1)$ of the equivariant cohomology theory. Thus the coefficients of the (q,z)-series L are polynomials in one variable, which we prefer to denote $\hbar$. With this notation, $Euler(LV_d) = 5p(5p-\hbar)(5p-2\hbar)\cdots(5p-5d\hbar)$.

THEOREM A ([**10**]). (a)

$$L(q,z) = \langle I(qe^{\hbar z}, \hbar^{-1}), I(q, -\hbar^{-1})\rangle,$$

where

$$\langle \phi, \psi\rangle := \int_{\mathbb{C}P^4} \phi\ \psi\ Euler(\mathcal{O}(5)) = \frac{1}{2\pi i}\oint \phi(P)\psi(P)\frac{5PdP}{P^5}$$

is the intersection pairing in the even cohomology algebra $\mathbb{Q}[P]/(P^4)$ *of the quintic, and*

$$I(q,\hbar^{-1}) := e^{P\ln q/\hbar} \sum_{d=0}^{\infty} q^d \frac{(5P+\hbar)(5P+2\hbar)\cdots(5P+5d\hbar)}{(P+\hbar)^5(P+2\hbar)^5\cdots(P+d\hbar)^5}$$

is a formal vector-valued function with coefficients in this algebra.

(b) *Components of the vector-valued function form a fundamental solution to the linear differential equation* (1).

PROOF. (a) Compute the coefficients of the series L explicitly by the Duistermaat–Heckman formula,

$$\int_{LX_d} \Phi(p,\hbar) = \frac{1}{2\pi i}\oint \frac{\Phi(p,\hbar)dp}{p^5(p-\hbar)^5\cdots(p-d\hbar)^5},$$

and change the order of summation. (b) Substitute I in (1). □

The non-linear sigma-model. The definition of the numbers n_d accepted in [**9**] was given by Kontsevich [**15**] on the basis of intersection theory in moduli spaces of stable maps to X. Let $X_{n,d}$ denote the moduli orbifold of degree d genus 0 stable maps with n marked points. For any such map $f : (\Sigma, \varepsilon_1, \ldots, \varepsilon_n) \to X$, the section Q of the degree 5 line bundle over X defines an element in the $5d+1$-dimensional space $H^0(\Sigma, f^*\mathcal{O}(5))$, hence a section of the orbibundle V_d over $X_{n,d}$ formed by these spaces. According to Kontsevich's definition, the Euler class of this bundle capped with the fundamental class of the orbifold is taken in the role of the virtual fundamental class in the corresponding space of stable maps to the quintic. Taking into account the multiple cover formula [**1**] proved in [**18**], one gives the following recursive definition of the virtual numbers of degree d spheres in quintic threefolds:

$$\sum_{m|d} \frac{n_{d/m}}{m^3} := \int_{X_{0,d}} Euler(V_d).$$

The first step in our proof of the conjecture extracting these numbers from the differential equation (1) consists in mimicking Theorem A within the framework of moduli spaces.

Introduce the *graph space* $GX_{n,d}$ as the moduli orbifold of stable maps to $GX = \mathbb{C}P^1 \times X$ of degree d in projection on X and degree 1 in projection on $\mathbb{C}P^1$. Let GV_d be the orbibundle with the fiber $H^0(\Sigma, f^* \mathcal{O}_{\mathbb{C}P^1} \otimes \mathcal{O}_X(5))$ of dimension $5d+1$. The space and the bundle inherit the S^1-action from the Möbius transformation group on $\mathbb{C}P^1$.

There is a natural equivariant birational isomorphism (see the Main Lemma in [**9**])

$$\mu : GX_{0,d} \to LX_d, \tag{2}$$

defined as follows.

A stable map $f : \Sigma \to \mathbb{C}P^1 \times X$ of bi-degree $(1, d)$ is represented by the graph of a map $f_0 : \mathbb{C}P^1 \to X$ of some degree $d_0 \leq d$ and several "vertical" curves $f_i : (\Sigma_i, \varepsilon_i) \to \{\zeta_i\} \times X$ of bi-degrees $(0, d_i)$, $i = 1, \dots, r$, attached to the graph at the points $f_i(\varepsilon_i)$. The number r of such vertical curves can vary from 0 to d, and their total degree $d_1 + \dots + d_r$ equals $d - d_0$.

The graph component f_0 is described by five mutually prime binary forms of degree d_0 uniquely up to a common constant factor (we record them by five polynomials $(x_1'(\zeta), \dots, x_5'(\zeta))$. Multiply these polynomials by the same binary form of degree $d - d_0$ with the roots at $\zeta_1, \dots, \zeta_r$ of multiplicity $d_1, \dots, d_r$ respectively (it is encoded by the polynomial $x(\zeta) = (\zeta - \zeta_1)^{d_1} \cdots (\zeta - \zeta_r)^{d_r}$ with the obvious convention about the roots at $\zeta = \infty$). Then the polynomials $(x_1(\zeta), \dots, x_5(\zeta)) := (x(\zeta)x_1'(\zeta), \dots, x(\zeta)x_5'(\zeta))$ represent the image $\mu(f)$ in the space LX_d.

Obviously, the map μ is equivariant with respect to the natural actions of the automorphism group $PSL_2(\mathbb{C}) \times PSL_5(\mathbb{C})$ of $\mathbb{C}P^1 \times X$ on $GX_{0,d}$ and $LX_d = \text{Proj}(H^0(\mathbb{C}P^1, \mathcal{O}_{\mathbb{C}P^1}(d)) \otimes H^0(X, \mathcal{O}_X(1))^*)$, and is therefore independent of the choice of the coordinate systems $(x_1 : \dots : x_5)$ and ζ on X and $\mathbb{C}P^1$. In order to check that μ is regular, let us assume for the moment that the map f is transverse to the five coordinate hyperplanes $\mathbb{C}P^1 \times x_i^{-1}(0)$, to $\infty \times X$, and that the intersections of $f(\Sigma)$ with the coordinate hyperplanes are away from $\infty \times X$. Then these intersections are all simple (even if f is a multiple cover, they are simple distinct non-singular points on Σ); their projections to $\mathbb{C}P^1$ are away from ∞ and determine the roots of the polynomials $x_1(\zeta), \dots, x_5(\zeta)$. The projective coordinates of the intersection point $f(\Sigma) \cap (\infty \times X)$ determine the top coefficients of the polynomials $x_1(\zeta), \dots, x_5(\zeta)$ uniquely up to a non-zero scalar factor (we can fix it by requiring that the top coefficient of $x_1(\zeta)$ equals 1). Now consider the point $[f]$ represented by the map f in the moduli orbifold $GX_{0,d}$ and a local non-singular chart near this point. Due to the simplicity of the intersection points of $f(\Sigma)$ with the coordinate hyperplanes and with $\infty \times X$, the roots and top coefficients of the polynomials $x_i(\zeta)$ are regular functions of the map in a sufficiently small neighborhood of $[f]$ (if this is not obvious yet, consider the universal stable map $\mathrm{ev} : GX_{1,d} \to \mathbb{C}P^1 \times X$, and describe the intersection points as local sections of the forgetting map $\mathrm{ft} : GX_{1,d} \to GX_{0,d}$). Thus the map μ is regular at $[f]$, since top coefficients and roots uniquely determine the polynomials.

Choosing the coordinate systems on $\mathbb{C}P^1$ and X in general position with respect to a given f, we conclude that μ is regular everywhere.

The map μ allows one to compare the equivariant Euler classes of LV_d and GV_d. Let $-p$ denote the equivariant Chern class of the Hopf bundle over LX_d pulled-back to $GX_{0,d}$. Introduce the formal series

$$G(q,z) := \sum_{d=0}^{\infty} q^d \int_{GX_{0,d}} e^{pz} Euler(GV_d).$$

THEOREM B. *We have*

$$G(q,z) = \langle J(qe^{z\hbar}, \hbar^{-1}), J(q, -\hbar^{-1}) \rangle,$$

where the formal vector-valued function J with values in the cohomology algebra $\mathbb{Q}[P]/(P^4)$ is defined by

$$(3) \qquad J(q,\hbar^{-1}) := e^{(P \ln q)/\hbar} \Big[1 + \hbar^{-1} \sum_{d=1}^{\infty} q^d \, \mathrm{ev}_* \Big(\frac{Euler(V_d)}{\hbar - c} \Big) \Big],$$

c denotes the Chern class of the line orbibundle over $X_{1,d}$ formed by cotangent lines to the curves at the marked point, the map $\mathrm{ev} : X_{1,d} \to X$ is defined by evaluation of stable maps at the marked point, and the push-forward is well-defined by

$$\langle \phi, \mathrm{ev}_* \Psi \rangle = \int_{\mathbb{C}P^4} 5P\phi \, \mathrm{ev}_* \Psi := \int_{X_{1,d}} (\mathrm{ev}^* \phi) \Psi.$$

PROOF. (See [**9**], Section 6.) It consists in applying fixed point localization in equivariant cohomology. A fixed point of the S^1-action on $GX_{0,d}$ is represented by a stable map $\Sigma \to GX$ which consists of the graph of a constant map $\mathbb{C}P^1 \to X$ with two "vertical " curves of degrees $d - d'$ and d' mapped to the slices $\zeta = 0$ and $\zeta = \infty$ and attached to the graph. Thus the fixed point components are given by the diagonal constraint at the marked points in the Cartesian product $X^{(0)}_{1,d-d'} \times X^{(\infty)}_{1,d'}$, $d' = 1, \dots, d-1$ (and are isomorphic to $X^{(0)}_{1,d}$ and $X^{(\infty)}_{1,d}$ for $d' = 0, d$). The normal bundle to the fixed point component has the equivariant Euler class $(-\hbar)(-\hbar - c^{(0)})(\hbar)(\hbar - c^{(\infty)})$ (unless $d' = 0$ or d, in which case a half of the product should be taken). The Euler classes occur in the denominator of the localization formula. The bundle GV_d restricted to the fixed point set splits into $V'_{d-d'} \oplus V'_{d'} \oplus \mathrm{ev}^* \mathcal{O}(5)$, where $'$ symbolizes that the subspace in $H^0(\Sigma, f^*\mathcal{O}(5))$ consisting of sections vanishing at the marked point is taken. Due to the multiplicative property of Euler classes, $Euler(GV_d)$ factors correspondingly. Since the map μ sends the fixed point component in GX_d to the corresponding fixed point component in LX_d, the class p localizes to $\mathrm{ev}^*(P) + d'\hbar$. It remains only to rearrange the summation over d and d' as the double sum over d' and $d'' = d - d'$. □

The divisor equation. It is useful to figure out the place of the series J in the axiomatic structure of Gromov–Witten theory (see for instance Dubrovin's book [**7**]). Genus 0 Gromov–Witten invariants of a Kähler manifold Y define on $H^*(Y)$ the structure of a Frobenius manifold. It basically consists of the quantum cup-product in each tangent space and a family of flat connections $\nabla_\hbar$ on the tangent bundle defined by the quantum multiplication operators and depending on a parameter which we denote by $\hbar^{-1}$. Structural constants of the quantum cup-product are third partial derivatives $F_{\alpha\beta\gamma}$ of the *potential*

$$F(T) = \sum_{n,d} \frac{q^d}{n!} (T, \dots, T)_{n,d},$$

where T is a general cohomology class of Y,

$$(T, \dots, T)_{n,d} := \int_{[Y_{n,d}]^{\text{virt}}} \text{ev}_1^*(T) \cdots \text{ev}_n^*(T),$$

and the "foreign" formal parameter q is introduced in order to provide convergence of the d-sum in the l-adic topology at least. For instance, the numbers n_d for quintic threefolds are encoded by the *Yukawa coupling*

$$F_{ttt} = \sum_{n,d} \frac{q^d}{n!} (P, P, P, Pt, \dots, Pt)_{n+3,d},$$

which is obtained by differentiation of the potential restricted to the second cohomology space of the quintic and, as we will see soon, coincides with the formal Fourier series

$$K(qe^t) := \sum_{d=0}^{\infty} (P, P, P)_{3,d} q^d e^{dt} = 5 + \sum_{d=1}^{\infty} \frac{n_d d^3 q^d e^{dt}}{1 - q^d e^{dt}}.$$

The vector fields on $H^*(Y)$ flat with respect to the connections $\nabla_\hbar$ also represent Gromov–Witten invariants. Namely, the following matrix is the fundamental solution to the system of linear differential equations $\nabla_\hbar S = 0$:

$$S_{\psi\phi}(T, \hbar^{-1}) := \langle \psi, \phi \rangle + \sum_{n,d} \frac{q^d}{n!} \left(\psi, T, \dots, T, \frac{\phi}{\hbar - c} \right)_{n+2,d}, \tag{4}$$

where $\phi, \psi \in H^*(Y)$, and c is the Chern class of the universal cotangent line at the last marked point (as specified by the position of this class in the correlator).

Of course, all the correlators for quintic threefolds Y are defined by means of intersection theory in $X_{n,d}$ with the Euler classes of the bundles V_d pulled-back from $X_{0,d}$ by forgetting maps.

THEOREM C.

$$\langle J(qe^t, \hbar^{-1}), e^{-(P \ln q)/\hbar} \phi \rangle = \langle 1, \phi \rangle + \sum_{n,d} \frac{q^d}{n!} \left(1, Pt, \dots, Pt, \frac{\phi}{\hbar - c} \right)_{n+2,d}. \tag{5}$$

PROOF. ([**9**], Section 6). The proof consists in applying the *string equation* followed by an iterative application of the *divisor equation*. The string and divisor equations are, respectively,

$$\left(1, T, \dots, T, \frac{\phi}{\hbar - c} \right)_{n+2,d} = \hbar^{-1} \left(T, \dots, T, \frac{\phi}{\hbar - c} \right)_{n+1,d}, \tag{6}$$

(7)

$$\left(P, T, \dots, T, \frac{\phi}{\hbar - c} \right)_{n+2,d} = d \left(T, \dots, T, \frac{\phi}{\hbar - c} \right)_{n+1,d} + \hbar^{-1} \left(T, \dots, T, \frac{P\phi}{\hbar - c} \right)_{n+1,d},$$

where one should require $n \geq 2$ when $d = 0$. The equations allow us to push the classes $\text{ev}_i^*(1)$ and $\text{ev}_i^*(P)$ of degree ≤ 2 from $X_{2+n,d}$ to $X_{1,d}$ along the forgetting maps $X_{n+2,d} \to X_{n+1,d} \to \cdots \to X_{1,d}$. This transforms the sum of d-terms in the series (5) to $\hbar^{-1}(\phi \exp(t(P+d\hbar)/\hbar)/(\hbar - c))_{1,d}$ for $d > 0$ (and to $\langle \exp(tP/\hbar), \phi \rangle$ for $d = 0$, since $X_{3,0} = X$). □

A similar application of the divisor equation to the Yukawa coupling shows that $F_{ttt}(tP) = K(qe^t)$. On the other hand, by applying the string equation to the elements $S_{1,P}(tP, \hbar^{-1})$ of the matrix (4) we extract the Yukawa coupling as the coefficient at $\hbar^{-2}$ in the second derivative in t. The theorem therefore implies that

$$K(e^t) = \left\langle \hbar^2 \frac{d^2}{dt^2} J(e^t, \hbar^{-1}), P \right\rangle. \tag{8}$$

Thus in order to prove the quintic formula it suffices to identify J and I up to suitable mirror transformations.

We complete this subsection by including a proof of the string and divisor equations. The argument seems to be standard in Deligne–Mumford theory.

The maps $\mathrm{ft} : X_{k+1,d} \to X_k$ are defined [**15, 3**] by forgetting the first marked point ε_0 of a stable map $f : (\Sigma, \varepsilon_0, \dots, \varepsilon_k) \to X$ and producing a new stable map $\tilde{f} : (\tilde{\Sigma}, \varepsilon_1, \dots, \varepsilon_k) \to X$ by contracting those components of Σ which have become unstable. The fiber of the forgetting map is canonically identified with the quotient of the curve $(\tilde{\Sigma}, \varepsilon_1, \dots, \varepsilon_k)$ by the finite automorphism group of the map $\tilde{f}$. In particular the map ft has k canonical sections ε_i defined by the marked points in the fibers, and together with the evaluation map $\mathrm{ev}_0 : X_{k+1,d} \to X$ can be considered as the universal stable map with k marked points.

One derives the equations (6), (7) by comparing the Chern class c of the universal cotangent line bundle over $X_{1+k,d}$ with the pull-back $\tilde{c}$ of the corresponding class from $X_{k,d}$. The cotangent lines $T^*_{\varepsilon_k}\Sigma$ and $T^*_{\varepsilon_k}\tilde{\Sigma}$ are canonically identified *unless* ε_k is situated on the same irreducible component Σ_0 of Σ as ε_0, the map f restricted to Σ_0 is constant, and Σ_0 carries no other marked points and contains only one singular point of the curve Σ. Since $\Sigma_0 \simeq \mathbb{C}P^1$ carries 3 special points of different nature, the cotangent line $T^*_{\varepsilon_k}\Sigma$ is canonically trivialized in this case. In terms of the universal cotangent line bundles l_k and $\tilde{l}_k = \mathrm{ft}^* l_k$ over $X_{k,d}$ with Chern classes c and $\tilde{c}$, this means that $\tilde{l}_k^* \otimes l_k$ has a canonical section non-vanishing outside the divisor $\varepsilon_k(X_{k,d})$ defined by the universal marked point $\varepsilon_k : X_{k,d} \to X_{k+1,d}$, and that l_k restricted to this divisor is trivial. Since the restriction of $\tilde{l}_k$ to the divisor coincides with the conormal bundle to this divisor by the very definition of the universal cotangent line $T^*_{\varepsilon_k}\tilde{\Sigma}$ on $X_{k,d}$, one concludes that the section has the 1-st order zero along the divisor. Thus $\delta = c - \tilde{c}$ is Poincaré-dual in $X_{k+1,d}$ to the hypersurface $\varepsilon_k(X_{k,d})$, and $c\delta = 0$.

We have

$$(1, T, \dots, T, \tilde{c}^m \phi)_{k+1,d} = 0, \quad (P, T, \dots, T, \tilde{c}^m \phi)_{k+1,d} = d(T, \dots, T, c^m \phi)_{k,d},$$

since the integrals of the classes 1 and P over degree d curves are equal to 0 and d respectively. Finally, replacing $\tilde{c}$ by $c = \tilde{c} + \delta$ yields the extra-terms

$$(T, \dots, T, c^{m-1}\phi)_{k,d} \quad \text{and} \quad (T, \dots, T, c^{m-1}P\phi)_{k,d}.$$

This implies the string and divisor equations, since $1/(\hbar - c)$ is the eigenfunction of the operation $c^m \mapsto c^{m-1}$ with the eigenvalue $1/\hbar$. □

Torus action. A link between I and J can be established via a recursion relation satisfied by their equivariant perturbations I^{eq} and J^{eq}.

Consider the standard action of the 4-dimensional torus G on $X = \mathbb{C}P^4$. The equivariant cohomology algebra $H^*_G(X)$ is generated over the coefficient ring $\mathbb{Q}[\lambda] = H^*(BG)$ by the equivariant Chern class $-P$ of the Hopf line bundle, and satisfies the relation

$$(P-\lambda_1)(P-\lambda_2)(P-\lambda_3)(P-\lambda_4)(P-\lambda_5) = 0$$

(where we assume that $\lambda_1 + \cdots + \lambda_5 = 0$). Evaluation of an equivariant cohomology class on the invariant fundamental class can be computed via fixed point localization:

$$\int_{[X]} \phi(P,\lambda) = \sum_{\alpha=1}^{5} \frac{\phi(\lambda_\alpha,\lambda)}{e_\alpha} = \frac{1}{2\pi i}\oint \frac{\phi(P,\lambda)dP}{(P-\lambda_1)\cdots(P-\lambda_5)},$$

where $e_\alpha = \prod_{\beta\neq\alpha}(\lambda_\alpha - \lambda_\beta)$ is the equivariant Euler class of the tangent space to X at the fixed point r_α where P restricts to λ_α. The equivariant Chern class of the anti-canonical line bundle $\mathcal{O}_X(5)$ equals $5P$, and we will denote by $\langle\phi,\psi\rangle$ the equivariant intersection pairing on $H^*_G(X)$ with this Chern class:

$$\langle\phi,\psi\rangle := \int_{[X]} 5P\phi\psi.$$

The torus G acts on the moduli orbifolds $X_{n,d}$, allowing one to introduce the equivariant Gromov–Witten correlators, such as

$$(T,\dots,T)_{n,d} \quad\text{and}\quad (\psi,T,\dots,T,\phi c^k)_{n+2,d},$$

where c and $\phi,\psi,T \in H^*_G(X)$ are equivariant cohomology classes. The correlators take values in $\mathbb{Q}[\lambda]$ and turn into corresponding non-equivariant Gromov–Witten invariants when specialized to $\lambda = 0$.

Similarly, one can use $S^1\times G$-equivariant intersection theory in the spaces LX_d and $GX_{0,d}$, and carry over Theorems A(a), B, C to the equivariant setting. The proofs are identical to those given above, but the equivariant counterpart of the hypergeometric series I is now defined by the series

$$I^{eq} := e^{(P\ln q)/\hbar}\sum_{d=0}^{\infty} q^d \frac{\prod_{m=1}^{5d}(5P+m\hbar)}{\prod_{m=1}^{d}\prod_{\alpha=1}^{5}(P-\lambda_\alpha+m\hbar)}, \tag{9}$$

with coefficients in the *equivariant* cohomology algebra of X generated by P (and which represents a fundamental solution to some fifth order ODE, which is however irrelevant for our goal in this paper).

In order to describe the recursion relation satisfied by the hypergeometric series I^{eq} let us strip off the factor $\exp(P\ln q)/\hbar$, denote the remaining series by $Z^{(hg)}$, and denote by $Z^{(hg)}_\alpha$ its fixed point localizations:

$$Z^{(hg)}_\alpha(q,\hbar^{-1},\lambda) = \sum_{d=0}^{\infty} \frac{q^d}{d!\hbar^d}\frac{\prod_{m=1}^{5d}(5\lambda_\alpha+m\hbar)}{\prod_{m=1}^{d}\prod_{\beta\neq\alpha}(\lambda_\alpha-\lambda_\beta+m\hbar)}.$$

Coefficients of the formal q-series $Z^{(hg)}_\alpha$ are degree 0 rational functions in $\hbar$ with the first order pole at $\hbar = (\lambda_\beta-\lambda_\alpha)/m$, $\beta\neq\alpha$, $m = 1,\dots,d$, and a high order pole at $\hbar = 0$. Rewriting the rational functions as sums of elementary fractions, we arrive at the following recursion relation:

(10)

$$Z_\alpha(q, \hbar^{-1}, \lambda) = 1 + \sum_{d=1}^{\infty} q^d \frac{R_{\alpha,d}(\hbar, \lambda)}{\hbar^d} + \sum_{\beta \neq \alpha} \sum_{m=1}^{\infty} C_\alpha^\beta(m) \frac{q^m}{\lambda_\alpha - \lambda_\beta + m\hbar} Z_\beta\Big(q, \frac{m}{\lambda_\beta - \lambda_\alpha}, \lambda\Big),$$

where $R_{\alpha,d}$ are degree $\leq d$ polynomials in $\hbar$ with coefficients rational in λ, and $C_\alpha^\beta(m)$ are rational functions in λ which we will call the *recursion coefficients.* Since $Z_\beta^{(hg)} \equiv 1 \mod q$, the coefficient $C_\alpha^\beta(d)$ can be read off the q^d-term of the series $Z_\alpha^{(hg)}$ as the residue at the pole $\hbar = (\lambda_\beta - \lambda_\alpha)/d$.

We will refer to the sequence of polynomials $R_{\alpha,d}$ as the *initial condition*: given such an initial condition, the recursion relation allows us to recover the solution $\{Z_\alpha, \alpha = 1, \ldots, 5\}$ unambiguously.

Now, starting with the Gromov–Witten invariant J^{eq}, introduce the vector q-series $Z^{(GW)}(q, \hbar^{-1}, \lambda)$ with coefficients in the algebra $H_G^*(X, \mathbb{Q}(\hbar))$ by stripping off the factor $\exp(P \ln q)/\hbar$, and denote by $Z_\alpha^{(GW)}$ the localization of $Z^{(GW)}$ at the fixed r_α.

THEOREM D ([**9**], Section 11)). *The series* $\{Z_\alpha^{(GW)}, \alpha = 1, \ldots, 5\}$ *satisfy the recursion relation* (10) *with the same recursion coefficients* $C_\alpha^\beta(d)$ *(and with another initial condition).*

PROOF. ([**9**], Sections 9, 11). Our argument is based on localization to fixed points of the torus G action on the moduli orbifolds $X_{1,d}$. Consider a stable map $f : (\Sigma, \varepsilon) \to X$ representing such a fixed point. The combinatorial structure of the curve Σ is described by a tree of irreducible components (each isomorphic to $\mathbb{C}P^1$). Some components are mapped onto the straight lines in $X = \mathbb{C}P^4$ connecting the fixed points $r_\alpha, \alpha = 1, \ldots, 5$, of the torus action, and the map is a multiple cover $\zeta \mapsto \zeta^m$ in suitable affine coordinates on the source and target $\mathbb{C}P^1$, so that $\zeta = 0, \infty$ are mapped to the fixed points. The remaining irreducible components of Σ are mapped to the fixed points in X. The marked point ε must be mapped to one of the fixed points r_α.

The fixed point in $X_{1,d}$ represented by f does not contribute to $Z_\alpha^{(GW)}$ via localization formulas *unless* $f(\varepsilon) = r_\alpha$.

Suppose that ε is situated in an irreducible component of Σ mapped to r_α. Consider the whole connected component of the fixed point set $X_{1,d}^G$ in $X_{1,d}$ which contains the equivalence class $[f]$. We will show that this connected component contributes to $Z_\alpha^{(GW)}$ by a polynomial in $\hbar^{-1}$. Indeed, the component can be described as the (quotient by a finite group of the) product of some Deligne–Mumford spaces $\bar{\mathcal{M}}_{0,k}$ (why? — see [**15**], where the fixed point set is described). The universal cotangent line orbibundle over $X_{1,d}$ restricted to the connected component of $X_{1,d}^G$ coincides with the universal cotangent line at one of the marked points over one of the factors $\bar{\mathcal{M}}_{0,k}$. Thus its Chern class c is nilpotent on this component, and the geometric series $(\hbar - c)^{-1}$ reduces to a finite sum of terms $c^l/\hbar^{l+1}$. Notice that $l \leq \dim \bar{\mathcal{M}}_{0,k} = k - 3$ is bounded by the total degree of the map f.

The fixed point localization terms just discussed form the initial condition in (10). We will show that contributions of all other fixed points can be arranged as the recursive part of (10). The idea is to cut off the component of the curve $(\Sigma, \varepsilon) \to X$ carrying the marked point and to observe that the rest of the curve represents a torus-invariant curve of smaller degree.

In more detail, suppose that ε is situated at $\zeta = 0$ on a multiple cover $\zeta \mapsto \zeta^m$ of the line connecting r_α with r_β. Then the universal cotangent line orbibundle restricted to the connected component of $X^G_{1,d}$ is topologically trivial (since $(T^*_\varepsilon\Sigma)^{\otimes m}$ coincides in this case with the cotangent space $T^*_{r_\alpha}\mathbb{C}P^1$ to the line joining r_α and r_β), but it carries a nontrivial infinitesimal action of G given by the character $(\lambda_\alpha - \lambda_\beta)/m$. Thus the localization of $(\hbar - c)^{-1}$ at this fixed point component yields the simple fraction $mq^m(m\hbar + \lambda_\beta - \lambda_\alpha)^{-1}$. The factor m is eventually compensated by the order of the automorphism group of the map $\zeta \to \zeta^m$, which occurs in the denominator of localization formulas on orbifolds. The weight q^m counts the degree of this map as a curve in X.

The whole contribution of the fixed point component to $Z^{(GW)}_\alpha$ via localization formulas includes two more factors. Each of them takes in account the equivariant Euler classes of the orbibundle V'_d and of the normal orbibundle to $X^G_{1,d}$ which occurs in the denominator of the localization formula.

The first factor corresponds to the irreducible component $C = \mathbb{C}P^1$ of Σ carrying the marked point, and the second corresponds to the remaining part $\tilde{\Sigma}$ of the curve Σ. The map f restricted to $\tilde{\Sigma}$ has degree $d - m$ and represents a point in the space $X^G_{1,d-m}$. The fiber $H^0(\Sigma, f^*V')$ of V'_d contains the subspace of sections vanishing on C, which coincides with the fiber of V'_{d-m}. The normal spaces to $X^G_{1,d}$ split similarly into parts corresponding to C and $\tilde{\Sigma}$. The intersection point $\zeta = \infty$ of C with $\tilde{\Sigma}$ plays the role of the marked point $\tilde{\varepsilon}$ in $\tilde{\Sigma}$. The deformation of f corresponding to smoothing of the curve Σ at the double point $\tilde{\varepsilon}$ is represented in the tangent space to $X_{1,d}$ by the line $T_{\tilde{\varepsilon}}C \otimes T_{\tilde{\varepsilon}}\tilde{\Sigma}$ and contributes the factor $(\lambda_\beta - \lambda_\alpha)m^{-1} - \tilde{c}$ to the denominator of the localization formula. Thus the contribution of $\tilde{\Sigma}$ is correctly accounted by the factor $Z^{(GW)}_\beta(q, m/(\lambda_\beta - \lambda_\alpha), \lambda)$ in the recursion relation (10).

The remaining factor in the localization formula is a rational function of λ and can be computed explicitly as the ratio of two Euler classes in the case $m = d$ when $\tilde{\Sigma}$ is a point (in the example of quintics we are studying, the factor has actually been computed in [**15**]). It turns out to coincide with the recursion coefficient $C^\beta_\alpha(m)$. We recommend the reader to carry out this computation (it amounts to analyzing the torus action on spaces of holomorphic sections of $\mathcal{O}(5)$ and T_X lifted to the multiple cover $\zeta \mapsto \zeta^m$ of the line joining r_α and r_β) or at least to look at some details of this computation in [**15**]. □

A plausible argument in [**9, 11**] intended to explain the "miraculous" coincidence of the recursion coefficients has been formalized in [**17**].

Polynomiality. The recursion relation (10) has many more solutions with various initial conditions than the mirror transformations can handle. However, according to Theorem A and Theorem B, the solutions $Z^{(ng)}$ and $Z^{(GW)}$ have the following polynomiality property.

Let us call a solution $Z(q, \hbar^{-1}, \lambda)$ to the recursion relation (10) *polynomial* if the formal (q, z)-series

$$\langle Z(qe^{\hbar z}, \hbar^{-1}, \lambda), e^{Pz} Z(q, -\hbar^{-1}, \lambda)\rangle \tag{11}$$

has coefficients polynomial in $\hbar$.

The solution $Z^{(GW)}$, by the very definition (3), satisfies also the asymptotic condition

$$Z(q, \hbar^{-1}, \lambda) = 1 + o(\hbar^{-1}). \tag{12}$$

THEOREM E ([**9**], Proposition 11.5). *A polynomial solution to the recursion relation* (10) *satisfying the asymptotic condition* (12) *is unique.*

PROOF. We use perturbation theory. Let Z be a polynomial solution, and let $\delta R = R_d - R_d^{(GW)}$ denote the discrepancy in the initial conditions for Z and $Z^{(GW)}$ with minimal $d > 0$. Then Z and $Z^{(GW)}$ coincide modulo q^d due to the recursion relation. The polynomiality property for Z and $Z^{(GW)}$ modulo q^{d+1} translates into regularity at $\hbar = 0$ of

$$\langle \delta R(\hbar)\hbar^{-d}, e^{(P+d\hbar)z}\rangle + \langle \delta R(-\hbar)(-\hbar)^{-d}, e^{Pz}\rangle.$$

Localizations of this intersection index to fixed points in X are — for each power of $\hbar^{-1}$ — finite sums of monomials $z^l \exp(\lambda_\alpha z)$, $l = 0, 1, 2, \ldots$, $\alpha = 1, \ldots, 5$. Linear independence of such monomials for generic λ implies that the localizations

$$\delta R_\alpha(\hbar)\hbar^{-d}e^{\hbar z} + \delta R_\alpha(-\hbar)(-h)^{-d},$$

where $R_\alpha(\hbar)$ are some polynomials in $\hbar$ of degree $\leq d$, must themselves be regular at $\hbar = 0$.

Consider first $(A\hbar^{-2} + B\hbar^{-3})\exp(\hbar z) + (A\hbar^{-2} - B\hbar^{-3})$ modulo z^2. The regularity condition at $\hbar = 0$ implies $A = 0$ and then $B = 0$. Applying this argument inductively, we conclude that $\delta R_\alpha(\hbar)\hbar^{-d} = A_\alpha + B_\alpha \hbar^{-1}$, where A_α, B_α do not depend on $\hbar$.

Assuming now that Z also satisfies the asymptotic condition (12), we find that $\delta R = 0$. □

Mirror transformations. The hypergeometric series I^{eq} has the asymptotic expansion

$$I^{eq} = e^{(P \ln q)/\hbar}\Big(f_0(q) + f_1(q)\frac{P}{\hbar} + o(\hbar^{-1})\Big),$$

where the series

$$f_0 = \sum_{d=0} q^d \frac{(5d)!}{(d!)^5}, \qquad f_1 = \sum_{d=1} q^d \frac{(5d)!}{(d!)^5}\Big(\sum_{m=d+1}^{5d} \frac{5}{m}\Big)$$

are found from (9) (remember that $\lambda_1 + \cdots + \lambda_5 = 0$).

The *mirror transformations*, namely the division of I^{eq} by f_0 followed by the change of variable $\ln q \mapsto \ln q + f_1(q)/f_0(q)$, transform I^{eq} to a new vector-valued function with the same asymptotic behavior as

$$J^{eq} = e^{(P \ln q)/\hbar}(1 + o(\hbar^{-1})).$$

Therefore the following theorem guarantees that the transformed series coincides with J^{eq}. Passing to the non-equivariant limit $\lambda = 0$, we conclude that the same mirror transformations take I into J.

Thus the Yukawa coupling (8) is indeed extracted from the fundamental solution I to the differential equation (1) by the procedure conjectured in [**6**].

THEOREM F ([**9**], Propositions 11.3, 11.6). *The mirror transformations take polynomial solutions of the recursion relation* (10) *to polynomial solutions of the same recursion relation.*

PROOF. Straightforward (see [**9**]). The division operation does not change the form of the recursion relation, and also preserves the polynomiality property since the extra factor $f_0(q\exp(\hbar z))f_0(q)$ does not produce negative powers of $\hbar$ in the (q,z)-series (11). The change of the variables $\ln q \mapsto \ln q + g(q)$ transforms z in this series into $z + [g(q\exp(\hbar z)) - g(q)]/\hbar$. At $\hbar = 0$ the difference vanishes. It is therefore divisible by $\hbar$, and thus the change of variables preserves the polynomiality property too.

When applied to the recursion relation (10) literally, the change of variables modifies it to a new recursion relation. In the new form the elementary fraction

$$\frac{q^m}{m\hbar + \lambda_\alpha - \lambda_\beta} = \frac{q^m}{\hbar(\lambda_\alpha - \lambda_\beta)} \cdot \frac{1}{\hbar^{-1} + m/(\lambda_\alpha - \lambda_\beta)}$$

occurs with the extra factor

$$\begin{aligned} \delta &:= \exp[mg(q) + \lambda_\alpha g(q)\hbar^{-1} - \lambda_\beta g(q) m/(\lambda_\beta - \lambda_\alpha)] \\ &= \exp[\lambda_\alpha g(\hbar^{-1} + m/(\lambda_a - \lambda_\beta))]. \end{aligned}$$

Thus $\delta - 1$ is divisible by $\hbar^{-1} + m/(\lambda_\alpha - \lambda_\beta)$. Since $g(0) = 0$, the result of this division is a q-series with coefficients polynomial in $\hbar^{-1}$ at each power of q. Thus the transformation affects only the initial condition and takes a solution of the recursion relation into another solution. □

Updates.

Definitions. The definition of virtual numbers n_d in terms of Euler classes of the orbibundles V_d over $X_{0,d}$ should be considered as tentative, and has been replaced by a more universal construction, due to Li and Tian [**16**], of virtual fundamental classes $[Y_{0,d}]$ defined in intrinsic terms of the quintic 3-fold Y rather than in terms of the embedding $i : Y \subset X$. Thus in order to place the above proof of the quintic formula into the framework of contemporary definitions one needs to check that $i_*[Y_{0,d}]$ in $H_*(X_{0,d})$ equals the cap-product of the virtual fundamental class $[X_{0,d}]$ with $Euler(V_d)$. This is easy and can be done as follows.

Consider first a model problem studied in [**8**]: given a holomorphic section $s : B \to E$ of a vector bundle $E \to B$ over a compact complex manifold, construct a cycle in the zero locus $Z = s^{-1}(0)$ of the section which is Poincaré-dual to the Euler class of the bundle. The model problem is solved by the normal cone construction: the normal cone $C \subset E|_Z$ to the zero locus has pure dimension $\dim B$, and the homological intersection in $E|_Z$ of the fundamental cycle $[C]$ with that of the zero section represents in $H_*(Z)$ the required Euler class. The construction can be adjusted to the orbifold/orbibundle situation.

The virtual fundamental cycle construction in [**16**] is based on the observation that the normal cone $C \subset E|_Z$ is intrinsic with respect to the scheme structure of Z and the *tangent-obstruction* complex $ds : T_B|_Z \to E|_Z$ of vector bundles over Z defined by the differential of the section. The kernel of $ds|_{z\in Z}$ is the algebraic tangent space $T_z Z$. In the case when Z is a moduli space of stable maps to Y, Li and Tian exhibit a tangent-obstruction complex of orbibundles $T \to E$ with this property and by this define the intrinsic normal orbi-cone $C \subset E$ and the virtual fundamental class $[Z] := [C] \cap$ [zero section] in $H_*(Z, \mathbb{Q})$.

In our situation $Z = Y_{0,d}$ is given in the orbifold $X_{0,d}$ by a section s of $V_d : E_{0,d} \to X_{0,d}$. Using the exact sequence $0 \to T_Y \to T_X|_Y \to N_Y \to 0$, where N_Y is the normal bundle to Y in X, we obtain the exact sequence

$$0 \to H^0(\Sigma, f^*T_Y) \to H^0(\Sigma, f^*T_X) \to H^0(\Sigma, f^*N_Y) \to H^1(\Sigma, f^*T_Y) \to 0$$

for each stable map $f : \Sigma \to Y$. Since the fibers of $V_d|_Z$ coincide with $H^0(\Sigma, f^*N_Y)$, this implies (via the description [**15, 16**] of the algebraic tangent space $T_{[f]}Y_{0,d}$ in cohomological terms of deformation theory) that the complex $ds : T_{X_{0,d}}|_Z \to E_{0,d}|_Z$ can take on the role of the tangent-obstruction complex in the definition of the virtual fundamental cycle $[Y_{0,d}]$. Thus this cycle represents in $H_*(X_{0,d})$ the Euler class of V_d.

While this obvious argument shows that the GW-invariant of Y in question can be computed in terms of the GW-theory for the convex bundle $V : E \to X$, some other GW-invariants of Y cannot be interpreted in terms of the bundle. In order to distinguish the GW-theory of the bundle from the GW-theory of Y, we, following A. Schwarz, refer in [**11**] to the first one as the GW-theory of the *supermanifold* ΠE.

The map μ. Applications of our approach to complete intersections in toric manifolds more general than projective spaces show that some steps in the above proof are redundant. For the rest of this paper we discuss several such steps which can be simplified or avoided. The first of them is our use of the map $\mu : GX_{0,d} \to LX_d$.

The map was used to define the equivariant class p on $GX_{0,d}$ as a pull-back of the corresponding class on LX_d and thus assure the polynomiality property of the GW-invariant J^{eq}. Consider instead the following $S^1 \times G$-equivariant GW-invariant of $\mathbb{C}P^1 \times X$:

$$\sum_{d=0}^{\infty}\sum_{n=0}^{\infty} \frac{q^d}{n!}[T, \dots, T]_{n,d}, , \tag{13}$$

where $T = z(p \otimes P)$, p is the generator of the S^1-equivariant cohomology of $\mathbb{C}P^1$ satisfying $p(p - \hbar) = 0$, P is the generator of the G-equivariant algebra of X, and $[\cdots]_{n,d}$ is the equivariant GW-invariant defined by integration over $GX_{n,d}$ against the Euler class of $GV_{n,d}$. The series (13) is defined without the fixed point localization and thus is a (q, z)-series with coefficients polynomial in $\hbar$ and λ. On the other hand, applying localization to fixed points of the S^1-action on $GX_{n,d}$ as in the proof of Theorem B (notice that p localizes to 0 at $\zeta = 0$ and to $\hbar$ at $\zeta = \infty$) and then using the divisor equation for zP as in the proof of Theorem C, we will find that the series (13) coincides with $\langle J^{eq}(q \exp \hbar z, \hbar^{-1}), J^{eq}(q, -\hbar^{-1})\rangle$. This argument was mentioned in [**11**] and was used in [**12**].

Theorem F. The invariance of the recursion relation (10) under mirror transformations can be deduced from the string and divisor equations. Namely, consider the G-equivariant GW-invariant $\mathcal{J}^{eq}$ defined by integration over $X_{n+2,d}$ against the Euler classes of $V_{n+2,d}$:

$$\langle \mathcal{J}, \phi\rangle = \langle 1, \phi\rangle + \sum_{(n,d)\neq(0,0)} \frac{q^d}{n!}\Big(1, T, \dots, T, \frac{\phi e^{P\ln q/\hbar}}{\hbar - c}\Big)_{n+2,d} \tag{14}$$

with $T = a(q) + b(q)P$, where a, b are power q-series vanishing at $q = 0$. One can derive a recursion relation for $\mathcal{J}^{eq}$ in exactly the same way as we derived the recursion relation for J^{eq} in Theorem D. The recursion coefficients will be the same as in Theorem D, but the initial condition will depend now on a and b. On the other hand, the string and divisor equations show that

$$\mathcal{J}^{eq} = e^{a(q)} J^{eq}(qe^{b(q)}, \hbar^{-1})$$

and is therefore a result of a mirror transformation applied to J^{eq}. Also, using the argument from the previous subsection with a suitable function of a and b in the role of T, we can make sure that $\mathcal{J}^{eq}$ must *a priori* satisfy the polynomiality condition as well.

Mirror transformations. As it was shown in [**9**] (Section 12) and [**12**] (Section 5), there is a "non-linear Serre duality" equivalence between genus 0 equivariant GW-theory for a convex supermanifold ΠE and a similar theory for the non-compact total space E^* of the concave dual bundle $V^* : E^* \to X$. Namely, their genus 0 GW-invariants differ by a change of variables which can be explicitly described in terms of GW-invariants of either of them (see [**12**], Section 5).

On the other hand, the mirror formulas can be generalized, as was shown in [**17**] and [**12**] (Section 4), to include genus 0 equivariant GW-invariants of concave bundles E^*. The proof is completely parallel to the one given above. However, in the case when the bundle V^* is the direct sum of at least two line bundles, the GW-invariant $J^{eq}_{E^*}$ for E^* is *equal* to the corresponding hypergeometric series $I^{eq}_{E^*}$ which in this case happens to satisfy the asymptotic condition of the uniqueness Theorem E. With this observation, a proof of the quintic formula looks as follows (see Section 5 in [**12**]).

Describe quintic 3-folds by two equations in $\mathbb{C}P^5$ of degree 1 and 5. For the concave bundle $\mathcal{O}(-1) \oplus \mathcal{O}(-5)$ over $\mathbb{C}P^5$, prove the equality $J^{eq}_{E^*} = I^{eq}_{E^*}$ following the steps Theorem B — Theorem D — Theorem E as explained above. Now the mirror transformation between $J^{eq}_{\Pi E}$ and $I^{eq}_{\Pi E}$ emerges from the general formulas of "non-linear Serre duality".

Theorem D. The proof of mirror formulas given in [**17**] is based on a recursive property of the $S^1 \times G$-equivariant Euler classes of the bundles GV_d over $GX_{0,d}$ called *Eulerity.* In fact the property can be easily deduced from our recursion relation (10) as it is done in [**9**] (Proposition 11.4 (b)). The inverse implication is also immediate. However, the proof of the Eulerity property given in [**17**] is based on arguments similar to our proof of Theorem B (localization for the S^1-action) and thus completely eliminates the role of localization formulas for the G-action as a computational tool (and uses only the fact of their existence).

With this observation, the reduction of the quintic formula to the "non-linear Serre duality" theorem looks particularly short: prove Eulerity in the case of the concave bundle $\mathcal{O}(-1) \oplus \mathcal{O}(-5)$ over $\mathbb{C}P^5$ following the argument in [**17**], and use a uniqueness theorem (parallel to Theorem E above) in order to identify $J^{eq}_{E^*}$ with $I^{eq}_{E^*}$.

We see that the only new (after Kontsevich's paper [**15**]) geometrical construction which has survived so far through all variants of the proof of the quintic formula is the S^1-*equivariant* theory on the graph spaces $GX_{0,d}$, which originates from the loop space interpretation [**10**] (see also [**21**]) of Gromov–Witten invariants.

All other ingredients of the proof have a somewhat combinatorial character, and can be interchanged and simplified. This progress does not mean, however, that the mirror symmetry phenomenon has been adequately understood.

References

[1] P. Aspinwall and D. Morrison, *Topological field theory and rational curves*, Comm. Math. Phys. **151** (1993), 245–262.

[2] V. Batyrev, I. Ciocan-Fontanine, B. Kim, and D. van Straten, *Conifold transitions and mirror symmetry for Calabi–Yau complete intersections in Grassmannians*, Nuclear Phys. **B514** (1998), 640–666.

[3] K. Behrend and Yu. Manin, *Stacks of stable maps and Gromov–Witten invariants*, Duke Math. J. **85** (1996), 1–60.

[4] A. Bertram and M. Thaddeus, *On quantum cohomology of symmetric product of an algebraic curve*, Preprint `math. AG/9803026`.

[5] G. Bini, C. de Concini, M. Polito, and C. Procesi, (a) *Givental's work*, Preprint `math. AG/9805097`; (b) *On the work of Givental relative to mirror symmetry*, Notes of Courses Given by Teachers at the School, Scuola Normale Superiore, Pisa, 1998..

[6] P. Candelas, X. C. de la Ossa, P. S. Green, and L. Parkes, *A pair of Calabi–Yau manifolds as an exact soluble superconformal theory*, Nuclear Phys. **B359** (1991), 21–74.

[7] B. Dubrovin, *The geometry of 2D topological field theories*, Lecture Notes in Math., vol. 1620, Springer-Verlag, Berlin, 1996, pp. 120–348.

[8] W. Fulton, *Intersection theory*, Springer-Verlag, Berlin, 1984.

[9] A. Givental, *Equivariant Gromov–Witten invariants*, Internat. Math. Res. Notices **1996**, 613–663.

[10] ———, *Homological geometry.* **I:** *Projective hypersurfaces*, Selecta Math. (N. S.) **1** (1995), 325–345.

[11] ———, *A mirror theorem for toric complete intersections*, Preprint, `alg-geom/9701016`; Topological Field Theory, Primitive Forms, and Related Topics (M. Kashiwara et al., eds.), Birkhäuser, Boston, 1998, pp. 141–175.

[12] ———, *Elliptic Gromov–Witten invariants and the generalized mirror conjecture*, Preprint `math. AG/9803053`; Integrable Systems and Algebraic Geometry (Kobe/Kyoto, 1997; M.-H. Saito et al., eds.), World Sci. Publ., Singapore, 1998, pp. 107–155.

[13] B. Kim, *Quantum cohomology of flag manifolds G/B and quantum Toda lattices*, Ann. of Math. (2) **149** (1999), 129–148.

[14] ———, *Quantum hyperplane section theorem for homogeneous spaces*, Preprint `alg-geom/9712008`.

[15] M. Kontsevich, *Enumeration of rational curves via tori actions*, The Moduli Space of Curves (R. Dijkgraaf et al., eds.) Progress in Math., vol. 129, Birkhäuser, Boston, 1995, pp. 335–368.

[16] J. Li and G. Tian, *Virtual moduli cycles and Gromov–Witten invariants of algebraic varieties*, J. Amer. Math. Soc. **11** (1998), 119–179.

[17] B. Lian, K. Liu, and S.-T. Yau. *Mirror principle.* I, Asian J. Math, **1** (1997), 729–763.

[18] Yu. Manin. *Generating functions in algebraic geometry and summation over trees*, The Moduli Space of Curves (R. Dijkgraaf et al., eds.), Progress in Math., vol. 129, Birkhäuser, Boston, 1995, pp. 401–417.

[19] D. Morrison and R. Plesser, *Summing the instantons: Quantum cohomology and mirror symmetry in toric varieties*, Nuclear Phys. **B440** (1995), 279–354.

[20] R. Pandharipande, *Rational curves on hypersurfaces (after A. Givental)*, Sem. Bourbaki 1997/98, Astérisque, no. 252, Soc. Math. France, Paris, 1998, pp. 307–340.

[21] C. Voisin, *Symmétrie miroir*, Panoramas et Synthèses, vol. 2, Soc. Math. France, Paris, 1996.

Department of Mathemtics, University of California at Berkeley, Berkeley, CA 94720

Amer. Math. Soc. Transl.
(2) Vol. **196**, 1999

Stabilisation of Symplectic Inequalities and Applications

François Lalonde and Charles Pestieau

Abstract. In this short note, we show that the basic symplectic inequalities – the non-squeezing theorem and the energy-capacity inequality – are *stable* under product with any closed symplectic manifold, in the following sense. Let M^{2n} and W^{2m} be symplectic manifolds with W closed. Denote by $B^{2n}(a)$ the standard closed ball of $\mathbf{R}^{2n}$ of radius r satisfying $\pi r^2 = a$. We prove that any symplectic embedding of $B^{2n+2}(c) \times W$ in $B^2(\lambda) \times M \times W$ must satisfy $c \leq \lambda$. This evidently generalises the ordinary non-squeezing theorem, which we recover when W reduces to a point. Similarly, given any subset A of M, we prove that the disjunction energy $e(A \times W)$ of the cylinder $A \times W$ inside $M \times W$ must satisfy $e(A \times W) \geq \frac{1}{2}\mathrm{cap}(A)$, where $\mathrm{cap}(A)$ is the Gromov capacity of $A \subset M$. Hence $e(A \times W) \geq \frac{1}{2}\max(\mathrm{cap}(A \times W), \mathrm{cap}(A))$.

We then apply these stabilisations to prove a basic result on the extension of Hofer's geometry to all symplectic diffeomorphisms (not only the Hamiltonian ones), conjectured by Lalonde and Polterovich.

1. Introduction

1.1. Stabilisation of the non-squeezing theorem. Let M be any symplectic manifold (not necessarily compact) of dimension say $2n$. Lalonde and McDuff proved in [**LM**] the following general version of Gromov's non-squeezing theorem [**G**]: a symplectic embedding $B^2(c)$ in $B^2(\lambda) \times M$ is possible only if $c \leq \lambda$. In this paper, we consider the following generalisation of that theorem.

Theorem 1.1. *Let M and W be symplectic manifolds with W closed. If there is a symplectic embedding*

$$B^{2n+2}(c) \times W \to B^2(\lambda) \times M \times W,$$

then $c \leq \lambda$.

The proof of this theorem is based on J-holomorphic curves. Actually, it relies on the observation that a complete connected J-holomorphic curve with non-empty boundary passing through the fiber $\{0\} \times W$ of $B^{2n+2}(c) \times W$ must project to a J_0-holomorphic curve in $B^{2n+2}(c)$ passing through the center with non-empty boundary, provided that J decomposes as $J_0 \oplus J_W$ and W is closed. Thus the

---

1991 *Mathematics Subject Classification.* Primary 58F05.

The first author was partially supported by NSERC grant OGP 0092913 and FCAR grant ER-1199.

©1999 American Mathematical Society

argument used in the case of the ordinary non-squeezing theorem applies here as well, with the usual difficulty caused by J-holomorphic curves in general symplectic manifolds.

1.2. Stabilisation of the energy-capacity inequality. The second theorem is the stabilisation of the energy-capacity inequality in Hofer's geometry, which was proved in special cases in [**LP**], §4, using other methods. Let ϕ be any compactly supported Hamiltonian diffeomorphism of a symplectic manifold M, and define its Hofer norm $\|\phi\|_H$ as the infimum, over all smooth compactly supported Hamiltonian isotopies $\phi_t : M \times [0,1] \to M$ joining the identity to ϕ, of the quantity

$$\int_0^1 (\max_M H_t - \min_M H_t)\, dt,$$

where H_t is the Hamiltonian that generates the isotopy. Define then, for any subset $A \subset M$, the energy of disjunction $e(A)$ as the infimum of $\|\phi\|_H$ over all Hamiltonian diffeomorphisms ϕ that separate A from itself (i.e., $\phi(A) \cap A = \emptyset$). Obviously, the non-degeneracy of the Hofer norm is an immediate consequence of the following energy-capacity inequality:

$$e(A) \geq \frac{1}{2}\mathrm{cap}(A),$$

which was established in Euclidean space by Hofer [**H**] using variational techniques, and in [**LM**] in general using J-holomorphic techniques. We prove here the following stable version of this inequality.

THEOREM 1.2. *If A is a subset of a symplectic manifold M and W a closed symplectic manifold, then the energy of disjunction of $A \times W$, as a subset of $M \times W$, satisfies:*

$$e(A \times W) \geq \frac{1}{2}\mathrm{cap}(A).$$

Thus

$$e(A \times W) \geq \frac{1}{2}\max\{\mathrm{cap}(A), \mathrm{cap}(A \times W)\},$$

which means that our stable version of the inequality is especially useful when W has a small capacity with respect to A, since in this case the capacity of $A \times W$ is smaller that the capacity of A.

REMARK. It is expected that the sharper inequality without the factor $\frac{1}{2}$ holds in all cases. Indeed, one can prove this sharper inequality with mild hypotheses on the set A (see [**LM**] for the notion of *proper disjunction*). We will not consider this question here.

1.3. Application to bounded isometries. The group $\mathrm{Symp}(M)$ of all symplectic diffeomorphisms of a symplectic manifold M acts on its subgroup $\mathrm{Ham}_c(M)$ of compactly supported Hamiltonian diffeomorphisms by conjugation. This action preserves the Hofer norm; hence each symplectomorphism ϕ induces an isometry

$$C_\phi : \mathrm{Ham}_c(M) \to \mathrm{Ham}_c(M)$$

by $C_\phi(f) = \phi f \phi^{-1}$. Following Lalonde and Polterovich [**LP**], we say that ϕ is *bounded* if the corresponding isometry C_ϕ is bounded with respect to the Hofer norm on $\mathrm{Ham}_c(M)$, that is, if the expression

$$d(C_\phi(f), f) = \|f^{-1}\phi f \phi^{-1}\|_H = \|[f,\phi]\|_H = d_H(id, [f,\phi])$$

remains bounded as f runs over all elements of $\mathrm{Ham}_c(M)$.

For $\alpha \in (0, \infty]$, we define $r_\alpha(\phi) \in [0, \infty]$ as the supremum of $d(C_\phi(f), f)$ over the set of all Hamiltonian diffeomorphisms f in a ball of radius α centered at the identity. One easily shows ([**LP**]) that

$$r_\alpha(\phi\psi) \leq r_\alpha(\phi) + r_\alpha(\psi).$$

In particular, setting $r = r_\infty$, we have

$$r([\phi, \psi]) \leq 2r(\phi),$$

which we will need later.

We denote by $BI(M) \subset \mathrm{Symp}(M)$ the group of bounded isometries, and set

$$BI_0(M) = BI(M) \cap \mathrm{Symp}_0(M),$$

where $\mathrm{Symp}_0(M)$ denotes the identity component of $\mathrm{Symp}(M)$. The inequality $r([\phi, \psi]) \leq 2r(\phi)$ implies that $\mathrm{Ham}_c(M)$ is a (normal) subgroup of $BI_0(M)$ and that on this subgroup the new norm is bounded above by the Hofer norm (up to a constant). It was conjectured in [**LP**] that $BI_0(M)$ coincides with $\mathrm{Ham}_c(M)$. Our main result is the following theorem.

THEOREM 1.3. *If M can be expressed as a product $M = N \times W$ with N any product of closed surfaces and W any closed symplectic manifold with first real Betti number equal to zero, then the group $BI_0(M)$ coincides with* $\mathrm{Ham}_c(M)$.

This provides a metric characterisation of Hamiltonian diffeomorphisms in such manifolds. The proof of this theorem, given in the last section of this note, is a mild adaptation of the argument of [**LP**], §4, once one has the stable energy-capacity inequality.

2. Proof of the stable non-squeezing theorem

We will first give the general argument that easily generalises Gromov's geometric idea; this argument relies on the existence of some J-holomorphic cusp-curve in a general symplectic manifold. We will then indicate how one can show the existence of such a curve, using virtual stable moduli spaces.

REMARK. Note, however, that the existence of that curve can also be proved by a geometric argument similar to the one used by Lalonde and McDuff, which does not require the use of virtual stable moduli spaces. This latter proof will appear elsewhere (see [**P**]).

2.1. Gromov's argument revisited.

Let

$$\psi : B^{2n+2}(c) \times W \to D^2(\lambda) \times M \times W$$

be a symplectic embedding. We wish to prove that $c \leq \lambda$. Let ε be small and set $M' = S^2(\lambda + \varepsilon) \times M \times W$. Let J_W be a generic compatible almost complex structure on W, and J_0 the standard structure on the ball. Then extend the structure $\psi_*(J_0 \oplus J_W)$ to the whole manifold M', as an almost complex structure J' compatible with the form Ω on M'. We claim that there exists a J'-rational curve C passing through $\psi(0, w)$ (where $w \in W$ is any fixed point) which is not completely included in $\psi(B^{2n+2}(c) \times W)$, and whose area is bounded above by

$$\Omega([S^2 \times \{pt\} \times \{pt\}]).$$

Taking the inverse image by ψ, this gives a $(J_0 \oplus J_W)$-holomorphic curve C' in $B^{2n+2}(c) \times W$, passing through $\{0\} \times W$. That curve is defined on a closed smooth Riemann surface with *non-empty* boundary, the boundary being sent to the boundary of $B^{2n+2}(c) \times W$. Because W is closed, the J-curve has its boundary included in $(\partial B^{2n+2}(c)) \times W$. Thus its projection on the first factor is a J_0-holomorphic curve C_0 passing through 0 with non-empty boundary included in the boundary of $B^{2n+2}(c)$. Hence the minimality of that surface implies that its area is bounded below by c. Thus we get

$$c \leq \text{area}(C_0) = \int_{C_0} \omega_0 \leq \int_{C'} \omega_0 \oplus \omega_W = \int_C \Omega \leq \Omega([S^2 \times \{pt\} \times \{pt\}]) = \lambda + \varepsilon,$$

which holds for arbitrarily small ε. We conclude that $c \leq \lambda$, as required. □

Thus it remains to prove the existence of the J'-rational curve passing through $\psi(0, w)$ claimed above. This is the goal of the next subsection.

2.2. Virtual moduli spaces and the Gromov invariant. In this subsection, we recall the definition of stable virtual pseudo-holomorphic curves, following Kontsevich's idea. It is well-known that this definition should lead to well-defined Gromov invariants in all symplectic manifolds (not only the weakly monotone ones). The arguments have been written down by Ruan [**R**], Fukaya and Ono [**FO**], Liu and Tian [**LT**], and Li and Tian [**LiT**], and more recently by McDuff [**McD**] (in the case of a fibration over S^2, which applies to our situation). In our opinion, the latter is written in a quite satisfactory way. However, they all rely on versions of infinite dimensional orbibundles and generalised sections of them. Since that theory has not yet reached a full development, there is still some reservation about the results so obtained. This is why the second author wrote down a proof of the theorems ([**P**]) of this paper that does not rely on stable virtual moduli spaces, but follows instead the geometric construction of [**LM**].

We will restrict ourselves to the genus zero case, since this is enough for our purposes.

Let (V, Ω) be a symplectic manifold of dimension $2n$, and J an almost complex structure compatible with Ω. Let k be a non-negative integer, and β an element of $H_2(V, \mathbf{Z})$. A *stable map* of genus 0 with k marked points in the homology class β is a triple (Σ, z, u), where Σ is a compact complex variety of dimension 1 having at worst ordinary double points as singularities, with each component of genus 0, z is a k-tuple, $z = (z_1, \ldots, z_k)$, of distinct marked points on Σ different from the singular points, and $u : \Sigma \to V$ is a J-holomorphic map such that

(1) $u_*([\Sigma]) = \beta$, and
(2) for each component Σ_i of Σ, the map $u|_{\Sigma_i}$ is non-constant if $m_i < 3$ (here m_i is the number of points which are marked or singular).

Denote by $\bar{\mathcal{M}}_k(V, \beta)_p$ the space of such stable maps (p for *parametrised*) and by $\bar{\mathcal{M}}_k(V, \beta)$ the quotient by holomorphic reparametrisations; i.e., if $\phi : \Sigma \to \Sigma$ is a biholomorphic map, then (Σ, z, u) is identified with $(\Sigma, \phi^{-1}(z), u \circ \phi)$. The following theorem is then a mild adaptation of Gromov's compactness theorem:

THEOREM 2.1. *The space $\bar{\mathcal{M}}_k(V, \beta)$ is compact.*

The main result is then the following theorem.

THEOREM 2.2. *For generic compatible J, there is a perturbation of the equation $\bar{\partial}_J u = 0$ (defined by replacing the right hand side with a small non-vanishing term)* *such that the perturbed moduli space* $\mathcal{P}\bar{\mathcal{M}}_k(V,\beta)$ *carries a fundamantal class over* $\mathbf{Q}$. *Moreover, the evaluation map*

$$ev : \bar{\mathcal{M}}_k(V,\beta) \to V \times V \times \cdots \times V$$

defined by $ev(\Sigma, u) = (u(z_1), \ldots, u(z_k))$ *can also be defined on* $\mathcal{P}\bar{\mathcal{M}}_k(V,\beta)$. *It gives an element* $[\mathcal{P}\bar{\mathcal{M}}_k(V,\beta)]$ *of* $H_d(V^k;\mathbf{Q})$ *of dimension*

$$d = 2(c_1(\beta) + n + k - 3)$$

equal to the expected dimension, which is independent of the choice of J *and the perturbation. Given any piecewise smooth cycle in* V^k, *the perturbation can be chosen so that* $[\mathcal{P}\bar{\mathcal{M}}_k(V,\beta)]$ *is transversal to that cycle.*

In our situation, $V = M' = S^2 \times M \times W$, $\beta = \alpha = S^2 \times \{pt\} \times \{pt\}$ and $k = 1$. The perturbation is chosen so that the cycle $[\mathcal{P}\bar{\mathcal{M}}_1(M',\alpha)]$ is transversal to the 0-cycle consisting of the single point $\psi(0,w) \in M'$. Note that the dimension d of the cycle $[\mathcal{P}\bar{\mathcal{M}}_1(M',\alpha)]$ is equal to $2(2 + (1+n+m) + 1 - 3) = 2(1+n+m)$, which is the dimension of M'. The Gromov invariant is defined as the intersection number of these two cycles.

We can easily see that it is equal to 1. Indeed, for generic compatible $J_{M\times W}$ on $M \times W$, the structure $J_0 \oplus J_{M\times W}$ on M' and the unperturbed homogeneous equation $\bar{\partial}_{J_0 \oplus J_{M\times W}} u = 0$ are already generic: they lead to a smooth cycle of the expected dimension. But obviously the invariant is equal to 1 in this case, since the moduli space is just the one consisting of all flat S^2's of M' with one marked point. Thus the evaluation map sends this diffeomorphically onto M'.

Finally, the above theorem states that one gets the same value with a generic J on M' arbitrarily close to the J' defined in 2.1, and for an arbitrarily small perturbation of the right hand side of the $\bar{\partial}$-equation. This means that there is some stable cusp-curve Σ passing through $\psi(0,w)$ that satisfies the equation

$$\bar{\partial}_J u = \nu.$$

Let C be the component of Σ that passes through $\psi(0,w)$. The intersection of that component with $\psi(B^{2n}(c) \times W)$ is a complete curve with non-empty boundary included in the boundary of $\psi(B^{2n}(c) \times W)$ (this is because, by slightly perturbing the point $p = \psi(0,w)$, one can always assume that it belongs to a principal component of the cusp curve; thus we can avoid choosing a component that would lie entirely inside $\psi(B^{2n}(c) \times W)$). Now, letting J tend to J' and ν to 0, we get by Gromov's compactness theorem a J'-cusp-curve containing at least one component C such that (1) C passes through $\psi(0,w)$ and (2) C is not entirely included in $\psi(B^{2n}(c) \times W)$. But this is what we needed.

3. Proof of the stable energy-capacity inequality

In this section we prove Theorem 1.2. Basically, it can be derived from Theorem 1.1 in the same way as the ordinary energy-capacity inequality is derived from the non-squeezing theorem in [**LM**]. Let A be any subset of a symplectic manifold M, and W any closed manifold. We will show that if $f_t : W \times M \times [0,1] \to \mathbf{R}$ is a compactly supported Hamiltonian generating an isotopy ϕ_t such that $\phi_1(W \times A) \cap (W \times A) = \emptyset$, then

$$\int_0^1 (\max_M f_t - \min_M f_t)\, dt \;\geq\; \frac{1}{2}\mathrm{cap}(A).$$

We will call the quantity on the left e, the energy of f_t. Taking the infimum over all such Hamiltonians yields

$$e(A \times W) \geq \frac{1}{2}\text{cap}(A).$$

Let $B \subset A$ be the image of a standard ball of capacity c; we must show that $e \geq c/2$. Our method is to make use of f_t in order to construct a symplectic embedding

$$W \times B^{2n+2}(c) \to W \times M \times B^2(e + c/2 + 3\varepsilon),$$

of a type that we shall call wrapped. Gromov's non-squeezing theorem then states that $c \leq e + c/2 + 3\varepsilon$, which is the result we seek as ε goes to zero.

Let us now show how to construct the wrapped embedding. Picture a set $Y = P_1 \cup L \cup P_2 \subset \mathbf{R}^2$, comprised of two rectangles of area $c/2$ linked by a unit line. For instance, choose

$$P_1 = \{(u,v)|0 \leq u \leq 1/2, 0 \leq v \leq c\},$$
$$L = \{(u,v)|1/2 \leq u \leq 3/2, v = 0\},$$
$$P_2 = \{(u,v)|3/2 \leq u \leq 2, 0 \leq v \leq c\},$$

and call $Z = W \times B^{2n}(c) \times Y = W \times B^{2n}(c) \times (P_1 \cup L \cup P_2)$ the product set. First note that we can find a symplectic embedding

$$W^{2m} \times B^{2n+2}(c) \to W^{2m} \times B^{2n}(c) \times \mathcal{N}(Y),$$

for any neighborhood $\mathcal{N}(Y)$ of Y. Indeed it is easy to find an area preserving map $h : B^2(c) \to \mathcal{N}(Y)$ which splits $B^2(c)$ into two equal parts joined by a narrow band and mapping each part to a neighborhood of a rectangle P_i with the narrow band mapped to a neighborhood of L. We then lift this map to

$$W \times B^{2n+2}(c) \xrightarrow{i} W \times B^{2n}(c) \times B^2(c) \xrightarrow{(\text{id}\times\text{id}\times h)} W \times B^{2n}(c) \times \mathcal{N}(Y).$$

The second step is an intermediate unwrapped mapping

$$G : Z \to W \times M \times \mathbf{R}^2,$$

which we now describe. Let $g : W \times B^{2n}(c) \to W \times A \subset W \times M$ be the lift of the embedding $B^{2n}(c) \to A \subset M$. Define G as follows:

$$\begin{aligned} &G : W \times B^{2n}(c) \times Y \to W \times M \times \mathbf{R}^2, \\ &\forall\, (u,v) \in P_1, \quad (x,y,u,v) \mapsto (g(x,y), u, v), \\ &\forall\, (u,v) \in L, \quad (x,y,u,v) \mapsto (\phi_{u-1/2}(g(x,y)), u, f_{u-1/2}(\phi_{u-1/2}(g(x,y)))), \\ &\forall\, (u,v) \in P_2, \quad (x,y,u,v) \mapsto (\phi_1(g(x,y)), u, f_1(\phi_1(g(x,y)))), \end{aligned}$$

where $(u,v) \in Y \subset \mathbf{R}^2$, $(x,y) \in W \times B^{2n}(c)$ and $g(x,y) \in W \times M$. This embedding can be shown to be smooth if f_t is taken to be smooth for each t and equal to 0 in neighborhoods of $t = 0$ and $t = 1$. This may require choosing a Hamiltonian f_t of slightly greater energy $e + \varepsilon$. Moreover, it is easy to check that G is actually a symplectic map where we have chosen the standard symplectic structures on $Y \subset \mathbf{R}^2$ and $\mathbf{R}^2$.

In order to construct our wrapped embedding, observe that the image of G lies in the set $W \times M \times V$, where $V = P_1 \cup Q \cup P_2$, $Q = \{(u,v)|1/2 \leq u \leq 3/2, \min_M f_{u-1/2} \leq v(u) \leq \max_M f_{u-1/2} + \varepsilon\}$, so that the area of Q is $e + \varepsilon$, where e is the energy of the initial Hamiltonian. Now form an annulus X from the set

$P_1 \cup Q$ by identifying through the translation $u \mapsto u + 3/2$ the sides $u = 0$ of P_1 and $u = 3/2$ of Q, and benote by P' and Q' the images of P_1 and Q in X. We now define the wrapped embedding $W \times B^{2n}(c) \times Y \to W \times M \times X$, as follows:

$$W \times B^{2n}(c) \times P_1 \stackrel{G}{\to} W \times M \times P_1 \stackrel{(\mathrm{id}\times\mathrm{id}\times i)}{\to} W \times M \times P',$$

$$W \times B^{2n}(c) \times Q \stackrel{G}{\to} W \times M \times Q \stackrel{(\mathrm{id}\times\mathrm{id}\times i)}{\to} W \times M \times Q',$$

$$W \times B^{2n}(c) \times P_2 \stackrel{G}{\to} W \times M \times P_2 \stackrel{(\mathrm{id}\times\mathrm{id}\times i)}{\to} W \times M \times P',$$

where the first two inclusions $P_1 \stackrel{i}{\to} P'$ and $Q \stackrel{i}{\to} Q'$ are obvious and the last one is factored by the map $P_2 \to P_1$ given as $v \to v$ and $u \to u - 3/2$. Note that this map $(id \times id \times i) \circ G$ is injective. Indeed since ϕ_1 disjoins the set $W \times A$ we see that the images $(id \times id \times i) \circ G\ (W \times B^{2n}(c) \times P_1)$ and $(id \times id \times i) \circ G\ (W \times B^{2n}(c) \times P_2)$ do not overlap even though the images of P_1 and P_2 are overlaping in X. Further, the map $(id \times id \times i) \circ G$ is symplectic and smooth in the same way as G.

We now have a symplectic embedding

$$W \times B^{2n}(c) \times Y \to W \times M \times X(c/2 + e + \varepsilon),$$

and the symplectic neighborhhood theorem implies that it extends to a symplectic embedding $W \times B^{2n}(c) \times \mathcal{N}(Y) \to W \times M \times X(c/2 + e + 2\varepsilon)$ in a neighborhood $\mathcal{N}(Y)$ of Y. Let this map be preceded by our first embedding $W \times B^{2n+2}(c) \to W \times B^{2n}(c) \times \mathcal{N}(Y)$ given for any neighborhood $\mathcal{N}(Y)$ of Y, and let it be followed by the lift of an area preserving map $X(c/2+e+2\varepsilon) \to B(c/2+e+3\varepsilon)$. Combining these maps produces the required symplectic embedding

$$W \times B^{2n+2}(c) \to W \times M^{2n} \times B(c/2 + e + 3\varepsilon).$$

□

4. Application to bounded isometries

In this section we prove Theorem 1.3, which states that $BI_0(M) = \mathrm{Ham}_c(M)$ for a symplectic manifold $M = N \times W$, where N is a product of surfaces (of any genus) and W is a closed symplectic manifold whose first cohomology group is trivial. Given the results of the previous sections, this last theorem is an obvious generalisation of §§3 and 4 of [**LP**]. We recall the main steps for the sake of completeness.

EXAMPLE 4.1. The following stronger result is shown in Lalonde–Polterovich [**LP**]: $BI(S) = BI_0(S) = \mathrm{Ham}_c(S)$ if S is a closed surface.

The proof of the theorem is based on two lemmas.

LEMMA 4.2. *Let $M = N \times W$, where N is a product of surfaces of strictly positive genus and W is a closed manifold. If $\phi \in \mathrm{Symp}(N)$ is strongly unbounded, then $\phi \times \psi$ is an unbounded symplectomorphism of $N \times W$ for any $\psi \in \mathrm{Symp}(W)$.*

We define below what "strongly unbounded" means. To prove this lemma, we begin by following [**LP**] closely, and then apply our stable energy-capacity inequality. We first recall some definitions.

DEFINITION 4.3. Let $\tilde{N}$ be the universal cover of a symplectic manifold N (associated to a base point $p \in N$). Let $g \in \mathrm{Ham}_c(N)$ and $q = g(p)$. For each point $q_i \in \tilde{q} = \pi^{-1}(q) \subset \tilde{N}$, define a lift $\tilde{g} : \tilde{N} \to \tilde{N}$ of g by specifying $\tilde{g}(p) = q_i$. We will say that $\tilde{g}$ is *admissible* among these lifts if it is the lift of a Hamiltonian isotopy g_t from id to g.

Note that if $\tilde{g}$ is admissible then $\tilde{g} \in \mathrm{Ham}(\tilde{N})$. In addition we clearly have the following result:

PROPOSITION 4.4 (Proposition 1.5.A in [**LP**]). *Let $g \in \mathrm{Ham}_c(N)$ and suppose that each admissible lift $\tilde{g} : \tilde{N} \to \tilde{N}$ of g disjoins a symplectic ball of capacity c in $\tilde{N}$. Then $\|g\|_H \geq c/2$.*

DEFINITION 4.5. We say that a symplectomorphism $\phi \in \mathrm{Symp}(N)$ is *strongly unbounded* if for each $c > 0$ there is a Hamiltonian diffeomorphism $f : N \to N$ such that each admissible lift $\overline{[f, \phi]}$ of the commutator $[f, \phi]$ disjoins a symplectic ball of capacity c in $\tilde{N}$.

EXAMPLE 4.6. It is shown in [**LP**] that a non-Hamiltonian symplectomorphism ϕ of a surface of strictly positive genus which acts trivially on $H_1(S, \mathbf{R})$ is strongly unbounded. This becomes obvious once ϕ is composed whith an appropriate Hamiltonian diffeomorphism.

The following proof is the crucial step in proving Theorem 1.3.

PROOF OF LEMMA 4.2. For each arbitrarily large $c > 0$ let us choose a Hamiltonian diffeomorphism $f : N \to N$ as in Definition 4.5. Set $T = \phi \times \psi$ and notice that $[f \times id, \phi \times \psi] = [f, \phi] \times id = R$.

Let $\tilde{R} : \tilde{N} \times W \to \tilde{N} \times W$ be an admissible lift of R to the covering $\tilde{N} \times W$, where $\tilde{N}$ is the universal covering of N. Then $\tilde{R}$ is a Hamiltonian diffeomorphism which disjoins the set $B(c) \times W$. The stabilized energy-capacity inequality then tells us that $\|\tilde{R}\|_H \geq c/2$. On the other hand, since $\|R\|_H \geq \|\tilde{R}\|_H$, T must be unbounded. □

Recall that there is a homomorphism from the universal cover of $\mathrm{Symp}_0(M)$ to the first cohomology group $H^1(M, \mathbf{R})$. This homomorphism maps the path $\{\phi_t\}_{t \in [0,1]}$ to the cohomology class $\int_0^1 [\lambda_t] dt$, where λ_t is the closed 1-form generating the isotopy ϕ_t at time t. We call this the *flux homomorphism*, and the image of $\pi_1(\mathrm{Symp}_0(M), id)$ is called the *flux group*, denoted G. What is important here is to recognize that two symplectomorphisms in $\mathrm{Symp}_0(M)$ differ by a Hamiltonian diffeomorphism if and only if they have the same image modulo G.

Now here is the second lemma (cf. Theorem 1.3.C in [**LP**]).

LEMMA 4.7. *Let $\phi \times \psi$ be a bounded symplectomorphism of $S \times W$, where S is a surface of strictly positive genus and W is a closed symplectic manifold. Then $\phi \in \mathrm{Ham}(S)$.*

PROOF OF LEMMA 4.7. Suppose that ϕ is not Hamiltonian. We will reach a contradiction by showing that $\phi \times \psi$ is unbounded. Let us distinguish two cases.

Case 1. ϕ acts trivially on $H^1(S, \mathbf{R})$. According to Example 4.6, ϕ is strongly unbounded. This means we can apply Lemma 4.2, which shows that $\phi \times \psi$ is unbounded.

Case 2. ϕ acts non-trivially on $H^1(S, \mathbf{R})$. In this case, an argument from [**LP**] shows that there exists $\varphi \in \mathrm{Symp}_0(S)$ such that $[\phi, \varphi]$ is non-Hamiltonian. Indeed, let $v \in H^1(S\mathbf{R})$ be such that $v - \phi^*(v)$ does not belong to the flux group G of the surface S (such a v exists becouse G is countable). Choose any $\varphi \in \mathrm{Symp}_0(S)$ with flux equal to v mod G. Then the commutator $[\phi, \varphi]$ has flux $v - \phi^*(v)$ mod G and is therefore not Hamiltonian.

On the other hand, since $\varphi \in \mathrm{Symp}_0(S)$ we must have $[\phi, \varphi] \in \mathrm{Symp}_0(S)$, so that $[\phi, \varphi]$ acts trivially on $H^1(S, \mathbf{R})$. We may now return to case 1. Indeed, according to Example 4.6, $[\phi, \varphi]$ is strongly unbounded, and hence Lemma 4.2 tells us that $[\phi, \varphi] \times \mathrm{id} = [\phi \times \psi, \varphi \times id]$ is unbounded. This is a contradiction, because $r([\phi \times \psi, \varphi \times id]) \leq 2r(\phi \times \psi) < \infty$. □

PROOF OF THEOREM 1.3. Let $M = S_1 \times S_2 \times \cdots \times S_k \times W$ be given as in the statement of the theorem. Suppose that ψ belongs to $BI_0(M)$. It follows from the flux theory adove that we can write it as a composition of a Hamiltonian diffeomorphism f and a split symplectomorphism $\phi = \phi_1 \times \cdots \times \phi_k \times \phi_W$, which must be bounded too. If S_j is a surface of positive genus, then Lemma 4.7 tells us that ϕ_j is Hamiltonian. On the other hand, the first cohomology group is trivial for any factor of M which is not a surface of positive genus. In this case the flux theory tells us that the corresponding ϕ_j is automatically Hamiltonian. We conclude that $\phi = (\phi_1 \times \cdots \times \phi_k \times \phi_W)$ is Hamiltonian, and therefore so is $f \circ \phi$. □

References

[FO] K. Fukaya and K. Ono, *Arnold conjecture and Gromov–Witten invariants*, Topology **38** (1999), 933–1048.

[G] M. Gromov, *Pseudo-holomorphic curves in symplectic manifolds*, Invent. Math. **82** (1985), 307–347.

[H] H. Hofer, *On the topological properties of symplectic maps*, Proc. Roy. Soc. Edinburgh **A115** (1990), 25–38.

[LM] F. Lalonde and D. McDuff, *The geometry of symplectic energy*, Ann. of Math. (2) **141** (1995), 349–371.

[LP] F. Lalonde and L. Polterovich, *Symplectic diffeomorphisms as isometries of Hofer's norm*, Topology **36** (1997), 711–727.

[LiT] J. Li and and G. Tian, *Virtual moduli cycles and Gromov-Witten invariants of general symplectic manifolds*, Topics in Symplectic 4-Manifolds (Irvine, Ca, 1996; R. J. Stern, editor), Internat. Press, Cambridge, MA, 1998, pp. 47–83.

[LT] G. Liu and G. Tian, *Floer homology and Arnold conjecture* J. Diff. Geom. **49** (1998), 1–74.

[McD] D. McDuff, *Quantum homology of fibrations over S^2*, Preprint, 1998.

[P] C. Pestieau, *Stabilisation of two symplectic theorems* (to appear).

[R] Y. Ruan, *Virtual neighbourhoods and pseudoholomorphic curves*, Topics in Symplectic 4-Manifolds (Irvine, CA, 1996; R. J. Stern, editor), Internat. Press, Cambridge, MA, 1998, pp. 101–116.

UNIVERSITÉ DU QUÉBEC À MONTRÉAL

E-mail address: flalondemath.uqam.ca

E-mail address: pestieaumath.uqam.ca

Amer. Math. Soc. Transl.
(2) Vol. **196**, 1999

The Virtual Moduli Cycle

Dusa McDuff

This article is an attempt to describe one possible construction of the virtual moduli cycle that is used as a tool in the construction of Gromov–Witten invariants for a general symplectic manifold. There are many different versions of this construction. Here I will in the main follow Liu and Tian [**LiuT1**], [**LiuT2**], since their approach (when modified by an idea of Siebert's) seems to involve the least amount of analysis. However, it does involve quite a bit of topology, some of which they only outline. The aim here is to flesh out their picture and to explain the different ingredients that are needed to make the construction work. We do not try to give full proofs, nor do we work out all the ideas in full generality.

Liu and Tian work in the category of partially smooth spaces (spaces with two topologies), and represent the virtual moduli cycle by the zero set of a suitable multi-section of a multi-bundle. We show explicitly how to assemble this zero set into an object that we call a branched labelled pseudomanifold. Our theory is quite general, and suggests that every finite dimensional orbifold has a "resolution" of this form. An example is worked out in §4.4.

Other approaches to this question have been developed by Fukaya and Ono [**FO**], Li and Tian [**LiT**], Ruan [**R**], Siebert [**Sb**], and Hofer and Salamon [**HS**]. We shall restrict ourselves here to curves of genus 0, but similar considerations apply for curves of higher genus.

I wish to thank Salamon for explaining to me that the virtual moduli cycle should be thought of as a branched manifold, Polterovich for his many penetrating questions and useful ideas, Siebert, Tian, Fukaya, Liu and Ruan for informative discussions about general Gromov–Witten invariants, and Olga Buse and Haydee Herrera-Guzman for helpful comments on an earlier draft.

1. Gromov–Witten invariants

Given a symplectic manifold (X, ω) with compatible almost complex structure J, denote by $\mathcal{M}_{0,k}(X, J, A)$ the space of (unparametrized) J-holomorphic spheres in class A with k marked points. Elements of this space are equivalence classes $[h, z_1, \dots, z_k]$ of elements $(h, z_1, \dots, z_k)$, where $h : S^2 \to X$ is a somewhere injective (i.e. nonmultiply covered) J-holomorphic map and the z_i are distinct points in S^2.

1991 *Mathematics Subject Classification.* Primary 58Dxx; Secondary 58Exx.
Partially supported by NSF grant DMS 9704825.

©1999 American Mathematical Society

Elements $(h, z_1, \dots, z_k)$ and $(h', z'_1, \dots, z'_k)$ are equivalent if there is $\gamma \in \mathrm{PSL}(2, \mathbf{C})$ such that $h' = h \circ \gamma$, and $z_j = \gamma(z'_j)$ for all j.

When J is generic, $\mathcal{M}_{0,k}(X, J, A)$ is a manifold of dimension

$$\dim(X) + 2c_1(TX)(A) + 2k - 6.$$

It is usually not compact. However in many cases one can show that the evaluation map

$$ev : \mathcal{M}_{0,k}(X, J, A) \to X^k : \qquad [h, z_1, \dots, z_k] \to (h(z_1), \dots, h(z_k))$$

does represent a well-defined element in homology $H_*(X^k)$ that is independent of the choice of J and can be used to define the Gromov–Witten invariants by the following procedure.

Take H_* to mean integral homology modulo torsion. Then, given homology classes $v_1, \dots, v_k \in H_*(X)$ satisfying the dimension condition

$$\dim(X) + 2c_1(TX)(A) + 2k - 6 + \sum_j \dim(v_j) = k \dim(X),$$

the Gromov–Witten invariant $n_X(v_1, \dots, v_k; A)$ is defined to be the intersection number of the class represented by ev with the homology class $v_1 \times \cdots \times v_k$ in $H_*(X^k)$.

This approach is described by Ruan and Tian [**RT**] and by McDuff and Salamon [**MS**], for example, and basically works when X does not contain J-holomorphic spheres in classes B with $c_1(B) < 0$. The difficulty with it is that the presence of these "negative" spheres could in principle cause there to be so many elements "on the boundary" of $\mathcal{M}_{0,k}(X, J, A)$ that ev does not represent a homology class. (For a more detailed discussion of this, see [**S**].) To deal with this problem, one must first understand how to compactify $\mathcal{M}_{0,k}(X, J, A)$, and then try to construct from what might be a rather complicated compactification something that can represent a homology class. This class should of course be independent of all choices made and in particular of J. Here I will try to show how to construct from the compactification $\overline{\mathcal{M}}_{0,k}(X, J, A)$ a cycle (called the virtual moduli cycle) that represents the desired homology class. Sometime this cycle is called a *regularization* of the moduli space. Other authors, for example Siebert, just construct a homology class. But in all approaches, one constructs an element of *rational* rather than integral homology.

The paper is organized as follows. First I will describe the compactification of $\mathcal{M}_{0,k}(X, J, A)$, i.e. the space $\overline{\mathcal{M}}_{0,k}(X, J, A)$ of J-holomorphic stable maps. §2 describes its elements, and §3 shows how one can use the process of gluing in the domain to give it a topology. Roughly speaking, the idea now is that one can give an orbifold structure to a neighborhood $\mathcal{W}$ of $\overline{\mathcal{M}}$ in the space of all stable maps. The delbar operator $\overline{\partial}_J$ is a Fredholm section of a bundle over $\mathcal{W}$, and what one has to do is define a finite dimensional space R of allowed perturbations of $\overline{\partial}_J$. Then the zero set of a generic perturbation of $\overline{\partial}_J$ should represent the desired cycle.

However, it is not easy to put an orbifold structure on $\mathcal{W}$, at least in the category of Banach manifolds. Siebert [**Sb**] has succeeded in doing this, but all other authors make do with less structure. Nor is it easy to get one's hands on the perturbation space R: this can be defined locally on $\mathcal{W}$, but there is a problem with fitting the local perturbations together. What we do in §4 is describe explicitly how to use the weaker structure developed by Liu and Tian (of multi-bundles and multi-folds) to define a geometric object that represents a rational homology class. This object is called a branched pseudo-manifold, and is constructed from a Fredholm section of a multi-bundle over a multi-fold. Finally, in §5 we sketch Liu and Tian's arguments showing that the constructions described in §4 are indeed possible in the present context.

This whole process should be thought of as an existence result for these invariants. It is extremely difficult to use the definition given here for direct calculations. Rather one uses it to establish formal properties of the invariants which are then exploited to do calculations by other means. We sketch one nontrivial example in §5.3 below. See also [**Mc2**].

2. Stable maps and curves

It is a remarkable recent discovery (essentially due to Kontsevich) that the compactification $\overline{\mathcal{M}}_{0,k}(X,J,A)$ of $\mathcal{M}_{0,k}(X,J,A)$ can be described as the space of all J-holomorphic genus 0 stable maps. The elements of this space are equivalence classes $\tau = [\Sigma, h, z_1, \dots, z_k]$ of tuples $(\Sigma, h, z_1, \dots, z_k)$. Here the domain Σ is a connected union of components $\Sigma_i, i = 0, \dots, p$, each of which has a given identification with S^2. (Note that we consider Σ to be a topological space: the labelling of its components is a convenience and not part of the data.) The intersection pattern of the components can be described by a tree graph T with $p+1$ vertices, where each edge corresponds to an intersection point of the components represented by its vertices. No more than two components meet at any point. There are also k marked points $z_1, \dots, z_k$ placed anywhere on Σ except at an intersection point of two components. The positions of the marked points are often represented on the graph by labelled spokes emanating from the relevant vertex. These spokes are half-open intervals, i.e. they have no second endpoint.

The restriction h_i of the map h to Σ_i is assumed to be J-holomorphic, and we impose the *stability condition* that h_i is nonconstant (though possibly a multiple covering) unless Σ_i contains at least 3 special points. (By definition, *special points* are either points of intersection with other components or marked points.) Also $h_*([\Sigma]) = \sum_i (h_i)_*[\Sigma_i] = A$. Finally, we divide out by all holomorphic reparametrizations. In other words, if $\gamma : \Sigma \to \Sigma$ is a holomorphic map such that $\gamma(z_j') = z_j$ for all j, we consider $(\Sigma, h\circ\gamma, z_1', \dots, z_k')$ and $(\Sigma, h, z_1, \dots, z_k)$ to be equivalent. Note that when $k \le 2$ the group of reparametrizations for which $z_j' = z_j$ for all j has (real) dimension at least 2. (The other reparametrizations — those with $z_j' \neq z_j$ — in some sense do not count since they can be gotten rid of by normalizing the choice of z_j.) This defines the elements τ of $\overline{\mathcal{M}} = \overline{\mathcal{M}}_{0,k}(X,J,A)$. Components of Σ on which h is constant are often called *ghosts*, and those containing fewer than 3 special points are called *unstable*. The stability condition means that unstable components are never ghosts.

Each such τ has a finite automorphism group $\Gamma_\tau = \mathrm{Aut}([\Sigma, h, z_1, \dots, z_k])$. Given a representative $(\Sigma, h, z_1, \dots, z_k)$ of τ, we may identify Γ_τ with the group of all holomorphic maps $\gamma : \Sigma \to \Sigma$ such that $h \circ \gamma = h$, and $\gamma(z_j) = z_j$ for all j. For most elements this group is trivial, and it is always finite because of the stability condition. It may be nontrivial if any of the h_i are multiple coverings, or if two different h_i have the same image. (Note that γ might permute the components of Σ.) The presence of these automorphism groups means that in many cases the nicest structure that $\overline{\mathcal{M}}_{0,k}(X,J,A)$ can have is that of an orbifold: see Example 3.3.

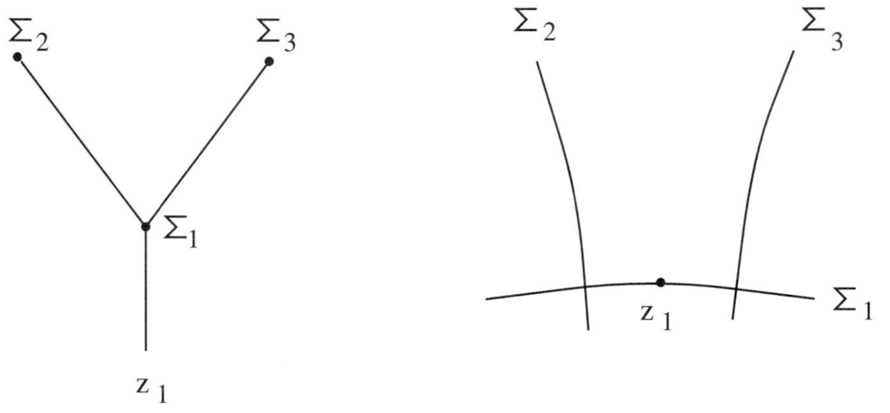

FIGURE 1

EXAMPLE 2.1. Let Σ have three components, with Σ_2 and Σ_3 both intersecting Σ_1, and let z_1 be a marked point on Σ_1. Then we can allow h_1 to be constant without violating stability. If in addition h_2, h_3 have the same image curve, there is an automorphism that interchanges Σ_2 and Σ_3. Since nearby stable maps do not have this extra symmetry, $\tau = [\Sigma, h, z_1]$ is a singular point in its moduli space. However, because marked points are labelled, there is no such automorphism if we put one marked point z_2 on Σ_2 and another z_3 at the corresponding point on Σ_3, i.e. so that $h_2(z_2) = h_3(z_3)$. One can also destroy this automorphism by adding just one marked point z_0 to $[\Sigma, h, z_1]$ anywhere on Σ_2 or Σ_3. Figure 1 gives two pictures of the domain of τ, one in terms of the corresponding tree T and the other a schematic picture of the 2-spheres in the domain with their marked points.

There is a special case in which X is a single point. All maps are now trivial, and the resulting space $\overline{\mathcal{M}}_{0,k}(pt, J, A)$ (usually written $\overline{\mathcal{M}}_{0,k}$) is known as the Mumford–Deligne compactification of the moduli space of marked spheres. Its elements are called *stable curves* of genus zero. The stability condition now says that every (nondegenerate) component of Σ contains at least 3 special points. When $k < 3$ the space $\overline{\mathcal{M}}_{0,k}$ reduces to a single point, which for consistency is taken to be represented by the degenerate Riemann surface consisting of a single point. Since the Möbius group PSL(2, $\mathbf{C}$) acts triply transitively on S^2, the space $\overline{\mathcal{M}}_{0,3}$ also is a single point, but now this is represented by a sphere with 3 fixed marked points $0, 1, \infty$.

EXAMPLE 2.2. $\overline{\mathcal{M}}_{0,4}$ is the compactification of the space $\mathcal{M}_{0,4}$ of 4-tuples $\{0, 1, \infty, w\}$ in S^2. The question here is to understand how to represent the limit of a sequence $\{0, 1, \infty, w_n\}$ as w_n converges to one of the other marked points, say 0. Since the marked points are assumed distinct, it is not allowed for w to be superimposed on 0. But these two points can occupy their own component of Σ. Thus the limit is $[\Sigma, z_1, z_2, z_3, z_4]$, where Σ is the one-point union of two copies of S^2, of which one contains the points z_1, z_4 and the other z_2, z_3. (The points are paired up this way because $z_1 = 0, z_4 = w$.) In §3.1 below we show how to use

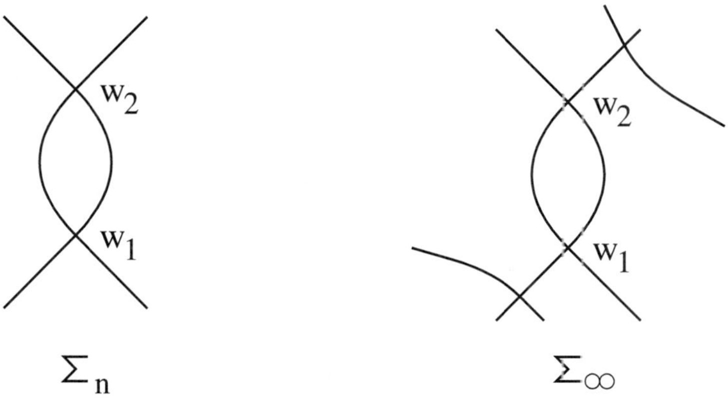

FIGURE 2

gluing to generate a neighborhood of this end in $\mathcal{M}_{0,4}$ starting from the limit point $[\Sigma, z_1, z_2, z_3, z_4]$. It will follow that $\overline{\mathcal{M}}_{0,4}$ is a manifold diffeomorphic to S^2.[1]

Observe that in this limit the pairs z_1, z_4 and z_2, z_3 appear symmetrically. This seems wrong at first glance because, after all, it is $w_n = z_4$ that is moving towards $0 = z_1$ while the other points are fixed. However, the set $\{0, 1, \infty, w_n\}$ is equivalent to $\{\gamma(0), \gamma(1), \gamma(\infty), \gamma(w_n)\}$ for any $\gamma \in \mathrm{PSL}(2, \mathbf{C})$, and if we choose γ_n so that $\gamma_n(0) = 0$, $\gamma_n(1) = 1$ and $\gamma_n(w_n) = \infty$, then $z_3 = \gamma_n(\infty)$ will converge to $z_2 = 1$. □

It is usually not hard to see how to parametrize the limit of a converging sequence of stable maps when these have genus 0. For example, the limit of a sequence $\tau_n = [\Sigma, h, z_1, z_{2,n}]$ in which $z_{2,n}$ is converging to z_1 (which is fixed) is a stable map with a new bubble component Σ_b attached to the point of Σ where z_1 was, and with the two marked points on Σ_b. Since h is independent of n, the new component is a ghost, i.e. h maps it to a single point.

Identifying the limit gets more tricky when one considers maps of higher genus. The following example in genus 1 is taken from Fukaya and Ono [**FO**], Example 10.5.

EXAMPLE 2.3. Consider a sequence of curves C_n in $\mathbf{CP}^2$ each made up of a conic and a line intersecting at points x_n, y_n, and suppose that x_n converges to y_n as $n \to \infty$. Each C_n is the image of a stable map $\tau_n = [\Sigma, h_n]$, where Σ has two components of genus 0 that intersect each other in two distinct (and fixed) points w_1, w_2. Since $x_n = h_n(w_1)$ and $y_n = h_n(w_2)$ have the same limit, there must be a point on each component that is different from w_1, w_2 at which the derivative of h_n is blowing up. In other words, a bubble forms on each component at some point other than the intersection points w_1, w_2. (For a description of how this happens, see [**MS**], §4.3.) Therefore the domain of the limit $[\Sigma_\infty, h]$ has 4 components, one mapping to the line, another (disjoint from the first) mapping to the conic, and two others that are ghosts that each intersect the other twice and meet one of the other components: see Figure 2.

[1]In fact, all the spaces $\overline{\mathcal{M}}_{0,k}$ are manifolds: see [**HS**] for example.

2.1. The space $\mathcal{W}$. The virtual moduli cycle is made up of stable maps that satisfy a perturbed Cauchy–Riemann equation. Therefore its elements belong not to $\overline{\mathcal{M}}$ itself but to a suitably small neighborhood $\mathcal{W}$ of $\overline{\mathcal{M}}$ in the space of all (not necessarily holomorphic) stable maps. Below we will define some open neighborhoods of the points τ in $\overline{\mathcal{M}}$, and $\mathcal{W}$ will be taken to be their union. Observe that $\mathcal{W}$ has a natural *coarse* stratification in which each stratum consists of elements $\tau = [\Sigma_\tau, h_\tau, z_1, \dots, z_k]$, where the domain has fixed topological type, where the marked points $z_1, \dots, z_k$ are on specified components of Σ_τ, and where the decomposition $A = \sum A_i$ is given. Such a stratum $\mathcal{W}^D$ can be described by the decorated tree D in which the vertices of the corresponding spoked tree are labelled with the classes A_i. We will also sometimes use the *fine* stratification, in which the isomorphism class of Γ_τ is also fixed on each stratum.

In order to put a topology on $\mathcal{W}^D$ we must choose a local Γ_τ-equivariant slice (or cross-section) for the action of the reparametrization group on the space of parametrized maps so that the elements of small subsets in $\mathcal{W}^D$ can be considered as parametrized maps. To this end, given a representative[2] $\tilde\tau = (\Sigma, \tilde h, z_1, \dots, z_k)$ of τ, choose a minimum number of extra points $w_1, \dots, w_\ell$ on Σ so that the domain $[\Sigma, z_1, \dots, z_k, w_1, \dots, w_\ell]$ is stable, i.e. so that each component of Σ has at least 3 special points, and then choose for each j a small open codimension-2 disc $\mathbf{H}_j$ in X that meets $\tilde h(\Sigma)$ transversally in $\tilde h(w_j)$. For this to be possible the w_j should be generic points on Σ. We also require that the set of added points $\{w_1, \dots, w_\ell\}$ be as invariant as possible under the action of the finite group Γ_τ. This requirement will be spelled out later. Then define

$$\widetilde U_\tau^D = \left\{ \tilde\tau' = (\Sigma, \tilde g, z_1', \dots, z_k') : \tilde g(w_j) \in \mathbf{H}_j, ||\tilde g - \tilde h|| + \sum_i d(z_i, z_i') < \varepsilon \right\},$$

where d is the distance function in Σ. The norm $||\cdot||$ used on the space of maps should be stronger than the C^1-topology, since we need the elements in $\widetilde U_\tau^D$ to be C^1-close to $\tilde h$ to ensure that they meet $\mathbf{H}_j$ transversally. For example, the Sobolev norm $L^{2,p}$ with $p > 2$ would do.[3] Moreover, we assume that ε is so small that $[\tilde g(\Sigma_j)] = A_j$ for all j.

If the group Γ_τ is trivial, then the projection $\pi_\tau : \widetilde U_\tau^D \to \mathcal{W}^D$ that takes the point $\tilde\tau'$ to its equivalence class τ' is injective (for small enough ε), so that τ has a neighborhood in $\mathcal{W}^D$ modelled on an open subset in a Banach space. If Γ_τ is nontrivial, we must extend its action to a linear action on $\widetilde U_\tau^D$ in such a way that a neighborhood U_τ of τ in $\mathcal{W}^D$ can be identified with the quotient $\widetilde U_\tau^D / \Gamma_\tau$.

First of all, consider the case when $\Sigma = S^2$ and there are two marked points z_1, z_2, so that we have to choose one point w_1. In this case, $\tilde h$ must factor through a holomorphic map $\tilde f : S^2 \to S^2$ of degree $n > 1$, and Γ_τ is a cyclic group generated by a rotation γ that fixes z_1, z_2 and is such that $\tilde f \circ \gamma = \tilde f$. Choose w_1 so that the set $\tilde f^{-1}(\tilde f(w_1))$ contains n distinct points, and choose disjoint little discs about each of these points that are permuted by γ. By construction, any element $\tilde g \in \widetilde U_\tau^D$ is such that $\tilde g(w_1) \in \mathbf{H}_1$. Moreover, for small enough ε, $\tilde g^{-1}(\mathbf{H}_i)$ is a collection of n points, one in each of the little discs. (cf. [**LiuT1**], Lemma 2.2). Therefore there

[2]For clarity we will here write $\tilde h : \Sigma \to X$ to denote a particular map, reserving h to denote an equivalence class of maps.

[3]One could get away with a weaker norm (e.g. the $L^{1,p}$ norm) if one used Siebert's integral method of fixing the parametrizations.

is a unique point w_1' in the little disc containing $\gamma(w_1)$ such that

$$g(w_1') \in \mathbf{H}_1,$$

and we define $\phi_g^\gamma \in \mathrm{PSL}(2,\mathbf{C})$ to be the unique element that fixes z_1, z_2 and takes w_1 to w_1'. Then set

$$\gamma \cdot \tilde{g} = \tilde{g} \circ \phi_g^\gamma.$$

It is not hard to check that

$$\gamma \cdot \tilde{g} \in \widetilde{U}_\tau^D, \qquad \phi_g^\gamma \circ \phi_{\gamma\cdot\tilde{g}}^\gamma = \phi_g^{\gamma^2},$$

so that this does give an action of Γ_τ. It is not hard to see that it is linear (for small enough ε.) Moreover, a neighborhood of τ in $\mathcal{W}^D$ can be identified with the quotient

$$U_\tau^D = \widetilde{U}_\tau^D / \Gamma_\tau.$$

A triple $(\widetilde{U}_\tau^D, \Gamma_\tau, \pi_\tau)$, where $\pi_\tau : \widetilde{U}_\tau^D \to U_\tau^D$ is the projection, is called a *local uniformizer* for τ in $\mathcal{W}^D$. Note that the automorphism group $\Gamma_{\tau'}$ of an element $\tau' \in U_\tau^D$ is isomorphic to the stabilizer subgroup $\mathrm{Stab}(\tilde{\tau}') \subset \Gamma_\tau$ of any lift $\tilde{\tau}'$ of τ' to $\widetilde{U}_\tau$.

A similar argument works if $\Sigma = S^2$ has one special point, since Γ_τ is again a rotation group: one just has to pick w_1, w_2 to be generic points on different Γ_τ-orbits. If there are no special points, Γ_τ need no longer be a rotation group, but it is not hard to adapt the argument. Again one picks the three points w_j to be generic points on different Γ_τ-orbits, chooses a Γ_τ-invariant set of little disjoint discs containing the w_j, and, for each $\gamma \in \Gamma_\tau$, chooses ϕ_g^γ to be an appropriate approximation to γ.

Finally, consider arbitrary τ. The group Γ_τ is the semidirect product of a permutation group with a subgroup $\Gamma_{\tau,0}$ that acts trivially on the set of components of Σ. Furthermore, $\Gamma_{\tau,0}$ is a product $\prod \Gamma_i$, where Γ_i is a group of reparametrizations of Σ_i. We have seen how to choose the points w_j on any component Σ_i with nontrivial Γ_i. If the permutation group is nontrivial, we require further that the set $\{w_j\}$ be invariant under this group. Thus, we have the following result: see [**LiuT1**], Lemma 2.5.

LEMMA 2.4. *There is a neighborhood $\mathcal{W}^D$ of $\overline{\mathcal{M}}^D$ in the space of all stable maps of type D that is covered by local uniformizers $(\widetilde{U}_\tau^D, \Gamma_\tau, \pi_\tau)$.*

In fact, these local uniformizers give $\mathcal{W}^D$ the structure of an orbifold: see §4.

3. Gluing

The problem now is to describe how the strata $\mathcal{W}^D$ fit together as the topological type of the domain changes. Here one must use gluing. If one forgets the maps and just considers the finite-dimensional spaces of stable curves, then a stratum $\mathcal{S}$ on which Σ has $k+1$ components has (real) codimension $2k$. We shall see in §3.1 that in this case there is a tubular neighborhood theorem that identifies a neighborhood of $\mathcal{S}$ with a neighborhood of the zero section in a certain orbibundle of gluing parameters $a_1, \dots, a_k$, where each $a_i \in \mathbf{C}$ is attached to one of the intersection points $\Sigma_j \cap \Sigma_k$ of the components of Σ. (There are precisely k of these intersection points, since the components of Σ are attached according to a tree graph.) We describe the orbifold structure of $\mathcal{W}$ in §3.2.

3.1. Gluing the domains. We now describe how to glue in the space of stable curves, i.e. we will only consider the domain Σ together with its marked points. The discussion here is essentially taken from Fukaya and Ono [**FO**], §9. They assume that each domain Σ is provided with a metric that is flat in some neighborhood of the double points and on each component is in the conformal class corresponding to the given complex structure.

First consider the simplest situation when Σ has two components with $x_1 \in \Sigma_1$ identified to $x_2 \in \Sigma_2$, and with two marked points on each component. Then the gluing parameter space $\mathbf{C}$ should be identified with the vector space $T_{x_1}\Sigma_1 \otimes T_{x_2}\Sigma_2$. Let us write $\tau_i, 1 \le i \le 3$, for the stable curve of this form with z_i and z_4 on Σ_2 and the other two points z_ℓ, z_m on Σ_1. As explained in Example 2.2, the three elements τ_i should correspond to the three ends of the open stratum $\mathcal{M}_{0,4} = S^2 - \{0, 1, \infty\}$ in $\overline{\mathcal{M}}_{0,4}$. Therefore, for each i what we must do is show how to identify a small deleted neighborhood $V - \{0\}$ of 0 in $\mathbf{C}$ with a neighborhood of the corresponding end in $\mathcal{M}_{0,4}$, i.e. our aim is to construct an injective map

$$\Psi_i : V - \{0\} \longrightarrow \mathcal{M}_{0,4} : \qquad a \mapsto [\Sigma_a, z_1^a, \ldots, z_4^a]$$

whose image is a neighborhood of the corresponding end of $\mathcal{M}_{0,4}$.

Given a gluing parameter $a \in V - \{0\} \subset \mathbf{C}$, set $r = \sqrt{|a|}$. For small r and $j = 1, 2$, let D_j be the 2-disc $D_{x_j}(r)$ with center x_j and radius r in Σ_j, and put $U_j = \Sigma_j - \operatorname{int} D_j$. Think of Σ_j as $D_j \cup_{\iota_j} U_j$, where D_j and U_j are positively oriented discs (i.e. their orientation as a subset of Σ_j agrees with their orientation as a disc) and ι_j is an orientation reversing identification of their boundaries. For each gluing parameter $a \in V - \{0\}$ the new domain Σ_a should be $U_1 \cup U_2$. In order to make this precise we have to describe the way ∂U_1 is identified with ∂U_2, and then put a suitable metric on Σ_a.

Since the given metric on Σ is flat near the double points x_j, we can use the exponential map to identify the discs D_j isometrically with the discs of radius r in the tangent spaces $T_{x_j}\Sigma_j$. The gluing map is then the restriction of a map

$$\Phi_a : T_{x_1}\Sigma_1 - \{0\} \longrightarrow T_{x_2}\Sigma_2 - \{0\}$$

that is defined by the requirement that

$$u \otimes \Phi_a(u) = a \in \mathbf{C} = T_{x_1}\Sigma_1 \otimes T_{x_2}\Sigma_2,$$

for all $u \in T_{x_1}\Sigma_1$. Thus in local coordinates Φ_a is the inversion $u \mapsto a/u$. This defines the space Σ_a, and the metric (and hence complex structure) is given as follows. Since

$$\Phi_a^*(|du|^2) = \left|\frac{a}{u^2}\right|^2 |du|^2,$$

$\Phi_a(|du|^2) = |du|^2$ on the circle $|u| = r$. Hence, we may choose a function $\chi_r : (0, \infty) \to (0, \infty)$ so that the metric $\chi_r(|u|)|du|^2$ is invariant by Ψ_a and so that $\chi_r(s) = 1$ when $s \ge 2r$, and then patch together the given metrics on $\Sigma_1 - D_{x_1}(2r)$ and $\Sigma_2 - D_{x_2}(2r)$ via $\chi_r(|u|)|du|^2$.

Finally, we must look at what happens to the marked points. Observe that Σ_a has four marked points, z_ℓ, z_m on U_1 and z_i, z_4 on U_2. We can identify the sets U_j with the disc D^2 so that $z_i = z_\ell = 0, z_4 = z_m = 1/2$. Our normalisation of $\mathcal{M}_4$ assumes that the points z_1, z_2, z_3 are fixed and equal to $0, 1, \infty$ respectively. For each a let $\gamma_a : \Sigma_a \to S^2$ be the unique holomorphic diffeomorphism that takes the marked points z_1, z_2, z_3 on Σ_a to $0, 1, \infty$. The question now is to understand what

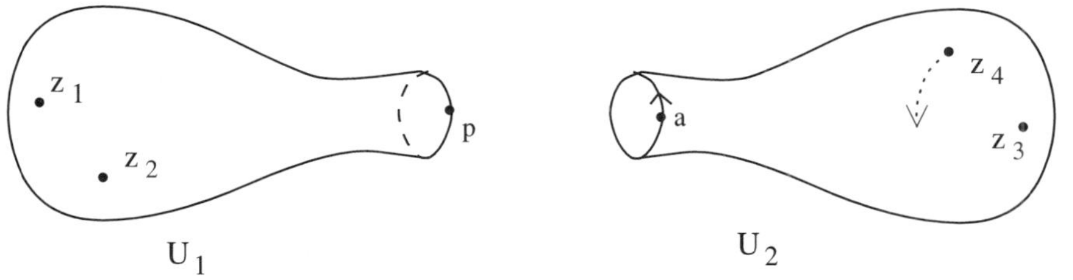

FIGURE 3. The dotted arrow shows the rotation of U_2 needed to match a to p as a rotates positively around ∂U_2.

happens to $\gamma_a(z_4)$ as a varies in a neighborhood of 0 in the gluing parameter space $\mathbf{C}$.

PROPOSITION 3.1. *The map* $a \mapsto \{0, 1, \infty, \gamma_a(z_4)\}$ *defines an orientation preserving diffeomorphism between a deleted neighborhood of* 0 *in* $\mathbf{C}$ *and a neighborhood of an end in* $\mathcal{M}_{0,4}$.

We will not attempt to prove this here, but will give some plausibility arguments. Consider the case when z_1 and z_4 are on the same component of Σ. As $|a| \to 0$ the metric on Σ_a has an increasingly pinched neck, which implies that $\gamma_a(z_4)$ tends to $\gamma(z_1) = 0$. Inspection shows that the image of this map is a full neighborhood of the end in $\mathcal{M}_{0,4} = S^2 - \{0, 1, \infty\}$ corresponding to the point 0. Hence all one has to check is that the map is injective and orientation preserving.

Here is an argument to show that orientation is preserved. For simplicity let us suppose that the induced metrics on U_j are radially symmetric and that $\ell = 1, m = 2, i = 3$. Then γ_a takes the points $0, 1/2$ on $U_1 = D^2$ to $0, 1$ on S^2 and the point 0 in $U_2 = D^2$ to ∞ on S^2, and is also radially symmetric. Let p be the point 1 on ∂U_1. This is identified to $a/p = a$ on ∂U_2. As a moves around 0 positively as seen from the deleted point x_2, it moves around ∂U_2 negatively with respect to the orientation of U_2. Therefore U_2 must be rotated positively to match a to the fixed point p. It follows that the point z_4 rotates positively round z_3, see Figure 3.

This shows how to glue two components together. One can obviously use the same method to glue any number of components.

PROPOSITION 3.2. *An element* $\tau = [\Sigma, z_1, \dots, z_k]$ *in* $\overline{\mathcal{M}}_{0,k}$ *whose domain has* $p+1$ *components has a neighborhood consisting of all stable curves obtained by first slightly moving the points* $z_1, \dots, z_k$ *and then gluing with an arbitrary small set of gluing parameters* $a_1, \dots, a_p$.

This is again a nontrivial result that we will not prove. Note that if any $a_j = 0$ one simply leaves that intersection alone. Therefore the induced stratification on a neighborhood of τ corresponds to the stratification of $\mathbf{C}^p$ given by the subspaces $V_I = \{a : a_j = 0, j \in I\}$, where $I \subset \{1, \dots, p\}$. Observe also that, because the automorphism group Γ_τ acts on the space of gluing parameters, $\overline{\mathcal{M}}_{0,k}$ has the structure of an orbifold.

In nice cases, the space $\overline{\mathcal{M}}_{0,k}(X, J, A)$ is also a finite dimensional orbifold. In the next example we show that the compactification $\overline{\mathcal{M}}_{0,0}(S^2, J, 2[S^2])$ of the space of self-maps of S^2 of degree 2 is a (degenerate) orbifold that is diffeomorphic to $\mathbf{C}P^2$ but all its points have automorphism group $\Gamma_\tau = \mathbf{Z}/2\mathbf{Z}$.

EXAMPLE 3.3. The space $\overline{\mathcal{M}}_0 = \overline{\mathcal{M}}_{0,0}(S^2, J, 2[S^2])$ has two strata. The first, $\mathcal{S}_1$, consists of self-maps of S^2 of degree 2, and the second, $\mathcal{S}_2$, consists of maps whose domain has two components, each taken into S^2 by a map of degree 1. The equivalence relation on each stratum is given by precomposition with a holomorphic self-map of the domain. It is not hard to check that each equivalence class of maps in $\mathcal{S}_1$ is uniquely determined by its two critical values (or branch points). Since these can be any pair of distinct points, $\mathcal{S}_1$ is diffeomorphic to the set of unordered pairs of distinct points in S^2. On the other hand there is one element σ_w of $\mathcal{S}_2$ for each point $w \in S^2$, the correspondence being given by taking w to be the image under h of the point of intersection of the two components.

If $\sigma_{\{x,y\}}$ denotes the element of $\mathcal{S}_1$ with critical values $\{x, y\}$, we claim that $\sigma_{\{x,y\}} \to \sigma_w$ when x, y both converge to w. To see this, let $h_{\{x,y\}} : S^2 \to S^2$ be a representative of $\sigma_{\{x,y\}}$ and let $\alpha_{\{x,y\}}$ be the shortest geodesic from x to y. (We assume that x, y are close to w.) Then $h^{-1}_{\{x,y\}}(\alpha_{\{x,y\}})$ is a circle $\gamma_{\{x,y\}}$ through the critical points of $h_{\{x,y\}}$. This is obvious if $h_{\{x,y\}}$ is chosen to have critical points at $0, \infty$ and if $x = 0, y = \infty$, since $h_{\{x,y\}}$ is then a map of the form $z \mapsto az^2$. It follows in the general case because Möbius transformations take circles to circles. Hence $h_{\{x,y\}}$ takes each component of $S^2 - \gamma_{\{x,y\}}$ onto $S^2 - \alpha_{\{x,y\}}$. If we now let x, y converge to w, we see that $\sigma_{\{x,y\}}$ converges to σ_w.

Hence $\overline{\mathcal{M}}_0$ is the quotient of $S^2 \times S^2$ by the involution $(x, y) \mapsto (y, x)$. This is well known to be $\mathbf{C}P^2$. Note that every element of $\mathcal{S}_1$ can be represented by a conjugate of $z \mapsto z^2$ and so has automorphism group of order 2. This remains true for elements of Σ_2: here the automorphism interchanges the two components of the domain.

Suppose now that one adds three marked points to the domain and looks at the subset of stable maps where these points z_0, z_1, z_∞ are taken to $\{0, 1, \infty\}$. This space, which we shall call $\overline{\mathcal{M}}_3$, is an 8-fold branched cover over $\overline{\mathcal{M}}_0$, where the inverse image of $\tau \in \overline{\mathcal{M}}_0$ is the set of ways of choosing which point in $h^{-1}(0)$, resp. $h^{-1}(1), h^{-1}(\infty)$, is called z_0, resp. z_1, z_∞. In [**Mc1**], Lemma 3.7, it is shown that this space can be identified with $\mathbf{C}P^2$, and that the branched covering map is

$$[x, y, z] \mapsto [x^2, y^2, z^2].$$

Moreover, the inverse image of the line $\mathcal{S}_2$ is a set of 4 lines in general position. These lines correspond to the 4 different ways in which 3 points can be arranged on the two components of the domain. Now all the groups Γ_τ are trivial. We will meet this space again in the example discussed in §5.3.

3.2. The topology on $\mathcal{W}$. The above process of gluing can also be used to describe a topology on $\mathcal{W}$. We have already seen how to define local uniformizers $(\widetilde{U}^D_\tau, \Gamma_\tau, \pi_\tau)$ for τ in its stratum $\mathcal{W}^D$. To deal with a full neighborhood, add a minimum number of marked points $(w_1, \dots, w_\ell)$ to Σ so that the curve $[\Sigma, z_1, \dots, z_k, w_1, \dots, w_\ell]$ is stable, and let V be a small neighborhood of 0 in the space of gluing parameters $\mathbf{C}^p$ for Σ as in Proposition 3.2. Then for each $a \in V$ there is a well-defined stable curve $[\Sigma_a, z_1^a, \dots, z_k^a]$ that is the result of gluing with parameters $a = (a_1, \dots, a_p)$, and a map

$$\psi_a : (\Sigma_a, z_1^a, \dots, z_k^a) \to (\Sigma, z_1, \dots, z_k)$$

that is injective except that it collapses the boundary of each disc ∂U_j to the intersection point of the relevant components in Σ. The neighborhood $\widetilde{U}_\tau$ of $\tilde{\tau} =$

$(\Sigma, \tilde{h}, z_1, \dots, z_k)$ now contains all points $\tilde{\tau}' = (\Sigma_a, \tilde{h}', z_1', \dots, z_k')$ where $a \in V$, $\tilde{h}'$ is close to $\tilde{h} \circ \psi_a$, and the $\psi_a(z_j')$ are close to z_j. This space has an obvious topology. It is now not hard to extend the action of the symmetry group Γ_τ to $\widetilde{U}_\tau$ and to show that the quotient

$$U_\tau = \widetilde{U}_\tau / \Gamma_\tau$$

can be identified with a subset of $\mathcal{W}$. The set $\mathcal{W}$ is defined to be the union of all the U_τ, and its topology is generated by these sets. Liu and Tian show that this topology is Hausdorff. Moreover, $\overline{\mathcal{M}}$ is a compact subset of $\mathcal{W}$: see [**LiuT1**], Proposition 4.1.

We will not try to put a smooth structure on $\mathcal{W}$ that is transverse to the strata. Instead we will, following Liu and Tian, consider $\mathcal{W}$ to be a space with two topologies (or a *partially smooth* space): see §4.1 below. Thus $\mathcal{W}$ is a (partially smooth) orbifold.

4. Branched pseudomanifolds

In this section we describe the topological nature of the virtual cycle. §4.1 discusses the partially smooth category and introduces the idea of a branched, labelled pseudomanifold. In §4.2 the concepts of multi-fold, multi-bundle and multi-section are defined and illustrated by examples. §4.3 shows how the zero set of a Fredholm multi-section can be assembled into a branched, labelled pseudomanifold. Though the main application is infinite dimensional, the definitions all apply in finite dimensions. We show in §4.4 how to calculate the "resolution" and the Euler number of the teardrop orbifold.

4.1. Spaces with two topologies. We will work in the category of *partially smooth* topological spaces Y. Thus Y is a Hausdorff topological space that is the image of a continous bijection

$$i_Y : Y_{sm} \to Y,$$

where Y_{sm} is a finite union of open disjoint subsets, each of which is a smooth Banach manifold. (Another way to think of Y is as a space with two topologies.) A morphism $f : Y \to X$ between two objects in this category is a continuous map $f : Y \to X$ such that the induced map $f : Y_{sm} \to X_{sm}$ is smooth. Usually, we will be interested in the case when X is a manifold, i.e. when i_X is a homeomorphism, so that X can be identified with the manifold X_{sm}. Morphisms $f : Y \to X$ will be called partially smooth maps, or, for short, simply maps. The path components of Y_{sm} are sometimes called *strata* and sometimes *components*. If Y is infinite dimensional, strata that are open subsets of Y are called top strata.

DEFINITION 4.1. A *closed oriented pseudomanifold* of dimension d is a compact Hausdorff partially smooth topological space Y such that one component of Y_{sm} is an oriented smooth d-dimensional manifold that is mapped by i_Y onto a dense open subset Y^{top} of Y. Moreover, all other components of Y_{sm} have dimension $\le d-2$. We write Y^{sing} for the image $Y - Y^{top}$ of these lower dimensional submanifolds.

It is easy to see that any partially smooth map f from such Y to a closed manifold X defines an element $f_*([Y])$ of $H_d(X, \mathbf{Q})$ by intersection. Indeed, since we are working rationally, the class $f_*([Y])$ is defined by its intersections with classes given by smooth maps $g : Z \to X$ from oriented closed manifolds Z into X. If $\dim Z + d = \dim X$, then we can jiggle g to make it meet each component of the image of Y_{sm} transversally. Hence it will be disjoint from Y^{sing} (for reasons of dimension), and will meet $f(Y^{top})$ transversally in a finite number of points. Moreover, if $g_0, g_1 : Z \to X$ are both transverse to f in this way, and are joined by a homotopy $G : Z \times [0,1] \to X$, then we can jiggle G (fixing $g_i = G|_{Z\times i}$) so that it is also transverse to f. The important point now is that the image of G still does not meet $f(Y^{sing})$, since Y^{sing} has codimension ≥ 2. It follows easily that the intersection number $f \cdot g$ of f with g is well defined and independent of choices. We now define $f_*([Y])$ to be the unique rational class such that

$$f_*([Y]) \cdot g_*([Z]) = f \cdot g, \quad \text{for all } g : Z \to X.$$

Thus, in the language of [MS1], the partially smooth map $f : Y \to X$ is a pseudo-cycle.

DEFINITION 4.2. A *branched pseudomanifold* Y of dimension d is a partially smooth space $Y_{sm} \to Y$ where the components of Y_{sm} have dimension at most d. The components of dimension d are called M_i, those of dimension $(d-1)$ are called B_j, and $Y_{\leq k}$ denotes the union of all components of Y_{sm} of dimension $\leq k$. Write

$$Y^{top} = \bigcup_i M_i, \quad B = \bigcup_j B_j, \quad Y^{sing} = Y - (Y^{top} \cup B) = Y_{\leq d-2}.$$

We assume that it is possible to give the set $M_i \cup_{j \in J_i} B_j$ the structure of a smooth manifold with boundary compatibly with its two topologies.[4] Here $j \in J_i$ if the closure $\overline{M}_i$ of M_i in Y meets B_j. Similarly, we assume $\overline{B_j} - B_j$ lies in Y^{sing}. The set B is called the *branch locus.*

One more ingredient is needed in order for a compact branched pseudomanifold to be a cycle, namely a suitable labelling of the top components. This is spelled out in the next definition.

DEFINITION 4.3. We say that a branched pseudomanifold Y of dimension d is *labelled* if Y is compact and if its top components M_i are oriented and have positive rational labels $\lambda_i \in \mathbf{Q}$, assigned so that the following condition holds:

> for each $x \in B$, pick an orientation of $T_x B$ and divide the components M_i that have x in their closure into two groups I^+, I^- according to whether the chosen orientation on $T_x B$ agrees with the boundary orientation. Then we require that
>
> $$\sum_{i \in I^+} \lambda_i = \sum_{i \in I^-} \lambda_i.$$

For example, if Y has dimension 0, it is just a collection of oriented labelled points (which means in practice that the labels can be both positive and negative), and there are no compatibility conditions at the branch locus since this is empty. However, even if one is considering the 0-dimensional case, one needs to look at 1-dimensional branched cobordisms between them in order to prove that the homology class represented by Y is independent of choices. This is nicely explained in [**S**], §5.4, but note that Siebert's point of view on branched manifolds is somewhat different from ours.

[4] By "compatibly" we mean that the inclusion of this set into Y is continuous, and that the inclusion of any stratum M_i or B_j into this manifold with boundary is smooth.

LEMMA 4.4. *Let Y be a (closed, oriented) branched and labelled pseudomanifold of dimension d. Then every partially smooth map f from Y to a closed manifold X defines a rational class $f_*([Y]) \in H_d(X)$.*

PROOF. As above, let Z be a smooth manifold of dimension complementary to Y, and let $g : Z \to X$ be a smooth map. Then we can jiggle g so that it does not meet $f(Y^{sing})$ and so that it meets $\bigcup f(M_i)$ transversally in a finite number of points. We then set

$$f \cdot g = \sum_i \varepsilon_i \lambda_i,$$

where $\varepsilon_i = \pm 1$ is the appropriate sign. It is not hard to check that the condition imposed on the labels in Definition 4.3 guarantees that this number is independent of the jiggling: cf. [**S**], Lemma 5.11. The class $f_*[Y]$ is then defined as before. □

Our basic aim is to use the minimum of structure for the purpose at hand. Nevertheless, the definition of branched pseudomanifold includes more information on the strata of codimension ≥ 2 than is strictly necessary for the preceding lemma to hold. We could get by with something more like the pseudocycles in [**MS**], §7.1, i.e. we could simply ask that Y^{sing} be the smooth image of a (probably disconnected) manifold of dimension $\leq d-2$, This definition has its own problems, since one would have to specify what "smooth" means. For instance, as is the case in our main example, there might be an evaluation map *ev* from Y to a smooth manifold X, and one would ask that the composite map be smooth. Since in [**Mc2**] we also need information on the strata of codimensions 2 and 3, we chose the present approach.

Another point is that, as much as is possible, we are trying to avoid specifying exactly how the strata of Y_{sm} fit together. If Y is branched then one does need some information of the normal structure to the codimension 1 components, but as is shown by the previous lemma, one can often get away without any other restrictions. However, later we will need to consider the intersection of two partially smooth spaces (see §4.3), and in order for this to be well behaved one does need more structure. This is the reason for the following definition.

DEFINITION 4.5. We will say that a partially smooth space Y has *normal cones* if every stratum S in Y has a neighborhood $\mathcal{N}(S)$ in Y whose induced stratification is that of a cone bundle over a link. More precisely, there is a commutative diagram

$$\begin{array}{ccc} \mathcal{N}(S)_{sm} & \to & \mathcal{N}(S) \\ {\scriptstyle \pi'} \downarrow & & {\scriptstyle \pi} \downarrow \\ S_{sm} & \to & S \end{array}$$

where the maps π' and π are oriented locally trivial fibrations with fiber equal to a cone over a link L. Here L is partially smooth, and the cone $C(L)$ is just the quotient $L \times [0,1]/L \times \{0\}$, stratifed so that it is the union of the vertex (the image of $L \times \{0\}$) with the product strata in $L_{sm} \times (0,1]$. Moreover each stratum in $\mathcal{N}(S)_{sm}$ projects onto the whole of S and so is a locally trivial bundle over S with fiber equal to a component of $C(S)_{sm}$. In particular, we identify S with the section of $\mathcal{N}(S)$ given by the vertices of the cones.

This definition implies that any stratum S' whose closure intersects S must contain the whole of S in its closure. In fact, near S it must look like the cone over some stratum of L_{sm}. Any finite dimensional Whitney stratified space has this normal structure. For example, any finite dimensional orbifold has a stratification such that the isomorphism class of the automorphism group Γ_x at the point x is constant as x varies over each stratum, and it is easy to see that this stratification has normal cones as defined here. It is also not hard to use arguments similar to those in §3.2 to show that the neighborhood $\mathcal{W}$ of $\overline{\mathcal{M}}$ in the space of all stable maps has normal cones with respect to the fine stratification. Finally, the process of gluing J-holomorphic maps shows that the finite dimensional local solution sets $\widetilde{Z}^{\nu}_{\tau}$ considered in Lemma 5.1 are stratified with normal cones.

4.2. Multi-folds and multi-sections. In the example of interest to us, the space $\mathcal{W}$ itself can be given the structure of an orbifold in the partially smooth category. However, it is hard to define a suitable global perturbation space, and hence the virtual moduli cycle, in this context. Different authors propose different solutions to this problem. Liu and Tian choose to minimize the needed analysis at the cost of introducing more topological complexity.

We begin by defining a generalization of the usual concept of orbifold that seems to be familiar to algebraic geometers as a "stack". We will call it a *multi-fold.* It is a certain kind of atlas (or covering) of a space $\mathcal{W}$ with two topologies that locally is an orbifold in the partially smooth category, in which the inclusions that relate one uniformizer to another in an orbifold are replaced by fiber products. We then show how to define multi-sections of a multi-bundle over a multi-fold. Our construction is taken from [LiuT1,2] but is actually fairly close to that of Fukaya and Ono: for example, they also consider something they call a multi-section. However, Liu and Tian's construction is topologically more transparent because they can restrict attention to multi-sections that are single-valued when pulled back to the desingularization of the multi-fold.

The theory described here could be generalized, but this would require quite a bit of work that we shall not do, since our aim is to understand the special case of the virtual moduli cycle. In this application, $\mathcal{W}$ is a neighborhood of the moduli space $\overline{\mathcal{M}}$ in the space of all stable maps. We assume that it is given the fine stratification, so that the isomorphism class of the group Γ_τ is fixed on each stratum. This means that the strata are manifolds. If instead one used the coarse stratification, then a stratum could have some orbifold singularities.

All maps, spaces and group actions considered below are in this partially smooth category. Branching will not be an issue until we construct the finite-dimensional virtual cycle in §4.3. Recall that a stratum (or component of Y_{sm}) is called a top stratum if it is an open subset of Y.

Multi-folds. Consider a space $\mathcal{W}$ that is a finite union of open sets $U_i, i = 1, \dots, k$, each with uniformizers $(\widetilde{U}_i, \Gamma_i, \pi_i)$ with the following properties. Each Γ_i is a finite group acting on $\widetilde{U}_i$, and the projection π_i is the composite of the quotient map $\widetilde{U}_i \to \widetilde{U}_i/\Gamma_i$ with an identification $\widetilde{U}_i/\Gamma_i = U_i$. The inverse image in $\widetilde{U}_i$ of each stratum in U_i is an open subset of a (complex) Banach space on which Γ_i acts complex linearly. For simplicity, we will assume that Γ_i acts freely on the points of the top strata in $\widetilde{U}_i$, and that the isomorphism class of the stabilizer subgroup $\mathrm{Stab}_i(\tilde{x}_i)$ of Γ_i is fixed as $\tilde{x}_i$ varies over a stratum of $\widetilde{U}_i$.[5]

[5]The stabilizer group $\mathrm{Stab}_i(\tilde{x}_i)$ of a point $\tilde{x}_i \in \widetilde{U}_i$ is the subgroup of Γ_i that fixes $\tilde{x}_i$.

To keep our construction as simple as possible, we will also assume that the set of local uniformizers $\{(\widetilde{U}_i, \Gamma_i, \pi_i)\}$ can be enlarged to an orbifold atlas. In an orbifold, the local uniformizers satisfy the following compatibility conditions:

(i) for every $x \in U_i \cap U_j$ there is an index ℓ such that $x \in U_\ell \subset U_i \cap U_j$;

(ii) for $m = i, j$, there are injections $\iota_{m\ell} : \widetilde{U}_\ell \to \widetilde{U}_m$ and group homomorphisms $\lambda_m : \Gamma_\ell \to \Gamma_m$ such that

$$\lambda_m(\gamma) \cdot \iota_{m\ell}(\tilde{y}) = \iota_{m\ell}(\gamma(\tilde{y})),$$

for all $\tilde{y} \in \widetilde{U}_\ell$ and $\gamma \in \Gamma_\ell$.

Thus, for example, we are not permitting a situation in which $U_i = \widetilde{U}_i = S^1$ for $i = 1, 2$ and $\pi_i : S^1 \to S^1$ quotients out by the action of $\mathbf{Z}/2\mathbf{Z}$ when $i = 1$ and by $\mathbf{Z}/3\mathbf{Z}$ when $i = 2$, though one could extend the theory to cover this case. Note that we do not assume that the cover $\{U_i\}_{1 \le i \le k}$ itself satisfies condition (i) above.

For each subset $I = \{i_1, \dots, i_p\}$ of $\{1, \dots, k\}$, set

$$U_I = \bigcap_{j \in I} U_j, \quad U_\emptyset = \emptyset.$$

Let $\mathcal{N}$ be the set of all I for which $U_I \ne \emptyset$. For each $I \in \mathcal{N}$ define the group Γ_I to be the product $\prod_{j \in I} \Gamma_j$, and then define $\widetilde{U}_I$ to be the fiber product

$$\widetilde{U}_I = \{\tilde{x}_I = (\tilde{x}_j)_{j \in I} : \pi_j(\tilde{x}_j) = \pi_\ell(\tilde{x}_\ell) \in U_I \text{ for all } j, \ell\} \subset \prod_{j \in I} \widetilde{U}_j,$$

with the two topologies induced from $\prod \widetilde{U}_j$. Clearly, Γ_I acts on $\widetilde{U}_I$ and the quotient $\widetilde{U}_I/\Gamma_I$ can be identified with U_I. Further, it is not hard to check that each $\widetilde{U}_I$ is a partially smooth space, with strata that project to strata in U_I. Again, the isomorphism class of the stabilizer subgroup

$$\mathrm{Stab}_I(\tilde{x}_I) = \prod_{i \in I} \mathrm{Stab}_i(\tilde{x}_i)$$

of $\tilde{x}_I$ in Γ_I is constant on each stratum, and trivial at points of top strata.

In the restricted case we are now considering, the topological structure of the fiber product $\widetilde{U}_I$ can best be understood in terms of the notion of "local component" that was introduced by Liu and Tian in [**LiuT2**]. Given a point $\tilde{x}_I \in \widetilde{U}_I$, choose $i_0 \in I$ and a small open neighborhood $\widetilde{C}$ of $\tilde{x}_{i_0}$ in $\widetilde{U}_{i_0}$. Then a neighborhood of $\tilde{x}_I$ in $\widetilde{U}_I$ can be identified with the set

$$\{(\gamma_j \circ \iota_{ji_0}(\tilde{y}))_{i \in I} \ : \ y \in \widetilde{C}, \gamma_{i_0} = id, \gamma_j \in \mathrm{Stab}_j(\iota_{ji_0}(\tilde{y})), j \ne i_0\}.$$

This is a finite union of sets $\widetilde{C}(\tilde{x}_I, \gamma)$, one for each element γ in the group

$$\mathrm{Stab}_{I - i_0}(\tilde{x}_I) = \prod_{j \in I - i_0} \mathrm{Stab}_j(\tilde{x}_j).$$

Each such set is homeomorphic (in the partially smooth category) to $\widetilde{C}$. It is also easy to check that the germs[6] at $\tilde{x}_I$ of the sets $\widetilde{C}_\gamma(\tilde{x}_I)$ are independent of the choices of i_0 and $\widetilde{C}$, as is the isomorphism class of the "reduced" group $\mathrm{Stab}'_I(\tilde{x}_I) = \mathrm{Stab}_{I - i_0}(\tilde{x}_I)$: see [**LiuT2**], Lemma 4.3. These germs are called the local components of $\widetilde{U}_I$ at $\tilde{x}_I$, and we will denote them by $\langle \widetilde{C}_\gamma(\tilde{x}_I) \rangle$. Note that

[6]Two subsets C_1, C_2 that both contain the point x are said to have the same germ at x if there is a neighborhood U of x such that $C_1 \cap U = C_2 \cap U$.

points $\tilde{x}_I$ in the top strata on $\widetilde{U}_I$ have a single local component, since their stabilizer groups are trivial. Liu and Tian define the *desingularization* $\widehat{U}_I$ of $\widetilde{U}_I$ to be the union of all such local components:

$$\widehat{U}_I = \{(\tilde{x}_I, \langle \widetilde{C}_\gamma(\tilde{x}_I)\rangle) \ : \ \tilde{x}_I \in \widetilde{U}_I, \gamma \in \mathrm{Stab}'_I(\tilde{x}_I)\}.$$

$\widehat{U}_I$ is topologized so that a germ of a neighborhood of $\widehat{U}_I$ at the point $(\tilde{x}_I, \langle \widetilde{C}_\gamma(\tilde{x}_I)\rangle)$ is homeomorphic to the local component $\langle \widetilde{C}_\gamma(\tilde{x}_I)\rangle$ itself. Thus the projection

$$proj \ : \ \widehat{U}_I \to \widetilde{U}_I$$

is locally a homeomorphism onto its image. The effect of considering $\widehat{U}_I$ is to get rid of the extra singularities of $\widetilde{U}_I$ that are introduced by passing to the fiber product: locally $\widehat{U}_I$ is homeomorphic to the initial sets $\widetilde{U}_j$.

EXAMPLE 4.6. For $j = 1, 2$ let $U_j = [0, \infty) = \mathcal{W}$ be uniformized by

$$(\widetilde{U}_j = \mathbf{R}, \Gamma_j = \mathbf{Z}/2\mathbf{Z}, \pi_j),$$

where $\Gamma_j = \mathbf{Z}/2\mathbf{Z}$ acts by multiplication by -1. Then, $\widetilde{U}_{12}$ can be identified with the union of the two lines $\{y = \pm x\}$ in $\mathbf{R}^2$, with the 5 strata

$$\{(\pm x, \pm x) : x > 0\}, \ \ \{(0, 0)\}.$$

Moreover, $\widehat{U}_{12}$ is the disjoint union of these two lines.

If $J \subset I$, there are projections

$$\pi^I_J : \widetilde{U}_I \to \widetilde{U}_J, \quad \lambda^I_J : \Gamma_I \to \Gamma_J,$$

where π^I_J quotients out by the action of the product group Γ_{I-J}. Observe that if $\pi^I_J(\tilde{y}_I) = \tilde{x}_J$ is in a top stratum, then the inverse image $(\pi^I_J)^{-1}(\tilde{x}_J)$ has $|\Gamma_{I-J}|$ points.

More generally, suppose that $\{V_I\}$ is a family of open sets that covers $\mathcal{W}$ and is such that $V_I \subset U_I$ for all I. (In the application V_I is in general much smaller than U_I, and the sets $V_j, j = 1, \dots, k$, no longer cover $\mathcal{W}$: see Example 4.10.) Then define $\widetilde{V}_I \subset \widetilde{U}_I$ to be the inverse image of V_I under the map $\pi : \widetilde{U}_I \to U_I$. Clearly Γ_I acts on each $\widetilde{V}_I$, and the map $\pi : \widetilde{V}_I \to V_I$ simply quotients out by this action. Further the projections $\pi^I_J : \widetilde{U}_I \to \widetilde{U}_J$ restrict to give partially defined maps $\widetilde{V}_I \to \widetilde{V}_J$ with domain $(\pi^I_J)^{-1}(\widetilde{V}_J)$. We will say that such a collection

$$\widetilde{\mathcal{V}} = \{(\widetilde{V}_I, \Gamma_I, \pi^I_J, \lambda^I_J) : I \in \mathcal{N}\}$$

is a *multi-fold atlas* for $\mathcal{W}$, or a multi-fold, for short. The motivation for considering the subcover $\{V_I\}$ will not be apparent until we start constructing multi-sections as in Example 4.7.

Multi-bundles. We now want to define a *multi-bundle* $\tilde{p} : \widetilde{\mathcal{E}} \to \widetilde{\mathcal{V}}$. The example of interest to us is induced from a collection of local orbibundles over the cover U_i, and so we will restrict to that case here, though one could make a more general definition that refers only to the atlas $\widetilde{\mathcal{V}}$. We start with a space

$$\mathcal{E} = \bigcup_i E_i$$

that has two topologies, and a map $p : \mathcal{E} \to \mathcal{W}$ with the property that each set $E_i = p^{-1}(U_i)$ has a local uniformizer $(\widetilde{E}_i, \Gamma_i, \Pi_i)$ such that the following diagram commutes:

$$\begin{array}{ccc} \widetilde{E}_i & \stackrel{\Pi_i}{\to} & E_i \\ {\scriptstyle \tilde{p}} \downarrow & & {\scriptstyle p} \downarrow \\ \widetilde{U}_i & \stackrel{\pi_i}{\to} & U_i. \end{array}$$

Here we require $\tilde{p} : \widetilde{E}_i \to \widetilde{U}_i$ to be a Γ_i-equivariant map that restricts over each stratum of $\widetilde{U}_i$ to a locally trivial vector bundle.[7] Thus the fiber $\widetilde{F}(\tilde{x}_i)$ of $\tilde{p}$ at each point $\tilde{x}_i$ is a vector space, but there is no natural way of identifying one with another if they lie over points in different strata. In the application, the points of $\widetilde{U}_i$ are parametrized stable maps $\tilde{\tau} = (\Sigma, \tilde{h})$ and the fiber $F(\tilde{\tau})$ is the space $L^{1,k}(\Lambda^{0,1}(\Sigma, \tilde{h}^*TP))$ of sections of a bundle over Σ. When $\tilde{\tau}$ moves from one stratum to another, the topological type of Σ may change, and so there is no easy way to identify these fibers.[8]

We assume further that the orbifold structure on $\mathcal{W}$ lifts to one on $\mathcal{E}$. Thus, $\mathcal{E}$ has the same local structure as $\mathcal{W}$, and we can define a multi-fold atlas $\widetilde{\mathcal{E}} = \{\widetilde{E}_I\}$ for it as above. By slight abuse of notation, we write $\widetilde{E}_I$ for the restriction to $\widetilde{V}_I$ of the fiber product of the $\widetilde{E}_i, i \in I$, over $\widetilde{U}_I$. More precisely, the fiber $\widetilde{F}(\tilde{x}_I)$ of $\widetilde{E}_I$ at $\tilde{x}_I = (\tilde{x}_i)_{i \in I} \in V_I$ has elements $(\tilde{x}_i, \tilde{v}_i)_{i \in I}$ where $\tilde{v}_i \in \widetilde{F}(\tilde{x}_i)$ and, for all i, j,

$$\Pi_i(\tilde{v}_i) = \Pi_j(\tilde{v}_j) \in \mathcal{E}.$$

Note that the desingularization $\widehat{E}_I \to \widehat{V}_I$ of $\widetilde{E}_I \to \widetilde{V}_I$ is an honest vector bundle, since locally it has the same structure as the maps $\widetilde{E}_{i_0} \to \widetilde{U}_{i_0}$. Thus $\widetilde{E}_I \to \widetilde{V}_I$ is the finite union of vector bundles, and its fibers are finite unions of vector spaces.

EXAMPLE 4.6 continued. Suppose that $\tilde{p} : \widetilde{E}_j \to \widetilde{U}_j$ is a trivial line bundle, on which Γ_j acts by $(x, v) \mapsto (-x, -v)$, and identify the quotients $\widetilde{E}_j/\Gamma_j = E_j$ in the obvious way. Then the fiber over a point (y, y) of $\widetilde{U}_{12}$ with $y \neq 0$ consists of the points $(y, v; y, v)$, while the fiber over $(y, -y)$ consists of the points $(y, v; -y, -v)$. When $y = 0$ the fiber is the union of the two lines $(0, v; 0, v)$ and $(0, v; 0, -v)$. Thus, the bundle $\widetilde{E}_{12} \to \widetilde{V}_{12}$ is the union of two line bundles, one over each line $y = \pm x$ in $\widetilde{U}_{12}$ with distinct fibers over the intersection point $(0, 0)$ of these lines that intersect only at the zero element. Observe also that $\mathrm{Stab}_{12}(0, 0) = \Gamma_{12} = \mathbf{Z}/2\mathbf{Z} \oplus \mathbf{Z}/2\mathbf{Z}$, and acts via

$$(\gamma_1, \gamma_2) \cdot (y, v; \pm y, \pm v) \mapsto (\gamma_1(y), \gamma_1(v), \gamma_2(\pm y), \gamma_2(\pm y))$$

Multi-sections. This is the heart of the construction. Everything we have done so far is aimed towards making it possible to define a large enough family of multi-sections to get transversality for the delbar operator. Intuitively a *multi-section* $\tilde{s}$ of $\widetilde{\mathcal{E}} \to \widetilde{\mathcal{V}}$ is a compatible collection $\{\tilde{s}_I\}$ of multi-valued sections. In other words, for each point $\tilde{x}_I$ of $\widetilde{V}_I$ we are given a nonempty finite subset $\tilde{s}_I(\tilde{x}_I)$ of $\widetilde{F}(\tilde{x}_I)$.[9] In the situation at hand, it is enough to consider multi-sections that are

---

[7]Observe that if the stabilizer $\mathrm{Stab}_i(\tilde{x}_i)$ of any point $\tilde{x}_i \in \widetilde{U}_i$ acts nontrivially on the fiber $\widetilde{F}(\tilde{x}_i)$, then the quotient object $E_i \to U_i$ is not a vector bundle but rather a (local) orbibundle.

[8]Siebert [**Sb**] actually makes such an identification, thereby simplifying the topology at the cost of some extra analysis.

[9]I am indebted to Polterovich for clarifying the definition of multi-bundles and multi-sections via Examples 4.6 and 4.7. The original presentation of Liu and Tian, which appeared in the early versions of [**LiuT1**] on which this paper is mostly based, was difficult to understand at this point

single-valued when lifted to the desingularization $\widehat{E}_I \to \widehat{V}_I$. Thus we will assume that there are sections $\widehat{s}_I : \widehat{V}_I \to \widehat{E}_I$ such that

$$\tilde{s}_I(\tilde{x}_I) = \{proj \widehat{s}_I(\tilde{x}_I, \langle \widetilde{C}_\gamma(\tilde{x}_I)\rangle) : \gamma \in \mathrm{Stab}'_I(\tilde{x}_I)\}.$$

In particular, when $I = \{j\}$, the section $\widehat{s}_j = \tilde{s}_j$ is a single-valued but not necessarily equivariant section of the bundle $\widetilde{E}_j \to \widetilde{U}_j$. Note also that the sets $\tilde{s}_I(\tilde{x}_I)$ contain a single element when $\tilde{x}_I$ is in a top stratum.

If $J \subset I$, we define the pullback $(\Pi^I_J)^{-1}(\tilde{s}_J)$ to be the multi-valued section that associates to $\tilde{x}_I$ the full inverse image $(\Pi^I_J)^{-1}(\tilde{s}_J(\tilde{x}_J))$, where $\tilde{x}_J = \pi^I_J(\tilde{x}_I)$. The compatibility condition is that $\tilde{s}_I$ restricts over $(\pi^I_J)^{-1}(\widetilde{V}_J)$ to the pullback of $\tilde{s}_J$.

As the following example shows, the compatibility conditions are quite strong. In fact, if the sets U_i overlap too much there are no interesting (i.e. nonequivariant) multi-sections, which is precisely the reason why we consider the subcover $\{V_I\}$ of the $\{U_I\}$.

EXAMPLE 4.7. Suppose that in the situation of the previous example we are given a section $\tilde{s}_1$ of $\widetilde{E}_1 \to \widetilde{U}_1$. Then, if $\tilde{s}_{12}$ denotes its pullback to $\widetilde{U}_{12}$,

$$\tilde{s}_{12}(y, y) = (y, \tilde{s}_1(y); y, \tilde{s}_1(y)), \quad \tilde{s}_{12}(y, -y) = (y, \tilde{s}_1(y); -y, \tilde{s}_1(-y)).$$

Thus $\tilde{s}_{12}$ is single-valued over the top strata in $\widetilde{U}_{12}$, since $y \neq 0$ there. Similarly, the pullback $\tilde{s}'_{12}$ of a section $\tilde{s}'_2$ over $\widetilde{U}_2$ has the form

$$\tilde{s}'_{12}(y, y) = (y, \tilde{s}'_2(y); y, \tilde{s}'_2(y)), \quad \tilde{s}'_{12}(-y, y) = (-y, -\tilde{s}'_2(y); y, \tilde{s}'_2(y)).$$

Therefore, in order for $\tilde{s}_{12}$ to be such a pullback, we need $\tilde{s}_1(y) = \tilde{s}'_2(y)$ and $-\tilde{s}_1(y) = \tilde{s}'_2(-y)$. This forces $\tilde{s}_1$ to be equivariant, i.e. $\tilde{s}_1(-y) = -\tilde{s}_1(y)$. In particular, $\tilde{s}_j(0) = 0$. Thus $\tilde{s}_1$ extends to a multi-section of $\widetilde{\mathcal{E}} \to \widetilde{\mathcal{U}}$ only under this equivariance condition.

Suppose now that $V_1 = [0, 2), V_2 = (4, \infty) = \widetilde{V}_2$ and $V_{12} = [0, 5)$, and let $\tilde{s}_1$ be any section of $\widetilde{E}_1 \to \widetilde{U}_1$ that has support in $(-3, 3)$. It is easy to see that now $\tilde{s}_1$ does extend to a multi-section over $\widetilde{\mathcal{V}}$, since $\tilde{s}_1(\tilde{y}) = 0$ if $\tilde{y}$ lies above the overlap $V_{12} \cap V_2$. Observe also that if $\tilde{s}_1(0) \neq 0$, then $\tilde{s}_{12}$ has two values at $(0, 0) \in \widetilde{V}_{12}$. □

REMARK 4.8. The graph $gr(\tilde{s}_I)$ is simply the union of the sets $\tilde{s}_I(\tilde{x}_I), \tilde{x}_I \in \widetilde{V}_I$, and it is often useful to impose geometric conditions on it, e.g. to require that both it and the projection map $gr(\tilde{s}_I) \to \widetilde{V}_I$ be an object in our category. For instance, such conditions are needed when discussing the intersection of the graph with the zero section as in Definition 4.11 below. This amounts to requiring that for each j the graphs $gr(\tilde{s}_j)$ have a stratification that is compatible with the projection to $\widetilde{V}_j$, is preserved by the action of the groups Γ_j, and is such that the rank of the stabilizer is constant on strata. Here one is allowed to refine the stratification of $\widetilde{V}_I$ provided that one does not subdivide its top strata. (Subdividing the top strata would introduce codimension 1 strata, which cause problems when one wants to define cycles.)

However, if one incorporates such geometric requirements into the definition of a multi-section, it is not possible to add an arbitrary pair of multi-sections since, for example, the sum $\tilde{s}_I + \tilde{s}'_I$ might equal zero on a very nasty set. Therefore we

---

since they did not make it clear that multi-sections could sometimes be multi-valued on lower dimensional strata. The exposition given here follows later versions of Liu andTian's work.

will not impose these conditions initially, though the sections that we will consider in practise will all satisfy such geometric conditions: see Example 4.10 below.

Constructing multi-sections. Any family of compatible Γ_i-equivariant sections $\tilde{s}_i$ of the local bundles $\tilde{p} : \widetilde{E}_i \to \widetilde{U}_i$ gives rise to a multi-section of $\widetilde{\mathcal{E}}$. For example, the Cauchy–Riemann operator $\overline{\partial}_J$ is a multi-section of this kind. However, one cannot regularize the moduli space just by considering equivariant perturbations of the Cauchy–Riemann equation: see §5.1. The whole point of the present construction is to give a way of extending nonequivariant sections of the bundles $\widetilde{E}_j \to \widetilde{U}_j$ to multi-sections of $\widetilde{\mathcal{E}} \to \widetilde{\mathcal{V}}$, so that they can form global perturbations of the Cauchy–Riemann operator. As is suggested by the previous example, such an extension is possible provided that the covering $\{V_I\}$ is suitably chosen. The next lemma is adapted from [**LiuT1**], Lemma 4.3. The notation $A \Subset B$ means that the closure $cl(A)$ of A is contained in B.

LEMMA 4.9. *Given any finite open covering $\{U_i\}$ of the compact subset $\overline{\mathcal{M}}$ of $\mathcal{W}$, there are open subsets $U_i^0 \Subset U_i$ and $V_I \subset U_I$ with the following properties:*

(i) $\overline{\mathcal{M}} \subset \bigcup_i U_i^0$ *and* $\overline{\mathcal{M}} \subset \bigcup_I V_I$.

(ii) $V_I \cap U_i^0 = \emptyset$ *if* $i \notin I$.

(iii) *If* $V_I \cap V_J \neq \emptyset$, *then one of the sets* I *or* J *contains the other.*

PROOF. If there are N sets U_i, choose for $n = 0, \dots, N$ open coverings $\{U_i^n\}$, $\{W_i^n\}$ of $\overline{\mathcal{M}}$ such that

$$U_i^0 \Subset W_i^1 \Subset U_i^1 \Subset W_i^2 \Subset \cdots \Subset U_i^N = U_i.$$

Then, if $\operatorname{card} I = |I| = \ell$, define

$$V_I = W_I^\ell - \bigcup_{J:|J|>\ell} cl(U_J^{\ell+1}).$$

It is easy to check that the required properties hold. □

The next example is taken from [**LiuT1**], Lemma 4.5, and explains how it is possible to construct a sufficiently large family of multi-sections.

EXAMPLE 4.10. Let the U_i^0 and V_I be as above, and (by shrinking $\mathcal{W}$) suppose that $\mathcal{W} = \bigcup_i U_i^0 = \bigcup_I V_I$. Consider a multi-bundle $\widetilde{\mathcal{E}} \to \widetilde{\mathcal{V}}$ as above, and suppose that for some j we are given a section $\sigma(j) : \widetilde{U}_j \to \widetilde{E}_j$ of the bundle $\widetilde{U}_j \to \widetilde{E}_j$ with support in $\widetilde{U}_j^0$. Then there is an induced multi-section $\tilde{s}(j)$ of $\widetilde{\mathcal{E}} \to \widetilde{\mathcal{V}}$ defined as follows:

> $\tilde{s}(j)_I$ is the zero section if $j \notin I$, and, if $j \in I$, is the restriction to $\widetilde{V}_I$ of the pullback to $\widetilde{U}_I$ of the graph of $\sigma(j)$.

Suppose further that $\sigma(j)$ is generic in the sense that, for all points $\tilde{x}_j \in \widetilde{U}_j$, no element of the stabilizer group $\operatorname{Stab}_j(\tilde{x}_j)$ fixes $\sigma(j)(\tilde{x}_j)$ unless $\sigma(j)(\tilde{x}_j) = 0$. We require also that the boundary of the support of $\sigma(j)$ is a union of strata. Then, it is not hard to see that its graph can be given a stratification compatible with the projections to the $\widetilde{V}_I$ as discussed in Remark 4.8. The essential point to check is that each set on which the number $|\tilde{s}(j)_I(\tilde{x}_I)|$ is constant is a union of strata. By construction this number is 1 over the top strata of $\widetilde{V}_I$, so there is no problem there. Note also that the genericity condition implies that, if $\pi_j^I(\tilde{x}_I)$ is not in the zero set of $\sigma(j)$, then $|\tilde{s}(j)_I(x_I)|$ is precisely equal to the number $|\mathrm{Stab}_{I-i_0}(\tilde{x}_I)|$ of local components at $\tilde{x}_I$ (i.e. is the maximum possible) and so is constant on strata.

The multi-sections that we will use in §5 have the form

$$\tilde{s} = \overline{\partial}_J + \sum_j \tilde{s}(j)$$

where $\overline{\partial}_J$ is the section defined by the Cauchy–Riemann operator. □

4.3. Construction of the virtual cycle. Finally, we need to see how multi-sections give rise to branched pseudomanifolds. From now on we will assume that the covering $\{V_I\}$ satisfies the conditions of Lemma 4.9. The following definition is very much *ad hoc*: we are gathering together a collection of useful properties of $\tilde{s}$ rather than formulating the best and most general definition. In particular, normally a Fredholm object has finite dimensional kernel and cokernel. Here we retain the information on the kernel plus the vestigial information on the cokernel that can be gleaned from condition (i).

DEFINITION 4.11. We say that $\tilde{s}$ is *Fredholm* of index d if the following conditions are satisfied.

(i) For each I, $gr(\tilde{s}_I)$ meets the zero section of $\widetilde{E}_I \to \widetilde{V}_I$ transversally in a d-dimensional (open) pseudomanifold $\widetilde{Z}_I$ whose top stratum coincides with its intersection with the top stratum of $\widetilde{V}_I$.

(ii) The top strata S of $\widetilde{Z}_I$ are provided with orientations that are preserved under the partially defined projections π^I_J.

(iii) The union $Z_{\mathcal{W}} = \bigcup_I \pi_I(\widetilde{Z}_I) \subset \mathcal{W}$ is compact.

Before showing how to assemble the local zero sets $\widetilde{Z}_I$ into a pseudomanifold, let us discuss the question of the existence of Fredholm sections. In the finite dimensional case, our assumption that $\widetilde{Z}_I$ has no codimension 1 strata implies that the sets $\widetilde{V}_I$ themselves have no codimension 1 strata (or at least no such strata that are anywhere near the zero set of $\tilde{s}$). In our main (infinite dimensional) example all bundles etc. are complex, and so all strata in $\widetilde{Z}_I$ have even codimension. Much of the work in [**LiuT1**] is taken up with showing that a suitable perturbation of the delbar operator does give rise to a Fredholm section. They construct a finite dimensional family of sections $\tilde{s}^\nu$, for $\nu \in R$, which are Fredholm in our sense for generic ν. This is discussed further in §5 below.

One would also hope that in the finite, even dimensional context, generic sections are Fredholm. Here $\widetilde{V}_I$ would have dimension $2n$, say, and the fibers of $\tilde{p} : \widetilde{E}_I \to \widetilde{V}_I$ would have dimension $2k$, so that $\tilde{s}$ would have index $d = 2n - 2k$. The idea would be to show that any multi-section can be perturbed so that it intersects the zero section of $\tilde{p} : \widetilde{E}_I \to \widetilde{V}_I$ transversally. In fact, the strata of $\widetilde{E}_I$ have the form $\tilde{p}^{-1}(S)$, where S is a stratum of $\widetilde{V}_I$ and, by construction, each stratum S' of $gr(\tilde{s}_I)$ lies inside a stratum $\tilde{p}^{-1}(S)$ in $\widetilde{E}_I$. Hence it would suffice to perturb $gr(\tilde{s}_I)$ so that the intersection of each S' with such S is transverse inside $\tilde{p}^{-1}(S)$. This can be arranged, but only in the presence of suitable normal cones.[10]

[10] The point here is that with stratified spaces one works inductively over the strata. If one perturbs $\tilde{s}_I$ over one stratum so that it is transverse to the zero section there, one needs to be able to extend this perturbation to nearby strata, and hence one needs information on how those strata fit together.

We now show how to glue the local pseudomanifolds $\widetilde{Z}_I$ together to get a closed branched labelled pseudomanifold Y that projects onto $Z_{\mathcal{W}} = \bigcup_I \pi_I(\widetilde{Z}_I)$. First of all we must replace the open sets $\widetilde{Z}_I$ by closed pseudomanifolds that we will call $\widetilde{Y}_I$. We will do this by choosing manifolds with boundary $\overline{V}'_I \subset V_I$ that cover $\mathcal{W}$, and then setting

$$\widetilde{Y}_I = \widetilde{Z}_I \cap \pi_I^{-1}(\overline{V}'_I).$$

Here $\widetilde{Y}_I$ is given the obvious first topology, and the components of $(\widetilde{Y}_I)_{sm}$ either lie on its boundary (i.e. in $\pi_I^{-1}(\partial \overline{V}'_I)$), or are the intersections of strata in $\widetilde{Z}_I$ with its interior. In order for there to be such a stratification, it suffices that the stratification of $\mathcal{W}$ can be refined so that each boundary $\partial \overline{V}'_I$ is a union of strata. Thus the boundaries $\partial \overline{V}'_I$ must intersect transversally and be in general position with respect to the original strata of $\mathcal{W}$. Strictly speaking, one can arrange this for general $\mathcal{W}$ only in the presence of suitable normal cones.

If one can find such sets $\widetilde{Y}_I$ we will say that they form a *shrinking* of the $\widetilde{Z}_I$. For now, we have stratified the $\widetilde{Y}_I$ so that their only codimension 1 strata lie in the boundary $\partial \overline{V}'_I$. These will form the branching locus of Y.

The next step is to construct a topological space Y such that there are continuous maps

$$\coprod_I \widetilde{Y}_I \longrightarrow Y \longrightarrow Z_{\mathcal{W}}.$$

Points of Y are equivalence classes $\langle z \rangle$ under the equivalence relation that is generated by setting

$$\tilde{y}_I \sim \tilde{z}_J, \qquad \text{if } J \subset I, \ \pi^I_J(\tilde{y}_I) = \tilde{z}_J.$$

The first topology on Y is the quotient topology, and the strata of its second (smooth) topology are formed by the images of strata in the $\widetilde{Y}_I$, subdivided as necessary. It will be convenient now to refine the stratifications of the $\widetilde{Y}_I$ so that the projections $q_Y : \widetilde{Y}_I \to Y$ take strata to strata. This introduces new codimension 1 strata in the $\widetilde{Y}_I$ coming from the boundaries of the $\widetilde{Y}_K$.

EXAMPLE 4.12. As an instructive example, consider the bundle $\widetilde{\mathcal{E}} \to \widetilde{\mathcal{V}}$ of Example 4.7 and suppose that $\widetilde{Z}_I$ is simply its zero section. Thus $\widetilde{Z}_I = \widetilde{V}_I$ for all i. Take $\overline{V}'_1 = [0,1], \overline{V}'_2 = [4.1, \infty)$ and $\overline{V}'_{12} = [.5, 4.5]$. Then Y projects to the base $\mathcal{W} = [0, \infty)$ by a map that is injective over $\overline{V}'_1 \cup \overline{V}'_2$ but is 2 to 1 elsewhere. In other words, the branching in Y comes from the properties of the cover $\overline{V}'_I$. Although this example does not have all the properties we require (e.g., it is not even-dimensional), this feature remains. Our multi-sections $\tilde{s}$ are single valued over top strata, so that $\widetilde{Z}_I$ has no branching. When one makes identifications to form Y, the branching does not come from properties of the multi-section $\tilde{s}$ that generates $\widetilde{Z}_I$ but from the properties of the cover $\{\overline{V}'_I\}$. □

It remains to define the labelling function. If S is a top stratum of Y lying in the image of $q_I : \widetilde{Y}_I \to Y$, we set

$$\lambda_I(y) = \frac{|\{q_I^{-1}(y)\}|}{|\Gamma_I|}, \qquad y \in S.$$

PROPOSITION 4.13. *Suppose that $\{V_I\}$ is as in Lemma 4.9, that $\tilde{s}$ is Fredholm, and that the zero sets $\widetilde{Z}_I$ have a shrinking $\widetilde{Y}_I$. Then the labelling function λ_I*

descends to a function λ on Y, and gives (Y, λ) the structure of a compact branched labelled pseudo-manifold.

PROOF. To prove the first statement, it suffices to check that $\lambda_I(y) = \lambda_J(y)$ if $J \subset I$ and $y \in q_J(\widetilde{Y}_J) \cap q_I(\widetilde{Y}_I)$. This holds because $\tilde{s}_I$ is the pullback of $\tilde{s}_J$ over $(\pi^I_J)^{-1}(\widetilde{V}_J)$, so that

$$|\{q_I^{-1}(y)\}| = |\{q_J^{-1}(y)\}| \cdot |\Gamma_{I-J}|, \quad \Gamma_I = \Gamma_J \cdot |\Gamma_{I-J}|.$$

To prove the second statement, we have to check the branching condition of Definition 4.3 on the codimension 1 components of Y. Thus we have to understand the effect of one of the generating equivalences $\tilde{y}_J \sim \tilde{z}_J$, where $J \subset I$, on this condition. Note first that a class $\langle y \rangle$ lies in a codimension 1 stratum B_ℓ in Y if for one K the representatives of $\langle y \rangle$ in $\widetilde{Y}_K$ lie on boundary strata B_K and if for all other J the representatives in $\widetilde{Y}_J$ lie in the interior of $\widetilde{Y}_J$.

We will call the side of B_K that meets the interior of $\widetilde{Y}_K$ the positive side, and will write $\{S^+_{K,\alpha}\}$ for the set of all top strata in $\widetilde{Y}_K$ whose closures contain representatives of $\langle y \rangle$. Similarly, for all other J we write $\{S^{\pm}_{J,\alpha}\}$ for the set of top strata in $\widetilde{Y}_J$ that contain representatives of $\langle y \rangle$, with signs $\pm$ assigned in the obvious way. By Lemma 4.9, the set of J in question have nonempty intersection I_y with K, and everything is essentially pulled back from $\widetilde{V}_{I_y}$. For example, two strata in $\{S^+_{J,\alpha}\}$ are identified in Y if and only if they have the same image in $\widetilde{V}_{I_y}$. Therefore, we only need to consider the effect of inclusions of the form $I_y \subset I$ on the branching condition. There are three cases, the first in which $K \neq I_y, I$, the second in which $K = I$, the third in which $K = I_y$. For instance, in Example 4.12 the codimension 1 strata of Y are the points $.5, 1, 4.1$ and 4.5. At the point $y = 4.1$, $K = \{2\} = I_y$ and $J = \{1, 2\}$, while at the point $y = 4.5$, $K = \{1, 2\}$ and $J = \{2\} = I_y$. Here inclusions of the first kind do not occur, inclusions of the second kind have no branching effect, while inclusions of the third kind give rise to the branching at the points $.5, 4.5$.

This illustrates what happens in general. Inclusions of the first kind are essentially neutral. For if J is either I_y or I there is a bijective correspondence between the components $\{S^+_{J,\alpha}\}$ and $\{S^-_{J,\alpha}\}$ on both sides that commutes with the identifications coning from $\sim$. So whatever happens is the same on both sides. In the second case, each stratum in $\{S^+_{K,\alpha}\}$ is mapped bijectively by $\pi^K_{I_y}$ onto a stratum in $\{S^+_{I_y,\alpha}\}$, and each of the latter strata is covered exactly $|\Gamma_{K-I_y}|$ times. Hence both sides of B in Y are the same, and no real branching occurs here either.

However, the situation is different if $K = I_y$. Then the components in $\{S^+_{I,\alpha}\}$ are identified via π^I_K with components in $\{S^+_{K,\alpha}\}$, while no identification is put on the components $\{S^+_{I,\alpha}\}$ on the other side. Since π^I_K is $|\Gamma_{I-K}|$ to 1, $|\Gamma_{I-K}|$ components on the negative side of B in Y will correspond to each component on the positive side. It is not hard to check that the sum of the labels on the two sides is the same. □

Obviously, one would now like to prove a statement to the effect that the homology class represented by the pseudo-manifold (Y_s, λ) defined by a Fredholm multisection $\tilde{s}$ is independent of the choice of multi-section and of coverings $\{U_j\}, \{V_J\}$ and so on. For such a statement to make sense, one first has to assume that there is a map $ev : \mathcal{W} \to X$, so that the classes one is comparing are given by the composite maps

$$Y_s \xrightarrow{\pi} \mathcal{W} \xrightarrow{ev} X.$$

Secondly, one has to set up a context in which there is a well-defined range of possible choices. This is no problem as far as the coverings are concerned, but does raise questions about what Fredholm multi-sections to consider. In the main example considered in §5 below, there is an evaluation map $ev : \mathcal{W} \to X$ and also a well-defined family of Fredholm sections, namely those of the form $\bar{\partial}_J + \sum_j \tilde{s}^\nu(j)$. Since this is a connected family, it is not too hard to show that the homology class defined by (Y, λ) is independent of choices.

Another natural framework to consider is the finite dimensional case. As an illustration of our method we show below how to calculate the Euler number of a teardrop T using multi-sections.

4.4. Finite dimensional orbifolds. Resolutions. As is shown by Example 4.12, the method used in the last section to piece the zero sets $\widetilde{Z}_I$ together into a branched, labelled pseudomanifold can be applied to the zero section of any finite dimensional multi-bundle over a finite dimensional multi-fold, provided that this is oriented. (Note that any finite dimensional oriented orbifold in the usual smooth category has no codimension 1 strata, since the strata come from points with nontrivial stabilizers and these always have codimension ≥ 2 in the oriented case.) In other words, as observed by Polterovich, one can construct from an oriented orbifold M a branched pseudocycle Y_M that maps to M and can be thought of as a *resolution* of M. Note that when constructing this we do not have to pass to the cover $\{\widetilde{V}_I\}$ but can simply use the $\{\widetilde{U}_I\}$, since the zero section is equivariant.

To illustrate this, let us look at the teardrop T. This is the simplest 2-dimensional orbifold. Topologically T is a 2-sphere, but it has one singular point at the north pole p_N where the group Γ_N is cyclic of order k. Therefore it can be covered by two open discs, U_1 containing p_N and U_2 containing the south pole p_S, that intersect in an open annulus U_{12}. Then $\widetilde{U}_2 = U_2$ and $\pi_1 : \widetilde{U}_1 \to U_1$ is given in polar coordinates (r, θ) by $(r, \theta) \mapsto (r, k\theta)$. The sets $\overline{V}'_I$ should be slightly smaller closed discs and annuli. The resulting branched pseudomanifold Y_T is the union of an open disc $D_2 = \mathrm{int}\,(\overline{V}'_2)$ with its bounding circle C and another disc $D_1 = U_1 - \mathrm{int}\,(\overline{V}'_2)$, whose boundary is attached to C by the k-fold covering $\theta \mapsto k\theta$. Points in D_1 have labels $1/k$, while those in D_2 have labels 1.

Calculating the Euler number. The general idea is this. Let M be a finite dimensional oriented orbifold with local uniformizers $(\widetilde{U}_j, \pi_j, \Gamma_j)$. Here we assume that $\widetilde{U}_j$ is an open subset of some oriented vector space and that Γ_j acts linearly, preserving orientation. It follows that any subspace with nontrivial stabilizer subgroup has codimension at least 2. Hence there are no codimension 1 strata and we are in a situation in which Fredholm sections can exist. (See the discussion after Definition 4.11.) Moreover, all strata have normal cones.

One can construct a multi-fold atlas $\mathcal{V}_M$ for M from any subcovering $\{V_I\}$ of the sets $\{U_I\}$ as in §4.2. The group Γ_j acts on the tangent bundle $\widetilde{E}_j = T\widetilde{U}_j \to \widetilde{U}_j$ of $\widetilde{U}_j$, and the quotient by these local actions is an orbibundle $\mathcal{E} \to M$.[11] From this we construct a multi-bundle $\widetilde{\mathcal{E}} \to \mathcal{V}_M$ as before. Given a suitable compactly

[11] As an exercise the reader might like to give a formal definition of an orbibundle (cf. [**FO**]). The point is to formulate exactly how the elements of the different local bundles E_i should be identified with each other.

supported section $\sigma(j)$ of $\widetilde{E}_j \to \widetilde{U}_j$, one can construct a multi-section $\tilde{s}(j)$ of $\widetilde{\mathcal{E}}$ as in Example 4.10. Then the section

$$\tilde{s} = \sum_j \tilde{s}(j)$$

should be well defined and Fredholm. Hence we can construct a 0-dimensional branched, labelled pseudomanifold $Y_{\tilde{s}}$ from its zero set. The Euler number of M is defined to be the number of points in $Y_{\tilde{s}}$, weighted by their labels. The claim is that this is independent of the choice of $\tilde{s}$. This looks very plausible: further details are left to the reader.

In the case of the teardrop, with U_1, U_2 as above, choose polar coordinates on U_1 so that the annulus $U_{12} = V_{12}$ is given by

$$(t, \theta) \;:\; t \in (1,4),\ \theta \in S^1 = \mathbf{R}/\mathbf{Z}.$$

Then let V_1, V_2 be smaller discs in U_1, U_2 respectively, such that

$$V_1 \cap V_{12} = (1,2) \times S^2, \quad V_2 \cap V_{12} = (3,4) \times S^1.$$

The tangent bundle $\widetilde{E}_1 = T\widetilde{U}_1 \to \widetilde{U}_1$ can be trivialised away from p_N by the vector fields $\tilde{\partial}_r, \tilde{\partial}_\theta$; and similarly, if (s, ϕ) are polar coordinates in U_2 centered at p_S, $\widetilde{E}_2$ can be trivialised by $\tilde{\partial}_s, \tilde{\partial}_\phi$. Note that the sections $\tilde{\partial}_r, \tilde{\partial}_\theta$ are invariant under the action of Γ_N and so descend to sections $\partial_r, \partial_\theta$ of $E_1 \to U_1 - p_N$. Moreover, on the overlap V_{12}, $\partial_r = -\partial_s$.

Let $\sigma(1) : \widetilde{U}_1 \to \widetilde{E}_1$ be a section of the form $\beta(r)\tilde{\partial}_r$, where the function β has an isolated zero at $r = 0$ and is everywhere ≤ 0, and let $\sigma(2) : \widetilde{U}_2 \to \widetilde{E}_2$ have the form $\gamma(s)\tilde{\partial}_s$, where the function γ has an isolated zero at $s = 0$ and is everywhere ≥ 0. Then it is not hard to see that the only zeros of $\tilde{s} = \tilde{s}(1) + \tilde{s}(2)$ are at p_N and p_S respectively. However, the point p_N is not in the top stratum of $\widetilde{U}_1$, and so condition (i) in Definition 4.11 is not satisfied. By slightly perturbing $\tilde{s}$ near p_N we can achieve transversality. Now there is one zero near p_N that counts with weight $1/k$ and one at p_S that counts with weight 1. Thus the Euler number is $1 + 1/k$.

One could try this with a different choice of $\sigma(1), \sigma(2)$. For example, suppose that we give U_2 rectangular coordinates (x, y), and let $\sigma(2) = \gamma(s)\partial_x$ for some cutoff function γ. Then $\sigma(2) = \gamma(s)(\cos(\phi)\partial_s - \sin(\phi)\partial_\phi)$, and so, in the coordinates $(r, \theta) = (-s, -\phi)$ of U_1 it looks like $\beta(r)(-\cos(\theta)\partial_r - \sin(\theta)\partial_\theta)$. Therefore it lifts to the section

$$\tilde{\sigma} = \beta(r)\left(-\cos(k\theta)\tilde{\partial}_r - \frac{\sin(k\theta)}{k}\tilde{\partial}_\theta\right)$$

of $\widetilde{E}_1$ over the circle $r = 4$. Now $\sigma(1)$ vanishes on this circle. Therefore, however we choose $\sigma(1)$, the only zeros of $\tilde{s} = \tilde{s}(1) + \tilde{s}(2)$ are going to be in $\widetilde{U}_1$, and they each will be weighted by $1/k$. Moreover the number of these zeros is the winding number of $\tilde{\sigma}$, where this is considered as a map $S^1 \to \mathbf{R}^2 - \{0\}$. But $\tilde{\sigma}$ is homotopic to the map

$$\theta \mapsto \cos(k\theta)\tilde{\partial}_r + \sin(k\theta)\tilde{\partial}_\theta = \cos((k+1)\theta)\tilde{\partial}_x + \sin((k+1)\theta)\tilde{\partial}_y,$$

and so has winding number $k + 1$. Hence we get $1 + 1/k$ as before.

We can check this by calculating the Euler number in a different way. Triangulate T, making p_N one of the vertices, and then take the sum of the weighted number of vertices and faces minus the weighted number of edges. Clearly, all edges, faces, and vertices other than p_N should have weight 1 as usual. However, intuitively it should be correct to count p_N with weight $1/k$. This method gives an Euler characteristic of $1 + 1/k$. To justify this, note that if there were 2 orbifold points, both of order k, then this calculation would give an Euler number of $2/k$. This is surely correct, since this second orbifold has a k-fold covering by the 2-sphere.

5. The virtual moduli cycle

In this section we briefly describe how to construct the virtual moduli cycle $\overline{\mathcal{M}}^\nu = \overline{\mathcal{M}}^\nu_{0,k}(X, J, A)$ as a branched pseudomanifold Y of dimension d, where $d = \dim X + 2c_1(A) + 2k - 6$. For simplicity, we will often ignore the marked points.

Here are the main features of the structure of Y. The elements of Y are parametrized stable maps $(\Sigma, \tilde{h})$ that each satisfy a perturbed Cauchy–Riemann equation

$$\overline{\partial}_J \tilde{h}(w) = \nu_{\tilde{h}}(w), \qquad w \in \Sigma,$$

where $\nu_{\tilde{h}}$ is a C^∞ smooth 1-form in the Sobolev space

$$\widetilde{L}_{\tilde{h}} = L^{1,p}\left(\Lambda^{0,1}(\Sigma, \tilde{h}^*(TX))\right),$$

consisting of all sections of the bundle $\Lambda^{0,1}(\Sigma, \tilde{h}^*(TX))$ that are $L^{1,p}$-smooth on each component. There is a finite-to-one map $Y \to \mathcal{W}$ that forgets the parametrization, such that each stratum in Y_{sm} of codimension $2k$ or $2k+1$ is taken to a stratum in $\mathcal{W}$ consisting of elements whose domains have $\leq k+1$ components. (We would have equality here if it were not for the extra strata that are introduced to deal with the singular points of the closure of the branching locus.) In particular, the top strata M_i and the branch components B_j consist of maps with domain a single sphere.

The evaluation map $ev : \overline{\mathcal{M}}^\nu \to X^k$ is given by composing the forgetful map $Y \to \mathcal{W}$ with the evaluation map $\mathcal{W} \to X^k$. $\overline{\mathcal{M}}^\nu$ is called a virtual moduli cycle because the class represented by $ev : \overline{\mathcal{M}}^\nu \to X^k$ can be used to calculate Gromov–Witten invariants. Sometimes $\overline{\mathcal{M}}^\nu$ is called a *regularization* of $\overline{\mathcal{M}}$.

5.1. The local construction. We saw in §3.2 that $\mathcal{W}$ has the local structure assumed in §4.2, i.e. it is a partially smooth orbifold. The next step is to define a bundle $\widetilde{\mathcal{L}}_\tau \to \widetilde{U}_\tau$ whose fiber at an element $\tilde{\tau}' = (\Sigma', \tilde{h}')$ is the space $\widetilde{L}_{\tilde{h}'}$ defined above. Note that the fiber changes when the topological type of the domain Σ' changes. But it is a locally trivial bundle over each stratum, and it is not hard to give it a global topology. Therefore it is a bundle in our category. Clearly, the action of Γ_τ by reparametrization lifts to $\widetilde{\mathcal{L}}_{\tilde{h}}$.

The Cauchy–Riemann operator $\overline{\partial}_J$ gives rise to a Γ_τ-equivariant section of $\widetilde{\mathcal{L}}_\tau \to \widetilde{U}_\tau$. Moreover, its linearization at $\tilde{\tau}'$ is[12]

$$D\tilde{h}' : L^{2,p}(\Sigma', (\tilde{h}')^*(TX)) \to \widetilde{L}_{\tilde{h}'},$$

where $p > 2$. Here, if Σ' has $\ell + 1$ components, the domain of $D\tilde{h}'$ consists of all $\ell + 1$-tuples

$$(\xi_0, \ldots, \xi_\ell), \qquad \xi_i \in L^{2,p}(\Sigma'_i, (\tilde{h}'_i)^*(TX),$$

such that

[12] If one used Siebert's way of fixing parametrizations, one could consider this as a map from $L^{1,p}$-sections to L^p-sections.

(i) $\xi_i(x) = \xi_j(x)$ at all intersection points x of the components Σ_i', Σ_j', and

(ii) $\xi_i(w_j) \in (\tilde{h}')^* T\mathbf{H}_j$ at all added points w_j.

Conditions (i) are needed since this domain is the tangent space of a space of maps $\Sigma' \to X$ with connected domain Σ', while conditions (ii) refer to the added marked points needed to stabilize the domain and correspond to the fact that we are restricting to a slice for the action of the reparametrization group. Observe that there are no similar compatibility conditions on the elements of the range $\widetilde{L}_{\tilde{h}'}$.

Now look at the "center" point τ of U_τ, and let $\tilde{\tau} = (\Sigma, \tilde{h}) \in \widetilde{U}_\tau$ be one of its lifts. Since $D\tilde{h}$ is elliptic, there is a finite dimensional subspace R_τ of the vector space $C^\infty\left(\Lambda^{0,1}(\Sigma \times X, pr^*TX)\right)$ such that the operator

$$D\tilde{h} \oplus e : L^{2,p}(\Sigma, \tilde{h}^*(TX)) \oplus R_\tau \to \widetilde{L}_{\tilde{h}} \tag{1}$$

is surjective, where e is given by restricting the sections in R_τ to the graph of $\tilde{h}$. In general the elements of R_τ cannot be chosen to be invariant under the automorphism group $\Gamma_{\tilde{\tau}}$ of $\tilde{\tau}$. However, we can enlarge R_τ by adding to it all its images under the elements of $\Gamma_{\tilde{\tau}}$, so that this group does act on it. It will be convenient to suppose that this action is free on the nonzero elements of R_τ. Clearly, this can be arranged by slightly perturbing R_τ.

The next step (see [**LiuT1**], §3) is to show that we can choose R_τ and a partially smooth family of embeddings $e_{\tilde{\tau}'} : R_\tau \to \widetilde{L}_{\tilde{h}'}$ so that the linearized operator

$$D\tilde{h}_{\tau'} \oplus e_{\tilde{\tau}'} : L^{2,p}(\tilde{h}_{\tau'}^*(TX)) \oplus R_\tau \to \widetilde{L}_{\tilde{h}'} \tag{2}$$

is surjective (with uniform estimates for the inverse) for all $\tilde{\tau}' \in \widetilde{U}_\tau$. Here we are allowed to shrink U_τ. The fact that this is possible is a deep result. If $\tilde{\tau}'$ lies in the same stratum as $\tilde{\tau}$, it holds because of the openness of the regularity condition. However, to prove this in general one has to show that there is a uniformly bounded family of right inverses to $D\tilde{h}_{\tau'} \oplus e_{\tilde{\tau}'}$ as $\tilde{\tau}'$ varies over a small enough neighborhood $\widetilde{U}_\tau$. This analytic fact is the basis of all gluing arguments: see, for example, [**RT**], [**LiuT1**], and [**MS**], Appendix A.

With this done, let $pr : \widetilde{U}_\tau \times R_\tau \to \widetilde{U}_\tau$ denote the projection and consider the pullback bundle $pr^*(\widetilde{\mathcal{L}}_\tau) \to \widetilde{U}_\tau \times R_\tau$. This bundle has a section s defined by

$$s(\tilde{\tau}', \nu) = \overline{\partial}_J(\tilde{h}_{\tau'}) + e_{\tilde{\tau}'}(\nu). \tag{3}$$

By construction, its linearization is surjective at all points $(\tilde{\tau}, 0)$ and so it remains surjective for $|\nu| \le \varepsilon$. One then shows by some variant of gluing that the intersection of s with the zero section is a (partially smooth) open pseudomanifold of dimension equal to $\operatorname{ind} D\tilde{h} + \dim R_\tau$, whose components are given by intersecting the solution space with the different strata in $\widetilde{U}_\tau$. In other words, s is Fredholm. Since R_τ is finite dimensional, the Sard–Smale theorem (cf. [**MS**], §3.4) now implies that for generic fixed $\nu \in R_\tau$ the corresponding section s^ν of $\widetilde{\mathcal{L}}_\tau \to \widetilde{U}_\tau$ is also Fredholm. Hence:

LEMMA 5.1. *For generic $\nu \in R_\tau$ the section s^ν is Fredholm, and its solution set $\widetilde{Z}_\tau^\nu$ has the structure of an open pseudomanifold of dimension $d = \operatorname{ind} Dh$.*

5.2. The global construction. Because $\overline{\mathcal{M}}$ is compact, it is covered by a finite number $U_j, j = 1, \dots, k$, of the sets of the form U_τ, and $\mathcal{W}$ is defined to be their union. We write $(\widetilde{U}_j, \Gamma_j, \pi_j)$ for the corresponding uniformizers, where

$$\pi_j : \widetilde{U}_j \to \widetilde{U}_j/\Gamma_j = U_j$$

is the obvious projection. Choose subcovers $\{U_i^0\}$ and $\{V_I\}$ as in Lemma 4.9, and then a partition of unity β_i on $\mathcal{W}$ that is subordinate to the covering $\{U_i^0\}$ and such that the boundaries of the sets where $\beta_i \neq 0$ are transversally intersecting submanifolds. Then define the section $\sigma(i)$ of the pullback bundle $pr^*(\widetilde{\mathcal{L}}_i) \to \widetilde{U}_i \times R_i$ by

$$\sigma(i)(\tilde{\tau}', \nu_i) = \beta_i(\tau') \, . \, e_{\tilde{\tau}'}(\nu_i),$$

as in Example 4.10. Next, choose $\varepsilon > 0$ so that, for all i, the linearization of $\overline{\partial}_J + \sigma(i)$ is surjective at all points where $\beta_i \neq 0$ and $|\nu| \leq \varepsilon$, and put

$$R = \bigoplus_i R_i, \quad R_\varepsilon = \{\nu = (\nu_i) \in R : |\nu_i| \leq \varepsilon, \text{ for all } i\}.$$

Let $\mathcal{W}_\varepsilon = \mathcal{W} \times R_\varepsilon$. The structure on $\widetilde{\mathcal{L}} \to \widetilde{\mathcal{V}}$ constructed in §4.2 can be pulled back to give a multi-bundle $pr^*(\widetilde{\mathcal{L}}) \to \widetilde{\mathcal{V}}_\varepsilon$. Hence this multi-bundle has a multi-section $\tilde{s}$ given by

$$\tilde{s}(\tilde{\tau}', \nu) = \overline{\partial}_J + \sum_i \tilde{s}(i)(\tilde{\tau}', \nu_i) = \overline{\partial}_J + \sum_i \beta_i(\tau') \, . \, e_{\tilde{\tau}'}(\nu_i).$$

Moreover, if the R_i are chosen so that the elements in $\bigcup_{i \in I} e_{\tilde{\tau}'}(R_i)$ are linearly independent for each $\tilde{\tau}' \in \widetilde{V}_I$, it is not hard to see that this section $\tilde{s}$ satisfies the geometric conditions of Example 4.10. The proof of Lemma 5.1 then adapts to show that $\tilde{s}$ is Fredholm. Hence the corresponding multi-section $\tilde{s}^\nu$ of $\widetilde{\mathcal{L}} \to \widetilde{\mathcal{V}}$ is Fredholm for generic $\nu \in R$. Therefore we can apply Proposition 4.13 to conclude:

PROPOSITION 5.2. *For generic $\nu \in R$ the zero sets $\widetilde{Z}_I^\nu$ of the multi-section $\tilde{s}^\nu$ of $\widetilde{\mathcal{L}} \to \widetilde{\mathcal{V}}$ fit together to give a compact branched and labelled pseudomanifold $\overline{\mathcal{M}}^\nu$ of dimension d.*

DEFINITION 5.3. Any d dimensional branched pseudomanifold $Y = \overline{\mathcal{M}}^\nu$ constructed as above from some covering of a neighborhood $\mathcal{W}$ of $\overline{\mathcal{M}}_{0,k}(X, J, A)$ will be called a *regularization* of $\overline{\mathcal{M}}_{0,k}(X, J, A)$.

By considering a common refinement of two covers, it is not hard to see that the rational bordism class of the map $ev : \overline{\mathcal{M}}^\nu \to X^k$ is independent of the choice of covers $\{U_j\}$ and $\{V_I\}$. Further, by considering a 1-parameter version of this construction, Liu and Tian show that, for any regularization $\overline{\mathcal{M}}^\nu$ of $\overline{\mathcal{M}}_{0,k}(X, J, A)$, this is independent of the choice of J and the perturbation term ν. In fact, it depends only on the deformation class of the symplectic form ω on X, on the homology class A and on the number of marked points k. The Gromov–Witten invariants are now defined by the method given in §1. It is not hard to show that, in all cases where the previous definition in terms of the uncompactified space $\mathcal{M}_{0,k}(X, J, A)$ makes sense, the new invariants equal the old. (On the other hand, I am not sure that anyone has yet checked that all the different ways of defining general Gromov–Witten invariants give the same result.)

Observe that by this method we construct a cycle that represents a *rational* homology class. Thus in general the Gromov–Witten invariants take rational values. In the case when the class A is primitive (i.e. not a multiple of any other integral class) one would expect that the invariant should be integral. However, this has not yet been proven.

REMARK 5.4. (i) Liu and Tian's definitions in [**LiuT1**] are actually slightly different from those given here. Their aim is to prove the Arnold conjecture, and so their basic domain, i.e. the domain of the uncompactified space $\mathcal{M}$, is a cylinder rather than a sphere. They consider this cylinder to be a sphere with two special points at the poles and a marked arc between them, and perturb the equation on this component by an appropriate Hamiltonian term so that the elements of $\mathcal{M}$ are Floer trajectories between periodic orbits. This means that they must replace the reparametrization group $\mathrm{PSL}(2, \mathbf{C})$ by a subgroup that respects this extra structure. Further, there are now two kinds of bubbling: the bubbling off of spheres and the lengthwise splitting of the cylinder into two or more pieces. Thus the stable curves that they consider have a more complicated structure than ours: there is a connected series of "principal components" that join the two poles — corresponding to a series of Floer trajectories — with attached bubble components. Moreover, the equation on all the principal components has a Hamiltonian perturbation term. But apart from these rather minor extra complications, their construction is exactly as described here.

(ii) Here we have chosen to equip the virtual moduli cycle with the minimum amount of structure needed for it to represent a homology class. However, one could keep track of more structure, if this is appropriate. For example, in [**Mc2**] we consider fibrations $P \to S^2$ with symplectic fibers M, and we need information on the strata of codimensions 3 and 4 in $\overline{\mathcal{M}}^\nu$ in order to relate the Gromov–Witten invariants of P to those of M. These strata correspond to stable maps whose domains have two components.

5.3. An example. So far there are rather few calculations of Gromov–Witten invariants using this general definition. Some results for manifolds that are symplectically fibered over S^2 are worked out in [**Mc2**]. We now work out an example in which the unperturbed moduli space $\overline{\mathcal{M}}$ is in fact a manifold, but it has the wrong dimension. In this situation one can often define a global obstruction bundle $E \to \overline{\mathcal{M}}$, and we will see that the cycle represented by the regularization $\overline{\mathcal{M}}^\nu$ is homologous to that represented by the zero set of a generic section of E. Hence one can calculate the desired invariants by investigating this zero set. The resulting global calculations are often best done in the context of algebraic geometry, because the topology of spaces of stable maps is so hard to understand.

Suppose that C is a symplectically embedded sphere in a 6-dimensional manifold X with $c_1(C) = 0$. (For example X might be a Calabi–Yau 3-fold.) We can choose J to be integrable near C, so that C is J-holomorphic and so that the normal bundle ν_C to C splits as the sum $\mathcal{O}(-1) \oplus \mathcal{O}(-1)$ of two line bundles each with Chern class -1. Then any J-holomorphic parametrization $h : S^2 \to X$ of C is regular, and C contributes $+1$ to the count of all J-holomorphic curves in class $A = [C]$. Let us work out the contribution of a k-fold cover of C to the invariant $n_X(H, H, H; kA)$, where H is a hypersurface such that $H \cdot A = 1$, at least in the case $k = 2$. According to Claire Voisin's proof in [**V**] of the Aspinwall–Morrison formula (cf. [**MS**], §9.3), this contribution is 1 for all k.[13]

Since $c_1(C) = 0$, the formal dimension of the moduli space $\mathcal{M}_{0,3}(X, J, kA)$ is 6 for all k. We are interested in counting the number of elements of this moduli space

[13]She uses a different compactification of the moduli space $\mathcal{M}$ which is easier to calculate with, but has to work to show that she really is calculating the same Gromov–Witten invariants as defined here.

whose three marked points meet representatives H_1, H_2, H_3 of the hypersurface H, where the H_i meet C in distinct points. Hence, if $\mathcal{M}_3(H, kA)$ denotes this cut-down moduli space, the formal dimension of $\mathcal{M}_3(H, kA)$ is 0. However, when $k > 0$ multiple covers of the form $h_k = h \circ f_k$ (where $f_k : S^2 \to S^2$ has degree k) are never regular. One can see this from the usual regularity criterion in [**MS**], Lemma 3.5.1, or by using the fact that the actual moduli space always has positive dimension, even after the intersection conditions are imposed. For example, when $k = 2$, it is diffeomorphic to the moduli space $\overline{\mathcal{M}}_3$ considered in Example 3.3, and so is diffeomorphic to $\mathbf{C}P^2$. Moreover, because a neighborhood of C in X is biholomorphically equivalent to a neighborhood of the zero section in the bundle $\mathcal{O}(-1) \oplus \mathcal{O}(-1)$ over S^2, there are no other curves in class $2A$ near double covers of C except these double covers themselves. Hence $\mathcal{M}_3(H, 2A)$ has a component $\overline{\mathcal{M}}(C)$ that can be identified with $\mathbf{C}P^2$, and we want to understand what the regularization $\overline{\mathcal{M}}^\nu(C)$ of this component is.

According to [**MS**], Lemma 3.5.1, for each $\tilde{\tau}$ in the open stratum $\mathcal{S}_1$ of $\overline{\mathcal{M}}(C)$ the cokernel of $D\tilde{h}$ can be identified with the space of holomorphic sections of the rank 2 bundle $\mathcal{O}(-2)^* \otimes K \oplus \mathcal{O}(-2)^* \otimes K$ over S^2, where K is the canonical bundle $T^*(S^2) = \mathcal{O}(-2)$. (In general for k-fold maps this cokernel would be the sum of two copies of $\mathcal{O}(-k)^* \otimes K$, since the normal line bundles $\mathcal{O}(-1)$ pull back to $\mathcal{O}(-k)$.) Hence this is a trivial rank 2 complex bundle with a 2-dimensional family of (constant) sections. Hence at each point $\tilde{\tau}$ of $\mathcal{S}_1$ the cokernel R_τ has 2 complex dimensions. Moreover, it clearly splits as the sum of two line bundles, one for each factor of ν_C.

Similarly, for $\tilde{\tau} \in \mathcal{S}_2$, the cokernel has two complex dimensions. Now it arises because of the compatibility condition $\xi_1(x) = \xi_2(x)$ put on the elements of the domain of $D\tilde{h}$ at the intersection point of the two components of Σ: see §5.1. Without this condition, the operator $D\tilde{h}$ would be surjective and with kernel equal to the tangent space to the reparametrization group G_R. The component of the equation that is tangential to C does not affect the image of $D\tilde{h}$, since one can alter the tangential component by adding to ξ_1 (or ξ_2) a vector field that is tangent to the action of G_R and hence in the kernel of $D\tilde{h}$. Hence the cokernel at $\tilde{\tau}$ can be identified with the normal space to C at $\tilde{h}(x)$. As τ varies over each line in the stratum $\mathcal{S}_2$, x varies over the curve C and so the cokernel bundle over this line is $\mathcal{O}(-1) \oplus \mathcal{O}(-1)$.

Altogether these fibers fit together to give a complex rank 2 bundle E over $\overline{\mathcal{M}}(C)$ that is called the obstruction bundle. We can use this bundle instead of R in the construction of the regularization. Because E is globally defined and all the groups Γ_τ are trivial, the sets $\widetilde{V}_I$ can be identified with the V_I, and multi-bundles and multi-sections are the same as usual bundles and sections. Hence a perturbation ν is simply a section of E, and the corresponding regularization $\overline{\mathcal{M}}^{\nu_E}(C)$ is just the zero set of this section (with no branching and all labels equal to 1.) Hence the Gromov–Witten invariant that we are trying to calculate is simply the Euler number of E.

Therefore the question boils down to understanding the topological type of this obstruction bundle E. Note that because the image curve C is fixed, the gluing of $\mathcal{S}_2$ to $\mathcal{S}_1$ really takes place in the domain, i.e. in the space of curves (Σ, z_1, z_2, z_3). It is therefore very plausible that E is the sum of line bundles $E = L \oplus L'$, i.e. the fact that the normal bundle to C splits as a sum of line bundles is reflected in the structure of E. To prove this one has to investigate the gluing process in detail. Usually this is done locally near a point $\tilde{\tau}$, and the arguments show that the cokernel of $D\tilde{h}$ can be identified with the cokernel $D\tilde{h}_a$ of the glued map. Here we need to look globally as $\tilde{\tau}$ varies over a line in $\mathcal{S}_2$ in order to understand the structure of E near this line. Similar questions are discussed in [**Mc1**], §4.2. In the notation used there, the cokernel bundle K over $\mathcal{S}_2$ is precisely $\mathcal{O}(-1) \oplus \mathcal{O}(-1)$, and it maps onto the cokernel of $D\tilde{h}_a$ for small a by the arguments given there. Moreover the map that identifies it with this cokernel can be chosen to respect the splitting of the cokernel bundle over $\mathcal{S}_1$. Hence E is the sum of line bundles $L \oplus L'$. It is then clear that $L = L'$ because of the local automorphism of X near C that interchanges the two factors of its normal bundle. Hence L restricts to $\mathcal{O}(-1)$ over each line, and so must be the universal line bundle over $\mathbf{C}P^2$. In particular the Euler number of E is 1.

These arguments should give the flavor of the topological questions that arise in this kind of context. Some related global calculations are carried out in [**Mc1**].

References

[FO] K. Fukaya and K. Ono, *Arnold conjecture and Gromov–Witten invariants*, Topology **38** (1999), 933–1048.

[HS] H. Hofer and D. A. Salamon, *Marked Riemann surfaces of genus* 0, preprint, 1997.

[LiT] Jun Li and Gang Tian, *Virtual moduli cycles and Gromov–Witten invariants for general symplectic manifolds*, Topics in Symplectic 4-Manifolds (Irvine, CA, 1996), Internat. Press, Cambridge, MA, 1998, pp. 47–83.

[LiuT1] Gang Liu and Gang Tian, *Floer homology and Arnold conjecture*, J. Diff. Geom., **49** (1998), 1–74.

[LiuT2] ———, *On the equivalence of multiplicative structures in Floer homology and quantum homology*, Acta Math. Sinica, **15** (1999).

[Mc1] D. McDuff, *Almost complex structures on* $S^2 \times S^2$, preprint `SG/9808008`.

[Mc2] ———, *Quantum homology of fibrations over* S^2, preprint, 1998.

[MS] D. McDuff and D. A. Salamon, *J-holomorphic curves and quantum cohomology*, University Lecture Notes, vol. 6, Amer. Math. Soc. Providence, RI, 1994.

[R] Y. Ruan, *Virtual neighborhoods and pseudoholomorphic curves*, Topics in Symplectic 4-Manifolds (Irvine, CA, 1996), Internat. Press, Cambridge, MA, 1998, pp. 101–116.

[RT] Y. Ruan and G. Tian, *A mathematical theory of quantum cohomology*, J. Diff. Geom. **42** (1995), 259–367.

[S] D. Salamon, *Lectures on Floer theory*, Symplectic Geometry and Topology (Proc. IAS/Park City Summer Institute, 1997; Ya. Eliashberg and L. Traynor, eds.), Amer. Math. Soc., Providence, RI, 1999, pp. 143–229.

[Sb] B. Siebert, *Gromov–Witten invariants for general symplectic manifolds*, preprint, 1997.

[V] C. Voisin, *A mathematical proof of a formula of Aspinwall and Morrison*, Compositio Math. **104** (1996), 135–151.

State University of New York at Stony Brook
E-mail address: `dusa@math.sunysb.edu`

Amer. Math. Soc. Transl.
(2) Vol. **196**, 1999

Engel Deformations and Contact Structures

Richard Montgomery

1. Introduction and Outline

An Engel structure is a maximally non-integrable two-plane field peculiar to dimension four. Engel structures are topologically stable in the same way that contact and symplectic structures are stable. Namely, any two such structures are locally diffeomorphic, and the typical perturbation of an arbitrary germ of a two-plane field on a four-manifold (read "one-form on an odd-dimensional manifold" or "two-form on an even-dimensional manifold") will be Engel (respectively "contact" or "symplectic"). The fact that any two Engel structures are locally diffeomorphic is contained in the the "Darboux theorem" for Engel structures, attributed to Engel by Cartan [**5**]. This normal form theorem asserts that any Engel structure is locally diffeomorphic to $\mathbb{R}^4$ with coordinates x, y, z, w and with its two-plane field defined by the vanishing conditions

$$dz - ydx = 0,$$

$$dy - wdx = 0.$$

If a k-plane field in n dimensions is stable in the above sense, then $k(n-k) \leq n$. (See [**19**] or [**16**].) The only positive integer solutions to this inequality are $k = 1$, n arbitrary, $k = n - 1$, n arbitrary, or $k = 2, n = 4$. The first possibility, $k = 1$, corresponds to line fields. The second, $k = n - 1$, is realized by contact structures when n is odd, and when n is even by an even-dimensional cousin of contact structures which we will call *even-contact structures*, following Ginzburg [**9**]. These are discussed in some detail in Section 2.3 below. The final possibility, $(k, n) = (2, 4)$, is realized by the case of interest to us, the Engel structures. The uniqueness of Engel structures within this list, and the special appearance of dimension four, are primary reasons for studying them.

In this paper we will study deformations of Engel structures. Our investigations have been guided by analogy with contact structures. A basic theorem for contact manifolds is the theorem of Gray [**11**]. This asserts that if two contact structures ξ_0, ξ_1 are homotopic through intermediate contact structures ξ_t, then the two structures are in fact diffeomorphic via an isotopy. We will see in Theorem 4 of the present paper that the analogous theorem is false for Engel structures.

1991 *Mathematics Subject Classification.* Primary 53C15.

©1999 American Mathematical Society

A major difference between Engel structures and contact structures is the existence of a *canonical flag*

$$L \subset \mathcal{D} \subset \mathcal{E} \subset TQ$$

of distributions attached to any Engel distribution. Here Q is the manifold, TQ its tangent bundle, and $\mathcal{D} \subset TQ$ is the Engel structure. $\mathcal{E}$ is the rank 3 subbundle of TQ which is formed (locally) by taking the span of the Lie bracket of any two linearly independent vector fields tangent to $\mathcal{D}$ and adding this to $\mathcal{D}$. L is a canonical line subbundle of $\mathcal{D}$ constructed as the kernel of some map based on the Lie bracket and $\mathcal{E}$. See Section 2.3 for details.

When we perturb $\mathcal{D}$, we will perturb its line field. But line fields are typically *not* structurally stable: small perturbations lead to nondiffeomorphic fields. Hence we expect that the analogue of Gray's theorem will fail in the Engel context. This accounts for Theorem 4. The idea just outlined can be found in Gershkovich [**8**], but the proof there appears incomplete and we have completed it. If one insists that the line field remain fixed during the Engel homotopy, then the analogue of Gray's theorem can be salvaged. Golubev [**10**] did this after the initial write-up of the present paper.

The main topic of the present paper is deformations of certain canonical Engel structures on four-manifolds of the form $M^3 \times S^1$ or $M^3 \times I$, where M^3 is a contact three-manifold. We will prove in Theorems 4 and 5 that the space of deformations is infinite dimensional. This is to be contrasted with the case of contact manifolds, where the deformation is discrete, according to Gray's theorem.

The central tool for obtaining this result is an an interplay between contact three-manifolds and Engel structures which is summarized by Theorems 1–3. These theorems are well known to a few experts, notably R. Bryant and others familiar with the works of E. Cartan. In one form or another they are all contained in the works of E. Cartan. They are central to our work. We could not find clear, modern statements of them, so we reformulate and prove them here. We do not make any claim to originality in Theorems 1–3 .

Theorem 1 constructs an Engel structure out of any contact three-manifold. Cartan called this construction 'prolongation' and applied it to general Pfaffian systems. Theorem 2 concerns the inverse of prolongation. This process allows us to associate a "local contact manifold" (more accurately, a transverse contact struture) to an Engel structure. This local manifold is the "local quotient" of the four-manifold by the integral curves of the Engel line field L. Through this description one also finds that the leaves of L inherit a real projective structure. Bryant and Hsu explicity described this real projective structure in their article [**4**]. Theorem 3 concerns the relationship between Engel and contact automorphisms. Theorems 1–3 may be summarized as saying that Engel geometry is a twisted product of three-dimensional contact and one-dimensional real projective geometries.

We end the paper with Theorem 6, where we construct an Engel manifold whose automorphism group is "very small" — essentially one-dimensional. This is to be contrasted with the contact case, where the automorphism group is always infinite-dimensional. This theorem is proved by combining Theorems 3 and 4.

When I began these investigations I had hoped that they might serve as a first step towards the development of an "Engel topology". Theorem 4 and 6 and the ideas around them have killed much of this hope. The Engel line field dominates global investigations. It appears that any deep global results on Engel structures will involve a strange mix of topology and dynamical systems.

2. Preliminaries

2.1. Definitions. Let Q be an n-dimensional manifold and TQ its tangent bundle. By a k-plane field or *distribution* on Q we mean a smooth linear subbundle $\mathcal{D} \subset TQ$. The integer k is the rank of the subbundle. We can think of $\mathcal{D}$ as a locally free sheaf of smooth vector fields on the manifold. Then $[\mathcal{D}, \mathcal{D}]$ will denote the sheaf generated by all Lie brackets $[X, Y]$ of sections X, Y of $\mathcal{D}$. Set $\mathcal{D}^2 = \mathcal{D} + [\mathcal{D}, \mathcal{D}]$ and continue to take Lie brackets, setting $\mathcal{D}^{j+1} = \mathcal{D}^j + [\mathcal{D}, \mathcal{D}^j]$. (One can check that $\mathcal{D}^j \subset [\mathcal{D}, \mathcal{D}^j]$ as sheaves, so that it is in fact unneccessary to add $\mathcal{D}^j$ in defining $\mathcal{D}^{j+1}$.) Write $\mathcal{D}^j(x) \subset T_xQ$ for the subspace obtained by evaluating all the vector fields in $\mathcal{D}^j$ at the point x. Write $n_j(x) = \dim(\mathcal{D}^j(x))$. We will also write $n_j(x; \mathcal{D})$ when we want to indicate the dependence of the integers on the distribution.

Definition 1. $\mathcal{D}$ is called *nonholonomic* if for all $x \in Q$ there is a $j = j(x)$ such that $\mathcal{D}^j(x) = T_xQ$.

$\mathcal{D}$ is called *maximally nonholonomic* if for each j we have

$$n_j(x; \mathcal{D}) = \max_{\mathcal{E}} n_j(x; \mathcal{E}),$$

where the maximum is taken over all rank k distributions $\mathcal{E}$ defined in some neighborhood of x.

Gershkovich and Vershik **[19]** call the list of integers $n_j(x)$ the *growth vector* of $\mathcal{D}$ at x. The list of integers $\max_{\mathcal{E}} n_j(x; \mathcal{E})$ is then called the *maximal growth vector* for the given k and n. The components of the maximal growth vector are of course independent of x. They equal the dimensions f_j of the graded components of the free Lie algebra on k generators, up until $f_j \geq n$. In case $(k, n) = (2, 4)$ the maximal growth vector is $(2, 3, 4)$.

Definition 2. An *Engel field*, or *Engel structure*, on a four-manifold Q is a rank two subbundle $\mathcal{D} \subset TQ$ whose growth vector is everywhere maximal, and consequently equals $(2, 3, 4)$. An *Engel manifold* is a four-manifold with an Engel structure.

Thus if $\mathcal{D}$ is an Engel structure, then $\mathcal{D}^2$ is a rank 3 distribution and $\mathcal{D}^3$ is the entire tangent bundle. More concretely, an Engel structure on a four-manifold is a two-plane field which is locally spanned by vector fields $\{X, W\}$, such that $\{X, W, [X, W], [X, [X, W]]\}$ frame the whole tangent bundle in the neighborhood in which two fields are defined.

Any two Engel structures are locally diffeomorphic, as mentioned at the beginning.

2.2. Prolongations. Let $\mathcal{E}$ be a distribution of k-planes on a manifold M. Imagine that we are interested in ℓ-dimensional submanifolds, $\ell < k$, of M which are tangent to $\mathcal{E}$. Such an study suggests forming the bundle $Q = Gr_\ell(\mathcal{E}) \to M$ whose fiber over $m \in M$ consists of the Grassmannian of all ℓ-dimensional linear subspaces $P \subset \mathcal{E}_m$. These ℓ-planes P represent the possible tangent spaces to the submanifolds. A curve in $Gr_\ell(\mathcal{E})$ consists of a moving pair $(m(t), P(t))$ with $P(t) \subset \mathcal{E}_{m(t)}$. Define a distribution on the manifold $Gr_\ell(\mathcal{E})$ by declaring that a curve $(m(t), P(t))$ is tangent to the distribution if and only if $\frac{dm}{dt} \in P(t)$. Alternatively, the distribution plane at the point (m, P) is $d\pi^{-1}_{(m,P)}(P)$, where $\pi : Gr_\ell(\mathcal{E}) \to M$ is the projection and $d\pi$ its differential.

DEFINITION 3. The total space of this Grassmann bundle, together with the distribution just constructed, is called a *prolongation* of $\mathcal{E}$.

This construction is due to Cartan [**5**]. See also the final section of [**3**], and the other references to Cartan therein.

EXAMPLE. Let us apply prolongation to the case where $\mathcal{E} = TM$ is the entire tangent bundle of the n-dimensional manifold M, and where $\ell = n - 1$, as would be the case if we were interested in studying hypersurfaces in M. The resulting prolongation of the tangent bundle is canonically isomorphic to $\mathbb{P}T^*M$, since the vanishing of a nonzero one-form $\theta_m \in T^*_m M$ defines a hyperplane in the tangent space at m. The distribution on $\mathbb{P}T^*M$ constructed by the prolongation process has rank $2n - 2$, compared with the dimension $2n - 1$ of $\mathbb{P}T^*M$. It is a contact structure. Indeed, this example explains the origin of the word "contact": a linear hyperplane within the tangent space at a point is called a "contact element" at that point, representing, as it does, contact with an alleged hypersurface passing through that point.

When we prolong with $\ell = 1$, the case corresponding to curves tangent to $\mathcal{E}$, then the total space of the Grassmann bundle is the projectivization $\mathbb{P}\mathcal{E}$ of $\mathcal{E}$. The fiber $\mathbb{P}\mathcal{E}_m$, $m \in M$, is the real projective space of dimension $k - 1$ consisting of the one-dimensional subspaces $\ell \subset \mathcal{E}_m$ of the vector space $\mathcal{E}_m$. Observe that if $\mathcal{E}$ has rank 2, then this process of prolonging with $\ell = 1$ increases the dimension of the space by 1, but keeps the rank of the distribution the same.

THEOREM 1. *The prolongation to $\mathbb{P}\xi$ of a contact structure $\xi \subset TM$ on a three-manifold yields an Engel structure on the four-manifold $\mathbb{P}\xi$.*

REMARK. Curves tangent to a contact structure on a three-manifold are called "Legendrian curves". So we might call $\mathbb{P}\xi$ the space of *Legendrian contact elements* for ξ.

PROOF. The Darboux theorem applied to the contact structure ξ yields coordinates (x, y, z) on M centered at any preassigned point $m \in M$ with the property that $\xi = \{dz - ydx = 0\}$ near m. The one-forms dx and dy are linear coordinates on each of the contact planes $\xi_{x,y,z}$ near m. Then $[dx : dy]$ form homogeneous coordinates on the corresponding projective line $\mathbb{P}\xi_{(x,y,z)}$. Write $dy/dx = w$ or $dy = wdx$ for the corresponding affine coordinate w. Then (x, y, z, w) are coordinates for the prolongation. The Pfaffian equations

$$\theta_1 := dz - ydx = 0$$

and

$$\theta_2 := dy - wdx = 0$$

define the prolonged distribution. Indeed, the first equation asserts that the moving point $m(t)$ with coordinates (x, y, z) is tangent to ξ, and the second equation asserts that this moving point is tangent to the line $\ell \subset \xi$ represented by the affine fiber coordinate w. *Note that the above representation is the normal form presented at the beginning of this article.*

The vector fields

$$W = \frac{\partial}{\partial w}$$

and

$$X = \frac{\partial}{\partial x} + y\frac{\partial}{\partial z} + w\frac{\partial}{\partial y}$$

span the prolonged distribution, as they are annihilated by θ^1 and θ^2. We compute $[W, X] = \frac{\partial}{\partial y}$ and $[X, [W, X]] = \frac{\partial}{\partial z}$. Now $W, X, [W, X]$ and $[X, [W, X]]$ span $\mathbb{R}^4$ everywhere in this coordinate neighborhood. It follows that the prolonged distribution on $\mathbb{P}(\xi)$ is indeed Engel, as defined by Definition 1, and as elucidated in the paragraph following that definition. □

As mentioned in the beginning, the *Engel normal form* asserts that in a neighborhood of any Engel manifold one can find coordinates (x, y, z, w), called *Engel coordinates*, with the above properties. Namely, the coordinate one-forms θ^1, θ^2 above define the distribution $\mathcal{D}$ by annihilating it. We call these one-forms the *canonical coframe*. The vector fields W, X written above and spanning $\mathcal{D}$ will be called *the canonical frame* in these coordinates. A modern derivation of the Engel normal form can be found in [**2**], where it is a special case of the "Goursat normal form".

COROLLARY 1. *If $Q^4 = M^3 \times S^1$ with M^3 oriented, then Q^4 admits an Engel structure.*

PROOF OF COROLLARY. Martinet [**14**] proved that any oriented M^3 admits a contact structure ξ. Lutz [**13**] extended Martinet's result by showing that ξ could be taken to represent any possible homotopy class of two-plane fields which arise within TM^3. (See Eliashberg [**7**] for more discussion.) In particular, the trivial class can be represented, in which case $\mathbb{P}\xi$ is is diffeomorphic to $M^3 \times S^1$. □

2.3. The characteristic line field and even-contact structures. An Engel distribution $\mathcal{D}$ determines a flag

$$L \subset \mathcal{D} \subset \mathcal{E}$$

of subbundles of the tangent bundle. Here

$$\mathcal{E} := \mathcal{D}^2 = [\mathcal{D}, \mathcal{D}]$$

is the rank three distribution which represents the '3' occuring in the growth vector $(2, 3, 4)$ in the definition of being Engel. $\mathcal{E}$ is spanned by $\mathcal{D}$ and the Lie bracket of any two vector fields which span $\mathcal{D}$. The line field L is defined by the condition

$$[L, \mathcal{E}] \subset \mathcal{E}.$$

It will be called the *characteristic line field*, or sometimes the *Engel line field.* Its integral curves will be called *characteristic leaves.*

In terms of Engel coordinates $\mathcal{E}$ is spanned by the canonical vector fields X, Y, W. L is spanned by W alone. $\mathcal{E}$ is annihilated by $\theta_1 = dz - ydx = 0$. L is annihilated by θ_1, $d\theta_1$, and θ_2.

EXAMPLE. If $Q = \mathbb{P}\xi$ is the prolongation of a contact three-manifold (M, ξ) with the Engel structure of Theorem 1, then its characteristic leaves are the circular fibers $\mathbb{P}(\xi_m)$, $m \in M$, of the fibration $Q \to M$. $\mathcal{E}$ is the pull-back of ξ by π.

$\mathcal{E}$, considered by itself, is an *even-contact structure.*

DEFINITION 4. An *even-contact* structure (sometimes called a "quasi-contact structure") is a field $E \subset TQ$ of hyperplanes on an *even-dimensional* manifold Q with the property that if θ is a one-form whose vanishing defines E (perhaps only locally), then the pointwise restrictions $d\theta|_{E_q}$ are two-forms of maximal rank on the E_q.

If we had insisted instead that the manifold were odd-dimensional, then we would have been giving the definition of a contact manifold.

We will now show that in our flag above, $\mathcal{E}$ determines L, without the knowledge of $\mathcal{D}$. A two-form on an odd-dimensional vector space always has a kernel. The even-contact condition is that the kernel of the $d\theta|_{E_q}$ on E_q be of minimal dimension, and consequently of dimension one. These kernels define a line field $L_q \subset E_q$, $q \in Q$, which we call the characteristic line for E. In dimension four the even-contact condition is equivalent to the condition $[E, E] = TQ$, so the $\mathcal{E} = \mathcal{D}^2$ of an Engel structure is indeed even-contact. Now any rank 2 subbundle F of an even-contact structure E with the property that $[F, F] \subset E$ must necessarily contain L. *From these observations it follows that the characteristic line field L of an Engel structure is determined completely by the even-contact structure $\mathcal{D}^2$.*

There is a Darboux theorem for even-contact structures. In dimension four it yields the normal form

$$\theta = dz - ydx$$

for the one-form whose vanishing defines the even-contact structure $\mathcal{E}$. (Note this is precisely the form θ_1 of the Engel normal form above.) The four-manifold coordinates are x, y, z, w, and the field $W = \frac{\partial}{\partial w}$ spans the characteristic line field L. This line field L is uniquely determined by the two conditions $\theta(L) = 0$, and $d\theta(L, \cdot) = 0$. Together, these equations say that the flow of any vector field tangent to L preserves E:

$$[L, E] \subset E.$$

For $U \subset Q$ an open set of an even-contact manifold, write U/L for the quotient space whose points are connected leaves of the characteristic line field L in U. Suppose that this quotient is "nice" in the sense that it is a three-manifold and that the natural projection $\pi : U \to U/L$ is a submersion. Then the quotient inherits a contact structure, denoted $\pi_* E$. Indeed, the flow along L leaves E invariant, so that we can push E down along L. In other words, the subspaces $d\pi_q E_q$ and $d\pi_p E_p$ are equal for q, p on the same leaf of L. This family of subspaces has dimension 2, since $L \subset E$ and we are dividing out by L. The condition $[E, E] = TU$ implies that $[\pi_* E, \pi_* E] = T(U/L)$. So the quotient distribution $\pi_* E$ is contact.

If U is the neighborhood for a canonical coordinate system (x, y, z, w), then (x, y, z) coordinatize the quotient space. The contact form for $\pi_* E$ has the same expression $dz - ydx$ as the even-contact form had on U.

If $\Sigma \subset M$ is a three-manifold transverse to L, then we may take $U = N(\Sigma)$ to be a tubular neighborhood of Σ swept out by flowing along L for short times. Then the quotient U/L is canonically isomorphic to Σ. And the contact structure on the quotient is identified in this way with $E \cap T\Sigma$.

These observations prove (a) and (b) of the following theorem.

THEOREM 2. *Let $\mathcal{E}$ be an even-contact structure on the four-manifold Q, and L its characteristic line field.*

(a) *If the leaf space Q/L is 'nice' in the above sense, then it inherits a contact structure $\pi_*\mathcal{E}$ from $\mathcal{E}$.*

(b) *If $\Sigma \subset Q$ is a three-manifold transverse to L, then $\mathcal{E}|_\Sigma := \mathcal{E} \cap T\Sigma$ is a contact structure on Σ.*

Suppose now, that $\mathcal{E} = \mathcal{D}^2$ *is the even-contact structure associated to an oriented Engel structure* $\mathcal{D}$. *Then* $\mathcal{D} \cap T\Sigma \subset \mathcal{E}_\Sigma$ *is a line subfield, and hence a global section of* $\mathbb{P}(\mathcal{E}|_\Sigma) \to \Sigma$. *We identify the image of this section with* Σ.

(c) *Any sufficiently small neighborhood* $N(\Sigma)$ *of* Σ *in* Q *is diffeomorphic, as an Engel manifold, to some neighborhood of* Σ *in* $\mathbb{P}(\mathcal{E}|_\Sigma)$ *with its prolonged Engel structure as described in Theorem 1.*

EXAMPLE. Take $Q = \mathbb{P}\xi$ to be the prolongation of a contact three-manifold (M, ξ) as in Theorem 1. Then $Q/L = M$, and the induced contact structure is the original contact structure ξ.

Before we go on to the proof of the theorem, we will rephrase its assertions (a) and (b). They assert that an even-contact structure on a four-manifold induces a *transverse contact structure* on that manifold. Let us pause to define the notion of transverse structure generally. Let L be a foliation of an n-manifold by s-dimensional leaves. Let P denote some type of geometric structure, e.g., contact structure, symplectic structure, volume form, metric, . . . , which may exist on manifolds of the complementary dimension $n - s$.

DEFINITION 5. We say that the foliation L of Q has a *transverse geometric structure* of type P if the following two conditions hold.

i) Any sufficiently small submanifold Σ transverse to the leaves of L and of complementary dimension to L is endowed with a P-structure.
ii) The flows along the leaves of L preserve these P-structures. More precisely, if Σ_1 and Σ_2 are two such transverse submanifolds, and if X is any vector field tangent to L whose time t flow maps Σ_1 to Σ_2, then this flow defines a P-isomorphism $\Sigma_1 \to \Sigma_2$.

Suppose that T is a tensor on Q whose restriction to any transverse submanifold Σ induces a structure of type P on Σ. Then T defines a transverse P-structure on Q *provided* the restrictions to the Σs of all of the Lie derivatives $L_X T$, X tangent to L, are zero. This is precisely our situation in the even-contact case. Take T to be the one-form θ whose vanishing defines the even-contact structure.

2.3.1. *Twisting and the projective structure on leaves.* An Engel structure is sandwiched between its even-contact structure $\mathcal{E}$ and its characteristic line field L. Almost any two-plane field $\mathcal{D}$ sandwiched between an even-contact structure and its line field will be Engel. Indeed, the fact that $L \subset \mathcal{D}$ implies that $[\mathcal{D}, \mathcal{D}] \subset \mathcal{E}$. To make sure $\mathcal{D}$ is Engel, we must impose the single nondegeneracy condition $[\mathcal{D}, \mathcal{D}] = \mathcal{E}$. Since $\mathcal{D}$ is rank 2, this is equivalent to the *twisting condition*:

$$[L, \mathcal{D}] = \mathcal{E}.$$

An Engel structure is a two-plane field sandwiched between an even-contact structure and its line field, and which "twists" about the line field in the above sense.

2.3.2. *The development map.* Now let U be a 'nice' neighborhood for L, so that $U \to U/L$ is a submersion. Let $\xi = \pi_* \mathcal{E}$ be the induced contact structure. A point $m \in U/L$ represents an integral curve $\ell \subset U$ for L. At each point $q \in \ell$ we have the two-plane field $\mathcal{D}_q \subset \mathcal{E}_q$, with $L_q \subset \mathcal{D}_q \subset \mathcal{E}_q$. Consequently $d\pi_q(\mathcal{D}_q) \subset \pi_*(\mathcal{E})_m$ is a line, or one-dimensional subspace, of the two-plane ξ_m. The set of all such lines forms the real projective line $\mathbb{RP}^1 = \mathbb{P}(\xi_m)$. In this manner we obtain a smooth map $\ell \to \mathbb{P}(\xi_m)$. We call this map, or its lift to the corresponding universal covers, the development map along the leaf ℓ. The "twisting" discussion above implies that the condition that $\mathcal{D}$ be Engel is equivalent to the condition that these development maps be local diffeomorphisms for all leaves. In other words, as we climb up any given leaf ℓ, the lines $d\pi_q(\mathcal{D}_q) \cong \mathcal{D}_q/L_q$ always rotate, and in the same sense, within the circle $\mathbb{P}(\xi_m)$.

DEFINITION 6. The map $\ell \to \mathbb{P}(\xi_m)$ restricted to a leaf of L, or the union $U \to \mathbb{P}(\xi)$ of these maps over the leaves ℓ lying in some 'nice' neighborhood U, will be called the *development map* associated to the Engel structure.

2.3.3. *Proof of Theorem 2.* We had proved (a) and (b) just before stating the theorem. It remains to prove (c). In proving (b) we observed that the quotient $N(\Sigma)/L$ was diffeomorphic to Σ itself. The development map defines a map $N(\Sigma) \to \mathbb{P}(\xi)$, $\xi = \mathcal{E}|_\Sigma$. Indeed, if we use L to make $N(\Sigma)$ into a fiber bundle, with fibers equal to intervals, then this map is a bundle map over the projections to Σ, and its restriction to each leaf ℓ has nonzero derivative. Consequently, the development map is a diffeomorphism onto its image, provided $N(\Sigma)$ is taken small enough. The coordinate description of the quotient procedure, and the use of w as affine coordinate on the lines $\mathbb{P}(\xi_m)$, show that this map preserves Engel structures. □

2.4. Summary. The characteristic line field of an Engel manifold admits a transverse contact structure, and the individual leaves of this line field admit a real projective structure.

3. Automorphisms

How do contact and Engel automorphisms relate to each other?

THEOREM 3 (Automorphism Theorem). (a) *Let $\mathbb{P}\xi$ be the Engel structure obtained by prolonging the contact three-manifold (M,ξ) as in Theorem 1. Then every Engel automorphism of $\mathbb{P}\xi$ is induced by a contact automorphism of M.*

In (b) *and* (c) *below, we suppose that Q is an Engel manifold which admits a global slice Σ, by which we mean a three-manifold which intersects every leaf of the Engel line field L transversally.*

(b) *Any Engel automorphism $F : Q \to Q$ is uniquely determined by its restriction to Σ, this being a contact map from Σ to its image $F(\Sigma)$.*

(c) *If the Engel automorphism F is close to the identity, then by using the flow along L we can define a "Poincaré return map" $\phi_F : \Sigma \to \Sigma$ associated to F. This return map is a contact automorphism of Σ which uniquely determines F.*

PROOF. (a) Any contact diffeomorphism $\phi : M \to M$ induces an Engel diffeomorphism ϕ_* of $\mathbb{P}\xi$ by differentiation: $\phi_*(m,\ell) = (\phi(m), d\phi_m(\ell))$.

Now suppose that F is an arbitrary Engel automorphism. Since the construction of L and $\mathcal{E} = \mathcal{D}^2$ depend only on $\mathcal{D}$, the map F must also preserve L and $\mathcal{E}$. Consequently it induces a contact map ϕ on the leaf space $M = \mathbb{P}(\xi)/L$, and this in turn induces the Engel diffeomorphism $\Phi = \phi_*$. We claim that $F = \Phi$.

$G = F^{-1} \circ \Phi$ is an Engel automorphism which induces the identity automorphism on M. It suffices to show all such automorphisms are the identity on $\mathbb{P}(\xi)$. In Engel coordinates such a map G has the form $(x,y,z,w) \mapsto (x,y,z,g(x,y,z,w))$. For it to be an Engel automorphism we must have $G^*\theta_1 = 0 \bmod\{\theta_1,\theta_2\}$, and $G^*\theta_2 = 0 \bmod\{\theta_1,\theta_2\}$, where $\theta_1 = dz - ydx, \theta_2 = dy - wdx$ are the canonical forms whose vanishing defines $\mathcal{D}$. Thus

$$dy - gdx = a\theta_1 + b\theta_2$$

for some functions a, b. Since no dz's occur on the left hand side, we have $a = 0$. Now write $g = w + h$. We obtain the equation $\theta_2 - hdx = b\theta_2$. Considering this equation mod θ_2, we obtain $h = 0$, so that $g = w$ and G is the identity.

(b) Let $F, G : Q \to Q$ be two Engel automorphisms. The claim is that if $F(x) = G(x)$ for all $x \in \Sigma$, then $F = G$. Again, consider $F \circ G^{-1}$. This is an automorphism which is the identity on Σ. According to Theorem 2 we can identify a neighborhood of Σ with a neighborhood of $\mathbb{P}\xi$, where $\xi \subset T\Sigma$ is the induced contact structure on Σ. Now apply result (a) to conclude that $F \circ G^{-1}$ is the identity in some neighborhood of Σ.

To show that $F = G$ everywhere, we invoke the fundamental theorem of projective geometry. A projective transformation of the real projective line which fixes three points is the identity. In particular, if such a transformation is the identity on some interval of the line, then it is the identity. This last assertion is valid for any connected one-dimensional manifold with a real projective structure, and hence for the leaves ℓ of L. Being Engel, $F \circ G^{-1}$ preserves the real projective structure on each leaf ℓ. But we have seen that $F \circ G^{-1}$ is the identity on an interval of every leaf. Thus $F = G$ everywhere.

(c) We represent a tubular neighborhood of Σ by a neighborhood of the zero section in the normal bundle $N(\Sigma)$, in such a way that the fibers of the normal bundle correspond to the leaves ℓ of L. If F is sufficiently C^1-close to the identity, then $F(\Sigma)$ can be represented as the image of a section $s : \Sigma \to N(\Sigma)$. We can then "push" this image down along the fibers of $N(\Sigma)$, homotoping the image back to the zero section. This homotopy corresponds to flowing along the leaves of L, and hence to the restriction of an *even-contact* automorphism $\Phi : Q \to Q$ to $F(\Sigma)$. *Warning: This flow Φ is not an Engel automorphism.* Our 'Poincaré return map' is $\phi_F := \Phi^{-1} \circ F|_\Sigma$, a contact automorphism of Σ. According to (c) of Theorem 2, our tubular neighborhood is Engel-diffeomorphic to a neighborhood in $\mathbb{P}(\xi)$. According to (a) just proved, ϕ_F determines F in this neighborhood, and according to (b) just proved this determines F everywhere. □

3.1. Coordinate form of automorphisms. Let (x, y, z, w) be canonical Engel coordinates, and let X, Y, Z, W be the canonical frame as before. Write $X[f] = df(X)$ for the directional derivative of f in the direction of the vector field X. Then any infinitesimal Engel automorphism V has the form

$$V = Y[f]X - X[f]Y + fZ + X^2[f]W$$

for some smooth function $f = f(x, y, z)$ *independent of w*. This formula can be derived by expanding out V in terms of the frame, and computing the Lie derivative L_V of the defining one-forms θ_1 and θ_2. Note that the projection of the vector field V to the space of x, y, z variables is well-defined, and is an infinitesimal contact automorphism, as it should be.

4. Deformations

We are ready to move on to our main study, the study of deformations of standard Engel structures.

DEFINITION 7. A *deformation* of an Engel structure $\mathcal{D}$ is a one-parameter family of Engel structures $\mathcal{D}_t$, $t \in I \subset \mathbb{R}$, which passes through $\mathcal{D}$ at $t = 0$: $\mathcal{D}_0 = \mathcal{D}$. The $\mathcal{D}_t$ vary smoothly with t. The interval I of deformation parameters may be any interval containing 0 in its interior.

Fix a contact three-manifold (M, ξ), where ξ is a globally trivial contact structure. Split

$$\xi = V_0 \oplus V_1$$

into the sum of two Legendrian line fields, V_0, V_1. Then $\mathbb{P}\xi \cong M \times S^1$, where (m, θ) corresponds to the Legendrian line spanned by the vector

$$v(\theta, m) = \cos(\theta)v_0(m) + \sin(\theta)v_1(m),$$

where the v_i are nonvanishing vector fields spanning V_i. This circle S^1 has circumference π, since v and $-v$ represent the same line.

Consider the domain $\Omega = \Omega(V_0, V_1)$ corresponding to the constraint $0 \leq \theta \leq \pi/2$. This consists of all Legendrian lines lying "between" V_0 and V_1.

DEFINITION 8. We call $\Omega(V_0, V_1) \subset \mathbb{P}(\xi)$, with its induced Engel structure, the *standard domain* associated to the pair of Legendrian line fields (V_0, V_1).

$\Omega(V_0, V_1)$ is a four-dimensional submanifold with boundary. Its boundary has two components, one corresponding to V_0 and the other to V_1. Being sections, each boundary component is an embeddded copy of M. $\Omega(V_0, V_1)$ is diffeomorphic to $M \times I$, $I = [0, \pi/2]$.

THEOREM 4. (a) *Let $\mathcal{D}_t$ be any sufficiently small deformation of the canonical Engel structure on the standard domain $\Omega(V_0, V_1)$. Then there is a one-parameter family of Engel diffeomorphisms $\phi_t : (\Omega(V_0, V_1), \mathcal{D}_t) \to (\Omega(V_{0,t}, V_{1,t}), \mathcal{D}_0)$. In other words, we can straighten out the deformed Engel structure at the expense of deforming the boundaries $V_{0,t}, V_{1,t}$ of the domain.*

(b) *Let $\mathcal{D}_t$ be a small deformation of the Engel structure on the whole of $\mathbb{P}\xi \cong M \times S^1$, where ξ is a parallelizable contact structure on M^3. Let L_t be its deformed line field, and ξ_t the deformed contact structure on M, induced by viewing M as a slice in $\mathbb{P}\xi$. Then the Poincaré return map for L_t is a contact map $\phi_t : (M, \xi_t) \to (M, \xi_t)$.*

(c) *Any near-identity contact isotopy of (M, ξ) to itself can be realized, as in* (b), *by some Engel deformation $\mathcal{D}_t$.*

This theorem, combined with the previous one, allows us to explore the space of moduli of Engel deformations of $M \times I$.

DEFINITION 9. Two Engel deformations $\mathcal{D}_t$ and $\mathcal{E}_t$ of an Engel structure $\mathcal{D}$ on Q represent the same *deformation germ* of $\mathcal{D}$ if there is an isotopy $\psi_t : Q \to Q$ such that $\psi_t^* \mathcal{E}_t = \mathcal{D}_t$.

In general, when we say "deformation germ" we will mean the germ in t of some deformation or homotopy h_t of a global geometric structure. Deformation germs are local in t but global in $q \in Q$, and involve dividing out by the appropriate automorphisms.

THEOREM 5. (a) *The space of deformation germs of the standard domain $(\Omega(V_0, V_1), \mathcal{D})$ with its standard Engel structure is canonically isomorphic to the space of deformation germs $(V_{0,t}, V_{1,t})$ of (V_0, V_1) of pairs of Legendrian line fields on (Σ, ξ) modulo contact isotopies. This space is infinite-dimensional.*

(b) *The space of deformation germs of the standard Engel structure on $\mathbb{P}\xi$ is equal to the space of deformation germs of the identity through contact isotopies ϕ_t of (Σ, ξ) modulo t-dependent conjugation: $\phi_t \sim g_t \circ \phi_t \circ g_t^{-1}$.*

REMARKS. 1. Gershkovich [8] realized that the Engel deformation space for $M \times S^1$ was nontrivial, and very large. He based his observations on perturbations of the line field, as we have. His proof was incomplete.

2. Since the first drafts of the current paper were written, Golubev [10] proved his theorem, mentioned in the introduction. It implies, roughly speaking, that the deformation germ of an Engel field is completely determined by the corresponding deformation germ L_t of its characteristic line field.

4.1. Proof of Theorem 4. (a) As $\mathcal{D}$ deforms through $\mathcal{D}_t$, its line field L deforms through L_t, and its even-contact field $\mathcal{E}$ deforms through $\mathcal{E}_t$. Identify M with its image in $\mathbb{P}(\xi)$ under the section $V_0 : M \to \mathbb{P}\xi$. This copy of M is a global slice for L, since L is the vertical bundle for the fibration $\mathbb{P}(\xi) \to M$. Consequently this M is a global slice for the nearby fields L_t. M inherits the deformed contact structures $\xi_t = \mathcal{E}_t \cap TM$ (Theorem 2). Now Gray's theorem [11] yields a diffeomorphism $f_t : M \to M$ taking ξ_t to $\xi = \xi_0$. Thus $f_t^*\xi_t = \xi_0$. It follows that the differential of f_t induces an Engel isomorphism $F_t : \mathbb{P}(\xi_t) \to \mathbb{P}(\xi)$, where $\mathbb{P}(\xi_t)$ is given the structure defined by prolongation of ξ_t, as in Theorem 1.

It follows from the proof of (c) of Theorem 2 that we have an Engel embedding $G_t : (\Omega(V_0, V_1), \mathcal{D}_t) \to \mathbb{P}(\xi_t)$. This is because M is a global slice for L_t, and because the leaf structure of L, and hence of nearby L_t, on $\Omega(V_0, V_1)$ is so simple—namely, this foliation is diffeomorphic to the foliation by intervals $\{m\} \times I$ in $M \times I$. Composing these two Engel embeddings, we get an Engel embedding $(\Omega(V_0, V_1), \mathcal{D}_t) \to (\mathbb{P}(\xi), \mathcal{D}_0)$. This embedding is close to the identity; hence its image must be close to $\Omega(V_0, V_1)$, with boundary components transverse to the fibers of L_0. Consequently this image is of the form $\Omega(V_{0,t}, V_{1,t})$ for some pair $(V_{0,t}, V_{1,t})$ of Legendrian line fields which are perturbations of the pair (V_0, V_1).

(b) The deformation $\mathcal{D}_t$ is defined on all of $\mathbb{P}\xi$. Use the same global section $M = V_0(M)$ for the L_t. Define Poincaré maps $\phi_t : M \to M$ by starting at $m \in M$ and travelling along the L_t-line passing through m until its next intersection $\phi_t(m)$ with M. This requires L to be oriented; hence the assumption on parallelizability of ξ. The map ϕ_t is a contact automorphism for the contact structure $\xi_t = \mathcal{E}_t \cap TM$, according to the discussion in Section 2.3 around Theorem 2.

(c) We prove that we can obtain any near-identity contact automorphism ϕ of the fixed contact structure (M, ξ) as the "Poincaré section" associated to an Engel deformation of the canonical Engel structure. We may suppose that the contact automorphism is the time-dependent flow of some time-dependent vector field $X = X(m; \theta)$, where we also think of θ as the time variable. Specifically, ϕ can be taken to be the time $\theta = \pi$ flow for this X. (Again, the circles of $M \times S^1$ have circumference π.)

As before, write elements of $\mathbb{P}(\xi)$ as the span of vectors

$$\cos(\theta)v_0(m) + \sin(\theta)v_1(m),$$

where v_0, v_1 are a pair of vector fields framing ξ. This induces a global trivialization $\mathbb{P}(\xi) \cong M \times S^1$, and under this diffeomorphism the canonical Engel distribution is spanned by the vector fields $\frac{\partial}{\partial\theta}$ and $v(m, \theta) = \cos(\theta)v_0(m) + \sin(\theta)v_1(m)$ on $M \times S^1$. Of course, $\frac{\partial}{\partial\theta}$ spans its characteristic line field $L = L_0$ for $\mathcal{D} = \mathcal{D}_0$.

Define the perturbed Engel structure by

$$\tilde{\mathcal{D}} = \operatorname{span}\{v, W\}$$

with v as before but with

$$W = \frac{\partial}{\partial\theta} + X(m,\theta).$$

We will show that this $\tilde{\mathcal{D}}$ is indeed an Engel field and that its characteristic line field is spanned by W. To begin, note that v and

$$u := -\sin(\theta)v_0(m) + \cos(\theta)v_1(m)$$

span ξ everywhere. Since X is an infinitesimal contact symmetry for ξ, and since v and u span ξ, we have $[X, v] = fv + gu$ for some θ-dependent functions f, g. On the other hand, $[\frac{\partial}{\partial\theta}, v] = u$, when v and u are thought of as vector fields on $\mathbb{P}(\xi) = M \times S^1$. Thus $[W, v] = fv+(1+g)u$. Consequently $[\tilde{\mathcal{D}}, \tilde{\mathcal{D}}]$ is rank 3 wherever $g \neq -1$. This can be arranged to happen everywhere, provided X is sufficiently small. In this case $\tilde{\mathcal{D}}^2 = [\tilde{\mathcal{D}}, \tilde{D}]$ is spanned by W, u and v. Since ξ is contact, $[u, v]$ is a fourth vector field, linearly independent from W, u, v everywhere. This shows that $[\tilde{\mathcal{D}}^2, \tilde{\mathcal{D}}^2] = T(M \times S^1)$, and consequently $\tilde{\mathcal{D}}$ is indeed Engel. By the same computation as used in finding $[W, v]$ above, we see that $[W, u]$ is also a linear combination of v and u everywhere. Consequently $[W, \tilde{\mathcal{D}}^2] \subset \tilde{\mathcal{D}}^2$, so W is indeed the characteristic line field for $\tilde{\mathcal{D}}$.

By construction, the time-s flow of W is of the form $(m, 0) \mapsto (\phi_s(m), s)$, when applied to a point $(m, 0)$. Here ϕ_s is the time-s flow of the time-dependent contact vector field $X(\cdot, s)$. It follows that the return map for $\tilde{\mathcal{D}}$ is the desired contact map ϕ. □

4.2. Proof of Theorem 5. (a) Except for the sentence about the deformation space being infinite-dimensional, the proof is a straightforward application of Theorems 4 and 5.

Regarding the infinite-dimensionality, we refer the reader to Cartan [**6**], or to Arnold's description [**1**] of Cartan. Cartan constructs a bijection between the germs of pairs of Legendrian line fields in three-space modulo contact automorphisms and the germs of second order differential equations of the form

$$\frac{d^2y}{dx^2} = f\left(y, \frac{dy}{dx}, x\right)$$

modulo germs of diffeomorphisms of the xy plane. The latter space is known to be infinite-dimensional, and some functional moduli are described in these references.

(b) The return map ϕ_t for ξ_t of the proof of part (b) of Theorem 4, together with the Gray isotopy $g_t : (M, \xi_t) \to (M, \xi_0)$, yields the contact isotopy $g_t \circ \phi_t \circ g_t^{-1}$. We leave it for the reader to check that this conjugation corresponds to the correct equivalence relation in the definition of 'deformation germ'.

5. An Engel structure with few symmetries

We end with the following example.

THEOREM 6. *There exist a three-manifold M and an Engel structure on $M \times S^1$ with the property that the automorphism group for this structure contains exactly one one-parameter subgroup.*

Thus if we knew that the automorphism group were a Lie group then it would be one-dimensional. Contrast this state of affairs with that of global symplectic and contact structures, whose symmetries can always be localized, and consequently are always infinite-dimensional. The asymmetry in this example is due to the characteristic line field.

5.1. The construction. The idea is to find a contact transformation $\phi = \phi_t$ for some parallelizable ξ which has very few symmetries, and then construct the corresponding Engel structure following (c) of Theorem 4. Our particular ϕ will be part of a one-parameter subgroup, accounting for the one-parameter group of Engel automorphisms.

We begin by assuming the existence of a near-identity contact automorphism ϕ of (M, ξ), ξ parallelizable, with the following property. First, ϕ is the time ϵ flow of some contact vector field X. Second, if $\psi_t = \exp(tY)$ is any other one-parameter group of contact automorphims which commutes with ϕ: $\psi_t \circ \phi = \phi \circ \psi_t$, then $\psi_t = \exp(tcX)$ for some constant c. In other words, the commutator subgroup for ϕ is as small as possible, given that ϕ is part of a one-parameter subgroup. This ϕ will be constructed later.

Apply the construction of (c) of Theorem 4 to obtain an Engel structure $\tilde{\mathcal{D}}$ on $\mathbb{P}\xi \cong M \times S^1$ which is a deformation of the canonical Engel structure and which has ϕ as its return map. In this construction, the induced contact structure on the slice $M \subset \mathbb{P}\xi$ is the original contact structure ξ on M, even after the deformation.

Now let $\Psi_t = \exp(tZ)$ be any smooth one-parameter family of Engel automorphisms for this deformed structure. As proved in Theorem 3, Ψ_t is induced by some one-parameter family of contact automorphisms $\psi_t : (M, \xi) \to (M, \xi)$. Tracing through the constructions, we see that if the Ψ_t are to be Engel automorphisms then the ψ_t must commute with the return map ϕ. Consequently $\psi_t = \exp(ctX)$, for some constant c, by assumption, and the automorphisms Ψ_t must equal the automorphism induced by ψ_t via (b) and (c) of Theorem 3.

It remains to construct the map ϕ with the desired property. Let M be the unit tangent bundle of a compact Riemann surface S with constant negative curvature. M has a canonical contact structure for which the geodesic flow is a one-parameter group of contact automorphisms. This contact structure on M is parallelizable, as it is for the unit tangent bundle of any oriented surface. It is well-known that in this negatively curved case the geodesic flow g_t is ergodic on M. We will see that $\phi = g_\epsilon$ for appropriate ϵ has the desired property.

If $C \subset S$ is an immersed curve, then its lift to M is uniquely defined. Consequently we may safely identify a geodesic in S with the corresponding orbit of g in M obtained by following the unit tangent vector to the geodesic.

To construct ϕ, recall that the closed geodesics on S form a countable collection of curves dense in S and also in M. Also recall that for each free homotopy class on S there is a unique closed geodesic. Let l_i denote the lengths of these closed geodesics—a countable collection of real numbers. Choose ϵ to be any real number which is smaller than the cut length of S and which is not rationally related to this collection of lengths; in other words, for which it is impossible to find a finite list of rationals $r_1, \ldots, r_N$ and lengths $l_1, \ldots, l_N$ such that $\epsilon = \sum r_i l_i$. The existence of such an ϵ is guaranteed in any small interval of reals because the set of such sums forms a countable set of numbers. We set $\phi = g_\epsilon$. The restriction of ϕ to any closed geodesic C is then equivalent to an irrational rotation of the circle C. It follows that if $p \in M$ lies on such a geodesic then the closure of its ϕ-orbit is all of C. The converse is also true: if $C \subset M$ is any ϕ-invariant set which is diffeomorphic to an embedded circle, then C is a closed geodesic. To see this, let $p \in C$ and let $\gamma \subset M$

be the geodesic (g-orbit) through p. Since $\phi = g_\epsilon$, we have $C \subset \gamma$. But both γ and C are immersed curves, hence they are equal.

We now check that ϕ has the desired commutator property. Suppose that $\psi_t = \exp(tY)$ is a one-parameter group of diffeomorphisms which commutes with ϕ. We will show that $\psi_t = g_{ct}$ for some constant c. First observe that ψ_t must map closed ϕ-invariant embedded circles C of ϕ to other such circles. Since these Cs are precisely the closed geodesics, ϕ maps closed geodesics to closed geodesics. Since ψ_t is isotopic to the identity, $\psi_t(C)$ must be in the same free homotopy class as C, when thought of as a loop in S. Consequently $\psi_t(C) = C$ for each closed geodesic C. Since the C's are dense, it follows that the vector field Y generating ψ_t must satisfy $Y = fX$, where X is the generator of geodesic flow and f is some functon. But $\phi^*Y = Y$, since ϕ and the ψ_t commute; and $\phi^*X = X$, since $\phi = \exp(\epsilon X)$. Therefore $\phi^* f = f$. Since the restriction of ϕ to each closed geodesic C is an irrational rotation, it follows that the restriction of f to each such C is a possibly C-dependent constant. But again the closed geodesics are dense and f is continuous, so $f = c$, a constant, on all of M. So $\psi_t = g_{ct}$, as claimed. □

5.2. Acknowledgements. The author is grateful to Yasha Eliashberg, Viktor Ginzburg, Andrew Klingler, Alex Golubev, Misha Zhitomirskii, and above all to Robert Bryant for useful discussions. This work was sponsored in part by National Science Foundation Grant DMS 94000515 and by a faculty research grant from the University of California at Santa Cruz.

References

[1] V. I. Arnol'd, *Geometrical methods in the theory of ordinary differential equations*, Springer-Verlag, New York, 1993.

[2] R. L. Bryant, S. S. Chern, R. B. Gardner, H. L. Goldschmidt, and P. A. Griffiths, *Exterior differential systems*, MSRI Publications, vol. 18, Springer-Verlag, New York, 1991.

[3] R. Bryant and L. Hsu, *Rigidity of integral curves of rank two distributions*, Invent. Math. **114** (1993), 435-461.

[4] R. Bryant, *An introduction to Lie groups and symplectic geometry*, Geometry and Quantum Field Theory, Park City Math. School, vol. 1, Amer. Math. Soc., Providence, RI, 1991, pp. 5–181.

[5] E. Cartan, *Sur quelques quadratures dont l'element differentiel contient des fonctions arbitraires*, Bull. Soc. Math. France **29** (1901), 118–130; reprinted in his *Oeuvres Complètes* (2nd ed.), Partie II, Editions de CNRS, Paris, 1984, pp. 397–410.

[6] ———, *Sur les variétes à connexion projective*, Bull. Soc. Math. France **52** (1924), 205–241; reprinted in his *Oeuvres Complètes* (2nd ed.), Partie III, Editions de CNRS, Paris, 1984, pp. 825-862.

[7] Y. Eliashberg, *Contact 3-manifolds twenty years since J. Martinet's work*, Ann. Inst. Fourier (Grenoble) **42** (1992), fasc. 1–2, 165–192.

[8] V. Gershkovich, *Engel structures on four-dimensional manifolds*, preprint, Univ. of Melbourne, 1993.

[9] V. Ginzburg, *Calculation of contact and symplectic cobordism groups*, Topology **31** (1992) 767–773.

[10] A. Golubev, *On the global stability of maximally non-holonomic two-plane fields in four dimensions*, Internat. Math. Res. Notes **1997**, 523–529.

[11] J. W. Gray, *Some global properties of contact structures*, Ann. of Math. (2) **69** (1959), 421–450.

[12] W-S. Liu and H. J. Sussmann, *Shortest paths for subRiemannian metrics on rank two distributions*, Mem. Amer. Math. Soc. **118** (1995), no. 564.

[13] R. Lutz, *Structures de contact sur les fibres prinipaux en cercles de dimension 3*, Ann. Inst. Fourier (Grenoble) **23** (1977), fasc. 3, 1–15.

[14] J. Martinet, *Formes de contact sur les varietes de dimension* 3, Lecture Notes in Math., vol. 209, Springer-Verlag, New York, 1971, pp. 142–163.

[15] R. Montgomery, *Abnormal minimizers*, SIAM J. Control Optim. **32** (1994), 1605–1620.

[16] ———, *Generic distributions and finite dimensional Lie algebras*, J. Differential Equations **103** (1993), 387–393.

[17] ———, *A survey of singular curves in subRiemannian geometry*, J. Dynam. Control Systems **1** (1995), 41–90.

[18] R. Montgomery, B. Shapiro. and M. Kazarain, *Characteristic classes for the degenerations of two-plane fields in four dimensions*, Pacific J. Math. **179** (1997), 355–370.

[19] A. M. Vershik and V. Ya. Gerhskovich *Nonholonomic dynamical systems, geometry of distributions and variational problems*, Itogi Nauki i Tekhn.: Sovremennye Problemy Mat.: Fundamental′nye Napravleniya, Vol. 16, VINITI. Moscow, 1987, pp. 5–85; English transl. in Encyclopaedia of Math. Sci., vol. 16 (Dynamical Systems, VII), Springer-Verlag, New York, 1994.

Mathematics Deptartment, University of California at Santa Cruz, Santa Cruz, CA 95064

E-mail address: rmont@cats.ucsc.edu

Amer. Math. Soc. Transl.
(2) Vol. **196**, 1999

Floer Homology, Novikov Rings and Clean Intersections

Marcin Poźniak

1. Introduction

Let $f : M \to \mathbb{R}$ be a Morse function on a compact Riemannian manifold (M, g). The Morse inequalities provide lower estimates for the number of critical points of f using topological information about the manifold, namely its Betti numbers. In fact, it turns out that the critical points and the connecting trajectories of the gradient flow carry complete information about the topology of the manifold. In particular, it is possible to recover the homology of the manifold as the homology of a chain complex generated by the critical points. This idea goes back to Smale [**Sma60**] and, more recently, has been formulated by Witten [**Wit82**]. Consider the gradient flow generated by the equation

$$\dot{\gamma} = -\nabla f(\gamma).$$

If x is a nondegenerate critical point of f, then its stable and unstable manifolds $W^s(x)$ and $W^u(x)$ intersect transversally and $\dim W^u(x) = \mathrm{ind}(x)$. For each critical point x of f, choose an orientation $\langle x\rangle$ of the tangent space $E^u_x := T_x W^u(x)$ and define a graded $\mathbb{Z}$-module $C_*(M, g, f)$ by taking $C_k(M, g, f)$ to be the free $\mathbb{Z}$-module generated by the set of all $\langle x\rangle$, where x is a critical point of f of index k,

$$C_k = \bigoplus_{\mathrm{ind}(x)=k} \mathbb{Z}\langle x\rangle.$$

If we assume that the gradient flow is of Morse–Smale type, then $W^u(x)\cap W^s(y)$ is a manifold of dimension $\mathrm{ind}(x) - \mathrm{ind}(y)$. In particular, if $\mathrm{ind}(x) - \mathrm{ind}(y) = 1$, then $W^u(x)\cap W^s(y)$ consists of isolated orbits. In fact, since the manifold is compact and the function f decreases along the nonconstant orbits, the number of these orbits is finite. With each of them one can associate a sign determined by the choice of the orientations $\langle x\rangle$ and $\langle y\rangle$. Let $n(y, x)$ denote the number of orbits connecting x and y, counted with signs. We define the homomorphism $\partial^f_k : C_k \to C_{k-1}$ by

$$\partial^f_k\langle x\rangle = \sum_{\mathrm{ind}(y)=k-1} n(x, y)\langle y\rangle. \tag{1.1}$$

1991 *Mathematics Subject Classification*. Pimary 58F05.

©1999 American Mathematical Society

Then $(C_*(M,f,g),\partial_*^f)$ is a chain complex and

$$H_*(C_*(M,f)) \cong H_*(M). \tag{1.2}$$

An immediate consequence is that the number of critical points of a Morse function is greater than or equal to the sum of the Betti numbers of the manifold. In the original proof of (1.2) (see [**Mil63**], [**Flo89b**], [**Sal90**]) one constructs a filtration $\emptyset = N_{-1} \subset N_0 \subset N_1 \subset \cdots \subset N_m = M$ of the manifold M such that all critical points of index k belong to $N_k \setminus N_{k-1}$. The set of critical points of index k can be viewed as an isolated invariant set for the gradient flow, and then (N_k, N_{k-1}) form its Conley index pair. The homology groups $H_i(N_k, N_{k-1})$ do not depend on the choice of the filtration, and all of them vanish except for $H_k(N_k, N_{k-1})$, which is a free module whose generators can be identified with the critical points. Hence $H_k(N_k, N_{k-1}) \cong C_k(M,f,g)$, and moreover it can be shown that the boundary homomorphism $\Delta_k : H_k(N_k, N_{k-1}) \to H_k(N_{k-1}, N_{k-2})$ in the exact sequence of the triple (N_k, N_{k-1}, N_{k-2}) coincides with ∂_k^f. On the other hand, $\Delta_k \circ \Delta_{k+1} = 0$, and a standard algebraic argument shows that the homology of the complex $\{H_k(N_k, N_{k+1}), \Delta_k\}$ is isomorphic to that of the manifold M.

The isomorphism (1.2) can be proved without referring to the identification above, via Fredholm theory. Here, $W^u(x) \cap W^s(y)$ is identified with the space of connecting orbits $\mathcal{M}(x,y)$, which arises as the zero set of a Fredholm section of a Banach vector bundle over the manifold of paths connecting x and y. This approach was invented by Floer ([**Flo88b**], [**Flo89a**], [**Flo89b**]), who applied it to a symplectic action functional a_H on an infinite dimensional manifold of loops whose critical points are the periodic orbits of a Hamiltonian system on an underlying symplectic manifold. If (M,ω) is a symplectic manifold, then a critical point of f can be seen as a 1-periodic orbit of a Hamiltonian vector field associated with f. This is one of the reasons behind Arnold's conjecture which states that for any 1-periodic Hamiltonian $H : S^1 \times M \to \mathbb{R}$ the number of closed orbits is greater than or equal to the sum of the Betti numbers provided that all periodic solutions are nondegenerate. Floer proved the conjecture, developing the Morse theory for the functional

$$a_H(\gamma) = \int_{D^2} u^*\omega + \int_0^1 H_t(\gamma(t))\,dt.$$

Here $\gamma : \mathbb{R} \to M$, and $u : D^2 \to M$ extends γ over the disc D^2. Thus a_H is defined locally on the space of loops in M. The classical methods of Morse theory fail for this variational problem. The L^2-gradient of a_H does not define a global flow, and moreover the index of any critical point is not finite. However, the Fredholm approach does not make use of the gradient flow and detects only the "relative" Morse index between two critical points. Moreover, even though the ambient manifold is infinite-dimensional, the spaces of connecting orbits $\mathcal{M}(x,y)$ enjoy the compactness property modulo "bubbling off" of pseudoholomorphic spheres (a phenomenon described by Gromov [**Gro85**], which can be prevented by additional assumptions on the symplectic manifold). This way, Floer was able to construct a chain complex $CF_*(M,H_t)$, generated by the periodic orbits, such that its homology is a symplectic invariant of M. Moreover, there is a direct link with the finite dimensional Morse theory here. If the Hamiltonian is time-independent, then its Morse complex is a subcomplex of $CF_*(M,H)$.

There are many generalizations of Floer's original results. For example, [**SZ94**] exploits the Maslov index to prove the existence of infinitely many periodic orbits, [**HS92**] and [**Ono93**] cover the case of weakly monotone manifolds, and [**LO93**] covers the nonexact case.

This thesis consists of two separate parts evolving around Floer's theory. The first part (§2) describes the construction of Novikov homology—that is, the generalization of the Morse complex to the case of a non-exact closed form. Since in general one cannot expect the existence of a suitable filtration $\{N_i\}$, an alternative proof that $\partial^2 = 0$ is needed. Thus there is a good reason to give a presentation of Floer's ideas in the simplest finite dimensional case. This part has mainly expository character (see also [**Sal90**] and [**Sch93**] for a very detailed presentation of Morse homology). Then we show that the Novikov homology is isomorphic to the homology of the manifold with coefficients in a local system (Theorem 2.2.2), a result proved independently by V. Lê Hông and K. Ono ([**LO93**]).

The second part (§3) uses Floer homology in the context of the Lagrangian intersections. These homology groups have not been computed in general. Theorem 3.4.11 provides a method for computing them when the intersection of the Lagrangian submanifolds is connected and sufficiently regular. As an example we describe the intersection of linear tori in T^{2n}. Also, in this part we show that the Floer homology for a non-exact perturbation of the zero section in the cotangent bundle is isomorphic to the Novikov homology of its flux form.

2. Novikov homology

2.1. Introduction. In [**Nov81**], [**Nov82**] Novikov gave a generalization of Morse theory to the case of a general closed 1-form α. He considered a covering of M, a Morse function f whose derivative is a lift of α, and the cell complex built of the "surfaces of steepest descent" of f, i.e., the unstable manifolds. The "multiplicity" of the critical points and noncompactness of the covering are dealt with by introducing a special coefficient ring and restricting attention to the "regions of finite energy". Novikov's papers do not include the proof that this construction defines a chain complex. Hofer and Salamon [**HS92**] employed Novikov's ideas to generalize Floer homology to a wider class of symplectic manifolds. They also remarked that one could define the Witten variant of Novikov homology in the finite dimensional situation. This construction, described below, has also been carried out by V. Lê Hông and K. Ono in [**LO93**]. They showed the existence of an isomorphism between Novikov homology and the homology with local coefficients, using an analytical method which they developed to prove the Arnold conjecture in the non-exact case. In this thesis we present an independent proof by the methods of differential topology. It seems, however, to be less suitable for generalization to the infinite dimensional situation.

2.2. **Novikov complex.** Let α be a smooth closed 1-form on a compact Riemannian manifold (M, g). If $\nabla\alpha$ denotes the covariant derivative of α, then for any zero point x of α the *Hessian* $\nabla\alpha(x) : T_xM \otimes T_xM \to \mathbb{R}$ does not depend on the particular choice of the metric. A zero point x is called *nondegenerate* if and only if $\nabla\alpha(x)$ is nonsingular. Then the *index* $\mathrm{ind}(x)$ of x is defined as the index of $\nabla\alpha(x)$, i.e., the dimension of its maximal negative subspace. The form α is called *nondegenerate* if all zeros of α are nondegenerate.

We want to construct an analogue of the Morse complex for α. Let X_α be the vector field dual to α:

$$g(X_\alpha(x), \xi) = \alpha(x)\xi \quad \text{for } \xi \in T_xM,\ x \in M,$$

and consider the flow $\psi_\alpha : \mathbb{R} \times M \to M$ generated by the equation

$$\dot\gamma = -X_\alpha(\gamma), \tag{2.1}$$

i.e., $\psi_\alpha(\cdot, x) : \mathbb{R} \to M$ is a solution to (2.1) such that $\psi_\alpha(0, x) = x$. For a critical point x of (2.1) (a zero of α) define stable and unstable sets

$$W^s(x) = \{y \in M : \psi_\alpha(y,t) \to x \text{ as } t \to +\infty\},$$
$$W^u(x) = \{y \in M : \psi_\alpha(y,t) \to x \text{ as } t \to -\infty\}.$$

A flow is said to be of *Morse–Smale type* if and only if stable and unstable manifolds of any two critical points intersect transversally. This condition is met for a generic metric g if the form α is nondegenerate (Subsection 2.5). Then the sets $W^u(x^-) \cap W^s(x^+)$ are manifolds of dimension $\operatorname{ind}(x^-) - \operatorname{ind}(x^+)$. Moreover $\mathbb{R}$ acts freely on $W^u(x^-) \cap W^s(x^+)$ via ψ_α, and so for $\operatorname{ind}(x^-) - \operatorname{ind}(x^+) = 1$ the set $W^u(x^-) \cap W^s(x^+)/\mathbb{R}$ is discrete. However, it may be infinite if α is not exact. There is a ($\mathbb{R}$ equivariant) bijective correspondence between $W^u(x^-) \cap W^s(x^-)$ and the set $\mathcal{M}(x^-, x^+)$ of all orbits $\gamma : \mathbb{R} \to M$ of the gradient flow satisfying the limit conditions

$$\lim_{s \to -\infty} \gamma(s) = x^-, \qquad \lim_{s \to +\infty} \gamma(s) = x^+, \tag{2.2}$$

given by the evaluation map

$$\mathrm{ev} : \mathcal{M}(x^-, x^+) \ni \gamma \mapsto \gamma(0) \in W^u(x^-) \cap W^s(x^+) \subset M.$$

We denote $\widehat{\mathcal{M}} := \mathcal{M}/\mathbb{R}$. To ensure the finiteness property we impose some restriction on the orbits $\gamma \in \mathcal{M}(x^-, x^+)$. Define the "energy" of a differentiable path $\gamma : \mathbb{R} \to M$ by

$$\ell^2(\gamma) := \int_{-\infty}^{+\infty} |\dot\gamma(s)|^2\, ds.$$

In particular, if $\gamma \in \mathcal{M}(x^-, x^+)$, then

$$\ell^2(\gamma) = -\int_{-\infty}^{+\infty} \langle X_\alpha(\gamma), \dot\gamma\rangle\, ds = -\int_\gamma \alpha, \tag{2.3}$$

and so the energy is locally constant on $\mathcal{M}(x^-, x^+)$. We say that a sequence $\{\gamma_\nu\}_{\nu=1}^\infty \subset \mathcal{C}^\infty(\mathbb{R}, M)$ converges to a *split trajectory* $(\gamma^1, \dots, \gamma^m)$ if there are collections of critical points $x^+ = x_0, x_1, \dots x_m = x^-$, orbits $\gamma^k \in \mathcal{M}(x_k, x_{k-1})$, and sequences of time shifts $\{s_\nu^k\}_{\nu=1}^\infty$, $k = 1, \dots, m$, such that $\gamma_\nu(\cdot + s_\nu^k)$ converges with all its derivatives on compact sets to γ^k.

PROPOSITION 2.2.1. *For every sequence $\{\gamma'_\nu\}_{\nu=1}^\infty \subset \mathcal{M}(x^-, x^+)$ satisfying*

$$\sup_{\nu \in N} \ell^2(\gamma'_\nu) < \infty$$

there is a subsequence γ_ν converging to a split trajectory $(\gamma^1, \ldots, \gamma^m)$ with

$$\ell^2(\gamma_\nu) = \sum_{k=1}^{m} \ell^2(\gamma^k)$$

if ν is sufficiently large. If the flow is of Morse–Smale type, then $\operatorname{ind}(x^+) < \operatorname{ind}(x_1) < \cdots < \operatorname{ind}(x^-)$.

PROOF (cf. [**Sal90**], p. 136). Let $x_0 = x^+$. First note that it is enough to find s_ν^k and γ^k such that $\gamma_\nu(s_\nu^k) \to \gamma^k(0)$. Then the required convergence will follow from the differentiable dependence on initial conditions for solutions of o.d.e.'s. Take an $\varepsilon > 0$ such that $d(x,y) > 2\varepsilon$ for any two different critical points x, y of f, and set

$$s_\nu^1 := \sup\{s \in \mathbb{R} : d(\gamma_\nu(s), x_0) > \varepsilon\}.$$

Passing to a subsequence we may assume that $\gamma_\nu(s_\nu^1)$ converges, and let γ^1 be the solution to (2.1) such that $\gamma^1(0) = \lim_{\nu\to\infty} \gamma_\nu(s_\nu^1)$. Then $d(\gamma^1(s), x_0) \leq \varepsilon$ for $s > 0$ and $\ell^2(\gamma^1) < \infty$, which implies that $\gamma^1 \in \mathcal{M}(x_1, x_0)$ for some critical point x_1. Indeed, assume the contrary, i.e., that there exist an ε_1 and a sequence s_ν converging to $-\infty$ such that $d(\gamma^1(s_\nu), x) > 2\varepsilon_1$ for every critical point x of α. We may assume that $s_\nu - s_{v+1} > 1$. If we define $M_\varepsilon := \{y \in M : d(y,x) \geq \varepsilon$ for any critical point $x\}$, then $c := \inf_{M_{\varepsilon_1}} |X_\alpha| > 0$. Set $t_\nu = \inf\{t \in [s_\nu, s_\nu + 1] : \gamma^1(t) \notin M_\varepsilon\}$ if this set is nonempty, and $t_\nu = s_\nu + 1$ otherwise. Then $d(\gamma^1(s_\nu), \gamma^1(t_\nu)) > \varepsilon_1$ in the first case, and so

$$\int_{s_\nu}^{t_\nu} |\dot\gamma^1|^2 \geq c \int_{s_\nu}^{t_\nu} |\dot\gamma^1| \geq \min\{c^2, c\varepsilon_1\},$$

which yields a contradiction. Proceeding by induction, assume that we have found x_j, $\{s_\nu^j\}$ and $\gamma^j \in \mathcal{M}(x_j, x_{j-1})$ for $j = 1, \ldots, k$. If $x_k \neq x^-$, then $d(\gamma_\nu(s), x_k) > \varepsilon$ for large $-s$. On the other hand, there is an $s^* < 0$ such that $d(\gamma^k(s), x_k) \leq \frac{\varepsilon}{2}$ for $s \leq s^*$, and so $d(\gamma_\nu(s_\nu^k + s^*), x_k) < \varepsilon$ for sufficiently large ν. Hence the number

$$s_\nu^{k+1} := \inf\{s \in \mathbb{R} : s < s_\nu^k + s^* \text{ and } d(\gamma_\nu(\sigma), x_k) < \varepsilon \text{ for } s < \sigma < s_\nu^k + s^*\}$$

is finite, and the sequence $s_\nu^k + s^* - s_\nu^{k+1}$ converges to infinity since for any $T > 0$ and $\sigma' \in [s^* - T, s^*]$ we have $d(\gamma_\nu(s_\nu^k + \sigma'), \gamma^k(\sigma')) < \frac{\varepsilon}{2}$ if ν is large enough. Consequently, a subsequence of $\gamma_\nu(\cdot + s_\nu^{k+1})$ converges to an orbit $\gamma^{k+1} \in \mathcal{M}(x_{k+1}, x_k)$ for some critical point x_{k+1}, which completes the induction step. The energy identity follows immediately from (2.3). □

Now we are ready to construct the *Novikov complex* for the form α. Choose an orientation $\langle x \rangle$ of E_x^u for each critical point of α. Let $\pi : \widetilde{M} \to M$ be a covering such that $\pi_1(\widetilde{M}) = \ker[\alpha]$, i.e., $\pi^*\alpha$ is exact: $\pi^*\alpha = df_\alpha$, where $f_\alpha : \widetilde{M} \to \mathbb{R}$ is a Morse function. Let $Z_k(f_\alpha) \subset \widetilde{M}$ denote the set of all critical points of f_α of index k, and define the graded $\mathbb{Z}$-module $C_* = C_*(M, \alpha, g)$, where

$$C_k \subset \prod_{\tilde{x} \in Z_k(f_\alpha)} \mathbb{Z}\langle \tilde{x} \rangle$$

consists of all $\xi = \{\xi_{\tilde{x}}\}_{\tilde{x} \in Z_k(f_\alpha)}$ such that

$$\#\{\tilde{x} \in Z_k(f_\alpha) : f_\alpha(\tilde{x}) > c, \xi_{\tilde{x}} \neq 0\} < \infty \quad \text{for all } c \in \mathbb{R}.$$

We also require that $\langle \tilde{x} \rangle = \pi^* \langle \pi(\tilde{x}) \rangle$. The gradient flow of f_α with respect to the metric $\tilde{g} = \pi^* g$ satisfies the Morse–Smale condition, and clearly

$$\mathcal{M}(x,y) \cong \bigcup_{\tilde{y} \in \pi^{-1}(y)} \mathcal{M}(\tilde{x}, \tilde{y})$$

for any critical points x, y of α and $\tilde{x} \in \pi^{-1}(x)$. Moreover, if $\tilde{\gamma} \in \mathcal{M}(\tilde{x}, \tilde{y})$, then

$$\ell^2(\pi \circ \tilde{\gamma}) = -\int_{\pi \circ \tilde{\gamma}} \alpha = -\int_{\tilde{\gamma}} df_\alpha = f_\alpha(\tilde{x}) - f_\alpha(\tilde{y}).$$

It follows from Proposition 2.2.1 that if $\operatorname{ind}(\tilde{x}) - \operatorname{ind}(\tilde{y}) = 1$, then the number of orbits connecting $\tilde{x}$ and $\tilde{y}$ is finite. To each of them we assign a number $n_{\tilde{\gamma}} \in \{-1, +1\}$ in the following way. Since the manifold $W^u(x)$ is contractible, every vector bundle over $W^u(x)$ is trivial and so $\langle x \rangle$ induces an orientation in the tangent bundle $TW^u(x)$. Similarly, $\langle y \rangle$ induces an orientation of $T_yW^u(y) \cong T_yM/T_yW^s(y)$ and hence an orientation of the normal bundle $TM/TW^s(y)$ of $W^s(y)$. For any point $p \in \mathcal{M}(x,y) \cong W^u(x) \cap W^s(y)$ the transversality condition implies the isomorphism

$$T_pW^u(x)/T_pW^u(x) \cap T_pW^s(y) \cong T_pM/T_pW^s(y),$$

and we choose an orientation of $T_p\mathcal{M}(x,y) = T_pW^u(x) \cap T_pW^s(y)$ so that this isomorphism is orientation preserving. Then the number $n_{\tilde{\gamma}}$ is chosen so that $d\pi(n_{\tilde{\gamma}}\dot{\tilde{\gamma}})$ is a positively oriented basis of $T_{\pi\circ\tilde{\gamma}}\mathcal{M}(x,y)$. Denote by $n(\tilde{x}, \tilde{y})$ the sum of $n_{\tilde{\gamma}}$ over all $[\tilde{\gamma}] \in \widehat{\mathcal{M}}(\tilde{x}, \tilde{y})$, and define the operator $\partial : C_* \to C_*$ by

$$(\partial \xi)_{\tilde{y}} = \sum_{\operatorname{ind}(\tilde{x}) = \operatorname{ind}(\tilde{y})+1} n(\tilde{x}, \tilde{y}) \xi_{\tilde{x}} \quad \text{for } \xi \in C_k.$$

This sum is finite, since $n(\tilde{x}, \tilde{y}) = 0$ for all $\tilde{x}$ with $f_\alpha(\tilde{y}) \geq f_\alpha(\tilde{x})$ and

$$\# \{\tilde{x} \in Z_k : f_\alpha(\tilde{y}) < f_\alpha(\tilde{x}),\, \xi_{\tilde{y}} \neq 0\} < \infty.$$

For a similar reason, $\partial \xi \in C_{k+1}$.

The $\mathbb{Z}$-modules C_k are in general infinite dimensional, since there may be an infinite number of lifts of a critical point of α. We take this into account by introducing an appropriate coefficient ring. Set $\Gamma = \pi_1(M)/\ker[\alpha]$, i.e., Γ is the group of deck automorphisms for $\widetilde{M}$, and let $\chi_\alpha : \Gamma \to \mathbb{R}$ be the homomorphism induced by $[\alpha] : \pi_1(M) \to \mathbb{R}$. Then Γ is necessarily isomorphic to $\mathbb{Z}^m$, and there is a $v \in \mathbb{R}^m$ with v_i rationally independent such that $\chi_\alpha(A) = v \cdot A$ for $A = (A_1, \ldots, A_m) \in \Gamma$. Define a subgroup $\Lambda_\alpha = \Lambda_\alpha(\mathbb{Z})$ of the product $\prod_{A\in\Gamma} \mathbb{Z}$ which consists of elements $\lambda = \{\lambda_A\}_{A\in\Gamma}$ satisfying

$$\# \{A \in \Gamma : \chi_\alpha(A) < c; \lambda_A \neq 0\} < \infty \quad \text{for all } c \in \mathbb{R}. \tag{2.4}$$

It is a ring with the multiplication given by

$$(\lambda * \mu)_A = \sum_{B\in\Gamma} \lambda_B \mu_{B^{-1}A}$$

and a principal ideal domain ([**HS92**]). It can be identified with the ring of formal power series $\lambda(t) = \sum_{A\in\Gamma} \lambda_A t_1^{A_1} \cdots t_m^{A_m}$ with λ satisfying (2.4). The groups C_k have a structure of a Λ_α-module defined in the similar manner:

$$(\lambda * \xi)_{\tilde{x}} = \sum_{A\in\Gamma} \lambda_A \xi_{A^{-1}\tilde{x}} \quad \text{for } \lambda \in \Lambda_\alpha,\ \xi \in C_*.$$

This is well defined since $f_\alpha(A^{-1}\tilde{x}) = f_\alpha(\tilde{x}) - \chi_\alpha(A)$, and clearly C_k is isomorphic to a free Λ_α-module generated by the critical points of α of index k. An isomorphism is given by picking an element $\tilde{x} \in \pi^{-1}(x)$ for each critical point x. Moreover, $\partial : C_* \to C_*$ is a Λ_α-homomorphism. Indeed,

$$\begin{aligned}(\partial\lambda * \xi)_{\tilde{y}} &= \sum_{\operatorname{ind}(\tilde{x}) = \operatorname{ind}(\tilde{y})+1} n(\tilde{x}, \tilde{y}) \sum_{A\in\Gamma} \lambda_A \xi_{A^{-1}\tilde{x}} \\ &= \sum_{A\in\Gamma} \lambda_A \sum_{\operatorname{ind}(\tilde{x}) = \operatorname{ind}(\tilde{y})+1} n(A^{-1}\tilde{x}, A^{-1}\tilde{y}) \xi_{A^{-1}\tilde{x}} \\ &= \sum_{A\in\Gamma} \lambda_A (\partial\xi)_{A^{-1}\tilde{y}} = (\lambda * \partial\xi)_{\tilde{y}},\end{aligned}$$

where the second equality follows since the gradient flow of f_α is Γ-equivariant and $n_{A\tilde{\gamma}} = n_{\tilde{\gamma}}$.

The ring Λ_α contains the group ring $\mathbb{Z}\Gamma$ of Γ as a subring. Hence we have the representation

$$\pi_1(M) \to \Gamma \to \mathbb{Z}\Gamma \to \operatorname{Hom}(\Lambda_\alpha),$$

which in turn defines a local system $\mathcal{L}_\alpha$ on M (see e.g. [**Whi78**]). The aim of this section is to prove the following result.

THEOREM 2.2.2 (Novikov homology, [**Nov81**], [**Nov82**], [**HS92**], [**LO93**]). *Let α be a closed nondegenerate form on a compact Riemannian manifold (M, g) such that the flow generated by the dual vector field X_α satisfies the transversality condition. Then $(C_*(M, \alpha, g), \partial^*)$ is a Λ_α-chain complex and*

$$H_*(C_*(M, \alpha_0, g_0); \Lambda_{\alpha_0}) \cong H_*(C_*(M, \alpha_1, g_1); \Lambda_{\alpha_1})$$

for any two forms in the same de Rham cohomology class. We call it Novikov homology and denote it by $HN_*(M, [\alpha]; \mathbb{Z})$, where $[\alpha] \in H^1_{DR}(M, \mathbb{R})$. It is isomorphic to the homology of the manifold with coefficients in the local system $\mathcal{L}_\alpha$.

Let μ_k denote the number of the critical points of α of index k, and β_k the rank of $H_k(M, \mathcal{L}_\alpha)$. The standard algebraic argument (see, e.g., [**Hir70**]) gives

COROLLARY 2.2.3. *There exists a polynomial Q with nonnegative integral coefficients such that*

$$\sum_{i=1}^{n} \mu_i t^i = \sum_{i=1}^{n} \beta_i t^i + (1+t)Q(t).$$

In particular, for $k \le n$

$$\sum_{i=1}^{k} (-1)^i \mu_i \ge \sum_{i=1}^{k} (-1)^i \beta_i.$$

2.3. Connecting orbits and glueing.

THEOREM 2.3.1. *Let (M, g) be a compact Riemannian manifold and α a nondegenerate closed form on M such that the flow generated by (2.1) satisfies the Morse–Smale condition for critical points. Let $C_*(M, g, \alpha)$ be the graded $\Lambda_\alpha(\mathbb{Z})$ module defined as in the previous section together with the degree -1 homomorphism ∂. Then $\partial^2 = 0 (\mathrm{mod}\ 2)$.*

PROOF. If we write down the formula for ∂^2:

$$(\partial^2\xi)_{\tilde{x}} = \sum_{\substack{\operatorname{ind}(ty)=k+1 \\ \operatorname{ind}(\tilde{z})=k}} n(\tilde{x},\tilde{y})n(\tilde{y},\tilde{z})\xi_{\tilde{z}}, \qquad \xi \in C_k,$$

we see that the condition $\partial^2 = 0(\text{mod } 2)$ means that the number of pairs of orbits $([\gamma^-],[\gamma^+]) \in \bigcup_{\operatorname{ind}(\tilde{y})=k} \widehat{\mathcal{M}}(\tilde{x},\tilde{y}) \times \widehat{\mathcal{M}}(\tilde{y},\tilde{z})$ is even for any two critical points $\tilde{x}$, $\tilde{z}$ of f_α with $\operatorname{ind}(\tilde{x}) = \operatorname{ind}(\tilde{z}) + 2 = k+1$. In fact, we will prove that such pairs of orbits occur in pairs. If $\operatorname{ind}(x) - \operatorname{ind}(z) = 2$, then the quotient manifold $\widehat{\mathcal{M}}(x,z) = \mathcal{M}(x,z)/\mathbb{R}$ is one-dimensional, i.e., consists of a number of circles and open intervals. Since the energy $\ell^2(\gamma)$ is constant on a connected component of $\mathcal{M}(x,z)$, each "end" of such an interval converges to a split trajectory $(\gamma^-,\gamma^+) \in \mathcal{M}(x,y) \times \mathcal{M}(y,z)$ (strictly speaking, if $[\gamma_\nu] \in \widehat{\mathcal{M}}$ does not converge in $\widehat{\mathcal{M}}$, then γ_ν converges to (γ^-,γ^+)).

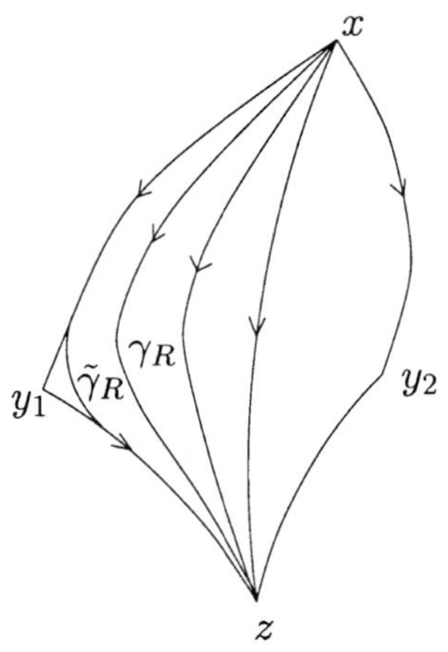

FIGURE 1

Hence in order to prove that $\partial^2 = 0$ it is enough to show that this is a one-to-one correspondence. For this sake we employ Floer's concept of glueing orbits, i.e., a family of orbits $\gamma_R \in \mathcal{M}(x,z)$, with $R \in \mathbb{R}$ and $R > R_0$, converging to (γ^-,γ^+). This family is unique, i.e., any orbit connecting x and z sufficiently close to γ^- and γ^+ belongs to the family γ_R.

We introduce the setup for the construction of γ_R. For any critical points x^- and x^+ of α we define the space $\mathcal{P} = \mathcal{P}(x^-,x^+)$:

$$\begin{aligned}\mathcal{P}(x^-,x^+) = \Big\{\gamma : \mathbb{R} \to M \text{ such that } & \gamma \in W^{1,2}_{\text{loc}}(M), \\ & \exists T > 0, \ \ \exists \xi_- \in W^{1,2}((-\infty,-T];T_{x^-}M), \\ & \exists \xi_+ \in W^{1,2}([T,\infty);T_{x^+}M) \\ & \text{such that } \gamma(s) = \exp_{x^\pm} \xi_\pm(s) \text{ for } \pm s \geq T\Big\}.\end{aligned}$$

Clearly any solution to (2.1) with limit conditions (2.2) belongs to $\mathcal{P}$. For $\gamma \in \mathcal{P}$ we will write $\mathcal{F}(\gamma) := \dot{\gamma} + X_\alpha(\gamma)$, i.e., $\mathcal{M}(x^-,x^+) = \mathcal{F}^{-1}(0)$. Define Banach spaces of vector fields along $\gamma \in \mathcal{P}(x^-, x^+)$:

$$L^2(\gamma) = \left\{\xi : \mathbb{R} \to TM : \xi(s) \in T_{\gamma(s)}M \text{ and } \int_{-\infty}^{+\infty} |\xi|^2 < \infty\right\},$$
$$W^{1,2}(\gamma) = \left\{\xi \in L^2(\gamma) : \nabla\xi \in L^2(\gamma)\right\},$$
$$W^{2,2}(\gamma) = \left\{\xi \in L^2(\gamma) : \nabla\xi \in W^{1,2}(\gamma)\right\},$$

where ∇ denotes the Levi–Civita connection on M. We define the operator $D_\gamma : W^{1,2}(\gamma) \to L^2(\gamma)$ by the formula $D_\gamma\xi = \nabla\xi + \nabla_\xi X_\alpha$. Let $D^* : W^{1,2}(\gamma) \to L^2(\gamma)$ be the formal adjoint of D. Since α is closed, ∇X_α is a symmetric tensor:

$$\langle \nabla_\xi X_\alpha, \zeta\rangle = \langle \xi, \nabla_\zeta X_\alpha\rangle \quad \text{for } \xi, \zeta \in T_xM,\ x \in M,$$

and so $D_\gamma^*\xi = -\nabla\xi + \nabla_\xi X_\alpha$.

THEOREM 2.3.2 ([**Sal90**]). *If x^- and x^+ are nondegenerate critical points of a closed form α and $\gamma : \mathbb{R} \to M$ is a smooth path satisfying (1.3), then D_γ is a Fredholm operator and*

$$\operatorname{ind} D_\gamma = \operatorname{ind}(x^-) - \operatorname{ind}(x^+).$$

If $\mathcal{F}(\gamma) = 0$, then D_γ is onto if and only if $W^u(x^-)$ and $W^s(x^+)$ intersect transversally along γ.

The Fredholm property allows us to formulate an "existence and uniqueness" result, a version of the implicit function theorem which is the main tool for the construction of glueing orbits.

THEOREM 2.3.3. *For any $c > 0$ there are positive constants c_1, ε_0 such that the following hold.*

a) *If $0 < \varepsilon < \varepsilon_0$ and $\gamma \in \mathcal{P}$ satisfies the conditions*

$$\|\dot\gamma\|_{L^2} \le c,$$
$$c\|D_\gamma^*\eta\|_{L^2} \ge \|\eta\|_{W^{1,2}} \quad \text{for any } \eta \in W^{1,2}(\gamma),$$
$$\|\mathcal{F}(\gamma)\|_{L^2} < \varepsilon,$$

then there is a unique vector field $\zeta \in D_\gamma^*(W^{2,2}(\gamma))$ such that $\|\zeta\|_{W^{1,2}} \le c_1\varepsilon$, the path $\gamma_0 = \exp_\gamma \zeta$ satisfies $\mathcal{F}(\gamma_0) = 0$, and

$$4c\|D_{\gamma_0}^*\eta\|_{L^2} \ge \|\eta\|_{W^{1,2}} \quad \text{for any } \eta \in W^{1,2}(\gamma_0).$$

b) *For γ as above, if $\mathcal{F}(\gamma) = 0$, then there is a C^∞-function ϕ defined on the neighborhood of zero in $\ker D_\gamma$, ϕ: $\ker D_\gamma \to D_\gamma^*(W^{1,2}(\gamma))$, such that for any $\zeta \in W^{1,2}(\gamma)$ we have $\zeta = \xi + \phi(\xi)$ for some ξ in the domain of ϕ if and only if $\mathcal{F}(\exp_\gamma \zeta) = 0$ and $\|\zeta\|_{W^{1,2}} < \varepsilon_0$.*

The proof of this theorem, based on Newton's method, is given in the next section.

Let x, y, z be critical points of α, let $\gamma^- \in \mathcal{M}(x, y), \gamma^+ \in \mathcal{M}(y, z)$, and let $\beta : \mathbb{R} \to [0, 1]$ be a cut-off function:

$$\beta(s) := \begin{cases} 0 & \text{for } s \le -1, \\ 1 & \text{for } s \ge 1. \end{cases}$$

For a positive $R \in \mathbb{R}$ define an approximate solution to (2.1):

$$\tilde{\gamma}_R(s) := \begin{cases} \gamma^-(s+R) & \text{for } s \leq -1, \\ \exp_y(\beta(s)\xi^+(s-R) + (1-\beta(s))\xi^-(s+R)) & \text{for } s \in [-1,1], \\ \gamma^+(s-R) & \text{for } s \geq 1, \end{cases}$$

where $\gamma^\pm(s) = \exp_y \xi^\pm(s)$ for $|s| > R-1$ and R sufficiently large.

PROPOSITION 2.3.4. *There are positive constants k, C, R_0 such that for $R > R_0$ the family defined by $\tilde{\gamma}_R$ satisfies the estimates*

$$\begin{aligned} \|\dot{\tilde{\gamma}}_R\|_{L^2} &\leq C, \\ C\|D^*_{\tilde{\gamma}_R}\eta\|_{L^2} &\geq \|\eta\|_{W^{1,2}} \quad \textit{for } \eta \in W^{1,2}(\gamma), \\ \|\mathcal{F}(\tilde{\gamma}_R)\|_{L^2} &\leq Ce^{-kR}. \end{aligned}$$

Thus $\tilde{\gamma}_R$ satisfies the assumptions of the part a) of Theorem 2.3.3 and leads to a family γ_R of glueing orbits. Then the uniform estimates on the norm of $D^*_{\gamma_R}$ yield the uniqueness result. The properties of γ_R are summarized below.

PROPOSITION 2.3.5. *The family γ_R of solutions of (2.1) converges to (γ^-, γ^+). Moreover, for any sequence $\{\gamma'_j\} \subset \mathcal{M}(x,z)$ converging to the pair (γ^-, γ^+) a curve γ'_j belongs to the same component of $\mathcal{M}$ as the family γ_R for j sufficiently large.*

In other words there exists precisely one end of a component of $\widehat{\mathcal{M}}(x,z)$ which converges to (γ^-, γ^+). This proves Theorem 2.3.1. □

2.3.1. *Proof of Theorem* 2.3.3. In the following we will — abusing notation — denote by K various constants depending only on the Riemannian metric and the function f.

Let ε_0 be less than the injectivity radius for the metric g. For $\gamma \in \mathcal{P}(x^-, x^+)$ define

$$U_\varepsilon(\gamma) := \{\xi \in W^{1,2}(\gamma) : \|\xi\|_{L^\infty} < \varepsilon\}.$$

Since $\|\xi\|_{L^\infty} \leq \|\xi\|_{W^{1,2}}$, the set $U_\varepsilon(\gamma)$ is open in $W^{1,2}(\gamma)$. We also denote $TM(\varepsilon_0) = \{v \in TM : |v| < \varepsilon\}$. The set $\mathcal{P}(x^-, x^+)$ is a Banach manifold modeled on $W^{1,2}(\gamma)$, where $\gamma \in C^\infty(M) \cap \mathcal{P}$ with the charts induced by the exponential map. The proof of this fact is essentially given in [**Kli78**], and we summarize it below for the sake of completeness. We need the following result.

LEMMA 2.3.6. *There is a positive constant K such that, for any interval $J \subset \mathbb{R}$ of length greater than 1, a path $\gamma_0 \in W^{1,2}_{\mathrm{loc}}(J; M)$, and a vector field $\xi_0 \in W^{1,2}(\gamma_0)$ with $\|\xi_0\|_{W^{1,2}} \leq \frac{\varepsilon_0}{2}$,*

$$\|\dot{\gamma}_1\|_{L^2} \leq \|\dot{\gamma}_0\|_{L^2} + K(1 + \|\dot{\gamma}_0\|_{L^2})\|\xi_0\|_{W^{1,2}}, \tag{2.5}$$

where $\gamma_1 = \exp_{\gamma_0} \xi_0$, and the following holds.

If $\xi_1 \in W^{1,2}(\exp_{\gamma_0} \xi_0)$, with $\|\xi_1\|_{W^{1,2}} \leq \frac{\varepsilon_0}{2}$, then there is a vector field $\xi_2 \in W^{1,2}(\gamma_0)$ such that

$$\exp_{\gamma_0} \xi_2 = \exp_{\exp_{\gamma_0} \xi_0} \xi_1 \tag{2.6}$$

and

$$\|\xi_2 - \xi_0\|_{W^{1,2}} \leq K(1 + \|\dot{\gamma}_0\|_{L^2} + \|\xi_0\|_{W^{1,2}})\|\xi_1\|_{W^{1,2}}.$$

Conversely, if $\xi_2 \in W^{1,2}(\gamma_0)$ *is such that* $\|\xi_2 - \xi_0\|_{W^{1,2}} \le \frac{\varepsilon_0}{2}$, *then there is a* $\xi_1 \in W^{1,2}(\exp_{\gamma_0}\xi_0)$ *satisfying (2.6) and*

$$\|\xi_1\|_{W^{1,2}} \le K(1 + \|\dot\gamma_0\|_{L^2} + \|\xi_0\|_{W^{1,2}})\|\xi_2 - \xi_0\|_{W^{1,2}}.$$

PROOF. We denote by $E(x,\xi)$ the representation of $\exp_x \xi$ in local coordinates x^i, ξ^i. Then

$$\dot\gamma_1 = \dot\gamma_0 + (E_x(\gamma_0,\xi_0) - \mathrm{id})\dot\gamma_0 + E_\xi(\gamma_0,\xi_0)\dot\xi,$$
$$E_x(\gamma_0,\xi_0) - \mathrm{id} = \int_0^1 E_{x\xi}(\gamma_0, t\xi_0)\xi_0\,dt,$$

and since $\nabla\xi = \dot\xi + \Gamma\xi\dot\gamma$, we have

$$|\dot\gamma_1| \le |\dot\gamma_0| + K(|\xi||\dot\gamma_0| + |\nabla\xi|),$$

where K bounds the derivatives of the exponential map and the Christoffel symbols Γ on $TM(\varepsilon_0)$. This yields (2.5).

We prove the first of the remaining inequalities. The vector field ξ_2 exists if the L^∞ norms of ξ_0 and ξ_1 are small, and clearly $\xi_2 \in W^{1,2}(\gamma_0)$. In normal coordinates at γ_0 we have $\xi_2 = E(\xi_0,\xi_1)$ and

$$\xi_2 - \xi_0 = E(\xi_0,\xi_1) - E(\xi_0,0) = \int_0^1 E_\xi(\xi_0, t\xi_1)\xi_1\,dt,$$
$$|\xi_2 - \xi_0| \le K|\xi_1|.$$

This implies the L^2 inequality. Differentiating (2.6) in local coordinates, we obtain

$$\begin{aligned} E_\xi(\gamma_0,\xi_2)(\dot\xi_2 - \dot\xi_0) &= E_\xi(E(\gamma_0,\xi_0),\xi_1)\dot\xi_1 \\ &\quad + (E_x(E(\gamma_0,\xi_0),\xi_1) - \mathrm{id})(E_\xi(\gamma_0,\xi_0)\dot\xi_0 + E_x(\gamma_0,\xi_0)\dot\gamma_0) \\ &\quad + (E_x(\gamma_0,\xi_2) - (E_x(\gamma_0,\xi_0))\dot\gamma_0 + (E_\xi(\gamma_0,\xi_0) - E_\xi(\gamma_0,\xi_2))\dot\xi_0. \end{aligned}$$

As E_ξ is invertible and $\|E_\xi^{-1}\|$ is bounded on $TM(\varepsilon_0)$, it is enough to estimate the right hand side in terms of ξ_1. We have

$$E_\xi(\gamma_0,\xi_0) - E_\xi(\gamma_0,\xi_2) = \int_0^1 E_{\xi\xi}(\gamma, \xi_0 + t(\xi_2 - \xi_0))(\xi_2 - \xi_0)\,dt,$$
$$\|E_\xi(\gamma_0,\xi_0) - E_\xi(\gamma_0,\xi_2)\| \le K|\xi_2 - \xi_0|.$$

Similarly

$$\|E_x(\gamma_0,\xi_2) - E_x(\gamma_0,\xi_0)\| \le K|\xi_2 - \xi_0|,$$
$$\|E_x(E(\gamma_0,\xi_0),\xi_1) - \mathrm{id}\| \le K|\xi_1|.$$

Putting these estimates together, we obtain

$$|\nabla\xi_2 - \nabla\xi_0| \le |\dot\xi_2 - \dot\xi_0| + |\Gamma\dot\gamma_0(\xi_2 - \xi_0)| \le K(|\nabla\xi_1| + |\xi_1|(|\dot\gamma_0| + |\xi_0| + |\nabla\xi_0|)),$$

where Γ are the Christoffel symbols. This yields the result. The second inequality is proved in a similar manner. ∎

From the lemma it follows that we can define the exponential map. For $\gamma \in \mathcal{C}^\infty \cap \mathcal{P}$ we have

$$\begin{aligned} \exp_\gamma &: U_{\varepsilon_0}(\gamma) \to \mathcal{P}(x^-, x^+), \\ \exp_\gamma \xi(s) &= \exp_{\gamma(s)} \xi(s). \end{aligned}$$

Indeed, if $\eta \in U_{\varepsilon_0}(\gamma)$ and $\gamma(s) = \exp_{x^\pm} \xi_\pm(s)$, then Lemma 2.3.6 guarantees the existence of $\eta_\pm \in W^{1,2}(J, T_{x^\pm}M)$ with $\exp_{x^\pm} \eta_\pm = \exp_\gamma \eta$. Also, if $\gamma_0, \gamma_1 \in \mathcal{C}^\infty \cap \mathcal{P}$, then

$$\exp_{\gamma^0}^{-1}(\exp_{\gamma_0}(U_{\varepsilon_0}(\gamma_0)) \cap \exp_{\gamma_1}(U_{\varepsilon_0}(\gamma_1)))$$

is open in $U_{\varepsilon_0}(\gamma_0)$. The transition map can be described as follows. Let U be an open subset of $\gamma^* TM$ defined by

$$U(s) = U \cap T_{\gamma_0(s)} M = \exp_{\gamma_0(s)}^{-1}(\exp_{\gamma_1(s)} T_{\gamma_1(s)} M(\varepsilon_0)).$$

Thus we have a bundle map $\Psi : U \to \gamma_1^* TM$, $\Psi = \exp_{\gamma_1}^{-1} \circ \exp_{\gamma_1}$, and the transition map is given by $\xi \mapsto \Psi \circ \xi$.

PROPOSITION 2.3.7 (cf. [**Kli78**]). *Let $\pi_0 : E_0 \to \mathbb{R}$, $\pi_1 : E_1 \to \mathbb{R}$ be two Riemannian vector bundles over $\mathbb{R}$, and $\Psi : U \to E_1$ a smooth bundle map defined in an open set $U \subset E_0$ with bounded derivatives in the vertical direction. Then the assignment $\xi \mapsto \hat{\Psi}(\xi) := \Psi \circ \xi$ defines a $\mathcal{C}^1$ map*

$$\hat{\Psi} : \hat{U} \to W^{1,2}(E_1)$$

on any open set $\hat{U} \subset \{\xi \in W^{1,2}(E_0) : \xi(s) \in U\}$.

PROOF. Let $D_v^k \Psi(x)$ denote the k-th derivative of $\Psi|_{\pi_0^{-1}(\pi(x))}$ at x. Then for $\xi, \eta \in \hat{U}$ we have $\nabla \Psi \circ \xi = D_v(\xi) \nabla \xi$ and

$$\begin{aligned} |\Psi \circ \xi - \Psi \circ \eta| &\le \sup_U |D_v \Psi| \cdot |\xi - \eta|, \\ |\nabla \Psi \circ \xi - \nabla \Psi \circ \eta| &\le |D_v(\xi) - D_v(\eta)| \cdot |\nabla \xi| + |D_v(\eta)| \cdot |\nabla \xi - \nabla \eta| \\ &\le \sup_U |D_v^2 \Psi| \cdot |\xi - \eta| \cdot |\nabla \xi| + \sup_U |D_v \Psi| \cdot |\nabla \xi - \nabla \eta|, \end{aligned}$$

which shows that $\hat{\Psi}$ is continuous. Similarly the bundle map $D_v \Psi : U \to L(E_0, E_1)$ induces a continuous map

$$\widehat{D_v \Psi} : \hat{U} \to W^{1,2}(L(E_0, E_1)).$$

If $A \in W^{1,2}(L(E_0, E_1))$, $\xi \in W^{1,2}(E_0)$ and $A\xi(s) = A(s)\xi(s)$, then

$$\begin{aligned} \|A\xi\|_{L^2} &\le \|A\|_{L^\infty} \cdot \|\xi\|_{L^2}, \\ \|\nabla(A\xi)\|_{L^2} &= \|\nabla A \xi + A \nabla \xi\|_{L^2} \le \|\nabla A\|_{L^2} \cdot \|\xi\|_{L^\infty} + \|A\|_{L^\infty} \cdot \|\nabla \xi\|_{L^2}, \end{aligned}$$

so that we have a linear continuous map

$$\# : W^{1,2}(L(E_0, E_1)) \to L(W^{1,2}(E_0), W^{1,2}(E_1))$$

with $\|\#\| \le 2$. We define the derivative of $\hat{\Psi}$ as $D\hat{\Psi} = \# \circ \widehat{D_v \Psi}$. By the Taylor formula,

$$\Psi(\xi(s)) - \Psi(\eta(s)) - D_v \Psi(\eta(s))(\xi(s) - \eta(s)) = r(\xi(s), \eta(s))(\xi(s) - \eta(s)),$$

where $r : U \times U \to L(E_0, E_1)$ is a bundle map with bounded derivatives. Hence

$$\hat{\Psi}(\xi) - \hat{\Psi}(\eta) - D\hat{\Psi}(\eta)(\xi - \eta) = \#\hat{r}(\xi, \eta)((\xi - \eta)$$

and $\|\#\hat{r}(\xi, \eta)\|_{W^{1,2}} \to 0$ as $\|\xi - \eta\|_{W^{1,2}} \to 0$. □

Thus the maps $\exp_\gamma$ introduce a differentiable structure on $\mathcal{P}$. Note that this structure is compatible for different metrics if they are C_1 close. Indeed, this reduces to the observation that for such metrics the $W^{1,2}$ norms are equivalent.

If $\gamma \in \mathcal{P}(x^-, x^+)$, then for large $|s|$ we have $\gamma(s) = \exp_{x^\pm} \xi_\pm(s)$ with $|\xi_\pm(s)| < \varepsilon_0$; and since X_α is locally Lipschitz, we have

$$|\mathcal{F}(\gamma)(s)| \le \|\mathrm{dexp}_{x^\pm}\xi_\pm(s)|(|\dot{\xi}_\pm| + |X_\alpha(\xi_\pm) - X_\alpha(0)|) \le K(|\dot{\xi}_\pm| + |\xi_\pm|)$$

and hence $\mathcal{F}(\gamma) \in L^2(\gamma)$. For $\xi \in B_{\varepsilon_0}(\gamma)$ denote by $\tau^\xi(t_0, t_1)$ the parallel transport along $\exp_\gamma t\xi$ from $\exp_\gamma t_0\xi$ to $\exp_\gamma t_1\xi$ and define the function $\mathcal{F}_{\gamma_0} : W^{1,2}(\gamma_0) \to L^2(\gamma_0)$ by the formula

$$\mathcal{F}_{\gamma_0}(\xi) = \tau^\xi(1,0)\mathcal{F}(\exp_{\gamma_0} \xi).$$

PROPOSITION 2.3.8. *The function $\mathcal{F}_{\gamma_0}$ is continuously differentiable in $B_{\varepsilon_0}(\gamma_0)$ with differential $d\mathcal{F}_{\gamma_0}(\xi) := D_\xi : W^{1,2}(\gamma_0) \to L^2(\gamma_0)$,*

$$D_\xi\eta := \tau^\xi(1,0)D_{\gamma_1} \circ \mathrm{dexp}(\xi)\eta + A_\xi(\eta, \mathcal{F}_{\gamma_0}(\xi)),$$

where $\gamma_1 = \exp_{\gamma_0} \xi$ and A is a bounded bilinear form on $TM(\varepsilon_0)$ and $A_0 = 0$. More specifically, there is a constant K, independent of γ_0, such that for every $\xi \in B_{\varepsilon_0}(\gamma_0)$ and $\eta \in W^{1,2}(\gamma_0)$

$$\|\mathcal{F}(\xi + \eta) - \mathcal{F}_{\gamma_0}(\xi) - D_\xi\eta\|_{L^2} \le K\|\eta\|_{W^{1,2}}\|\eta\|_{L^\infty}(\|\dot{\gamma}_0\|_{L^2} + 1)$$

and moreover

$$\|D_\xi\eta - D_0\eta\|_{L^2} \le K(\|\xi\|_{W^{1,2}} + \|\mathcal{F}_{\gamma_0}(\xi)\|_{L^2})\,\|\eta\|_{W^{1,2}}.$$

PROOF. Let $\xi, \eta \in W^{1,2}(\gamma_0)$ and set

$$u(r,t) = u(r,t,s) = \exp_{\gamma_0(s)} r(\xi(s) + t\eta(s)), \quad \gamma_r = u(r,0) \quad \text{and } \xi_t = \xi + t\eta.$$

Denote by τ^η the parallel transport along $u(1,t)$, and consider the difference

$$\begin{aligned} \mathcal{F}_{\gamma_0}(\xi + \eta) - \mathcal{F}_{\gamma_0}(\xi) &= \tau^{\xi_1}(1,0)\mathcal{F}(u(1,1)) - \tau^\xi(1,0)\mathcal{F}(u(1,0)) \\ &= \left(\tau^{\xi_1}(1,0) - \tau^\xi(1,0)\tau^\eta(1,0)\right)\mathcal{F}(u(1,1)) \\ &\quad - \tau^\xi(1,0)(\tau^\eta(1,0)\mathcal{F}(u(1,1)) - \mathcal{F}(u(1,0))). \end{aligned} \tag{2.7}$$

The first step is to obtain the estimate

(2.8)
$$\left\|\tau^\eta(1,0)\mathcal{F}(u(1,1)) - \mathcal{F}(u(1,0)) - D_{\gamma_1}\frac{\partial u}{\partial t}(1,0)\right\|_{L^2} \le K\|\eta\|_{W^{1,2}}\|\eta\|_{L^\infty}(\|\dot{\gamma}_0\|_{L^2} + 1).$$

Denote

$$\Phi_s(t) = \tau^\eta(t,0)\left(\frac{\partial u}{\partial s}(1,t) + X_\alpha(u)\right),$$

$$\Phi_s(0) = \mathcal{F}(\gamma)(s), \qquad \Phi_s(1) = \tau^\xi(1,0)\mathcal{F}(u(1,1))(s).$$

Since the derivative of the parallel transport of a vector field along a curve is the covariant derivative of the vector field,

$$\begin{aligned}\Phi'(t_0) &= \frac{d}{dt}|_{t=t_0}\tau^\eta(t_0,0)\tau^\eta(t,t_0)\left(\frac{\partial u}{\partial s}+X_\alpha(u)\right)\\ &= \tau^\eta(t_0,0)\nabla_t\left(\frac{\partial u}{\partial s}+X_\alpha(u)\right)=\tau^\eta(t_0,0)\left(\nabla_s\frac{\partial u}{\partial t}+\nabla_t X_\alpha(u)\right),\\ \Phi'(0) &= D_{\gamma_1}\frac{\partial u}{\partial t}(1,0).\end{aligned}$$

Here we use the fact that $\left[\frac{\partial u}{\partial t},\frac{\partial u}{\partial s}\right]=u^*\left[\frac{\partial}{\partial t},\frac{\partial}{\partial s}\right]=0$ and so $\nabla_s\frac{\partial u}{\partial t}=\nabla_t\frac{\partial u}{\partial s}$. To prove (2.8) we estimate the second derivative of Φ:

$$\Phi''(t)=\tau^\eta(t,0)\nabla_t\left(\nabla_s\frac{\partial u}{\partial t}+\nabla_t X_\alpha(u)\right).$$

We have

$$\begin{aligned}\nabla_t\nabla_s\frac{\partial u}{\partial t}-\nabla_s\nabla_t\frac{\partial u}{\partial t} &= R\left(\frac{\partial u}{\partial t},\frac{\partial u}{\partial s}\right)\frac{\partial u}{\partial t},\\ \nabla_t(\nabla_t X_\alpha(u)) &= \nabla_{\nabla_t\frac{\partial u}{\partial t}}X_\alpha(u)+\nabla_t(\nabla X_\alpha(u))\frac{\partial u}{\partial t}.\end{aligned}$$

Since both R and $\nabla(X_\alpha)$ are smooth sections of the appropriate tensor bundles and parallel transport is an isometry, it follows that

$$\left|\tau^\eta(t,0)R\left(\frac{\partial u}{\partial t},\frac{\partial u}{\partial s}\right)\frac{\partial u}{\partial t}\right|\le\|R\|_{L^\infty}\left|\frac{\partial u}{\partial t}\right|^2\left|\frac{\partial u}{\partial s}\right|, \tag{2.9}$$

$$\left|\tau^\eta(t,0)\nabla_t(\nabla X_\alpha(u))\frac{\partial u}{\partial t}\right|\le\|\nabla(X_\alpha)\|_{L^\infty}\left|\frac{\partial u}{\partial t}\right|^2, \tag{2.10}$$

and moreover

$$\left|\frac{\partial u}{\partial t}\right|\le\|\mathrm{dexp}(\xi+t\eta)\|_{L^\infty}|\eta|.$$

We estimate $\left|\frac{\partial u}{\partial s}\right|$ in local coordinates:

$$\left|\frac{\partial u}{\partial s}\right|=\left|\frac{d}{ds}tE(\gamma_1,\eta)\right|\le|E_x(\gamma_1,\eta)\dot\gamma_1|+|E_\xi(\gamma_1,\eta)\dot\eta|\le K(|\dot\gamma_1|+|\nabla\eta|+|\dot\gamma_1||\eta|)$$

since $|\dot\eta|\le|\nabla\eta|+|\dot\gamma_1||\eta||\Gamma|$, where Γ denote Christoffel symbols. Similarly

$$|\dot\gamma_1|\le K(|\dot\gamma_0|+|\nabla\xi|+|\dot\gamma_0||\xi|).$$

It remains to estimate $\nabla_t\frac{\partial u}{\partial t}(1,t)$ and $\nabla_s\nabla_t\frac{\partial u}{\partial t}(1,t)$. We have

$$\nabla_t\frac{\partial u}{\partial t}(1,t)=E_{\xi^i\xi^j}(\gamma,\xi+t\eta)\eta^i\eta^j+E_{\xi^i}(\gamma,\xi+t\eta)E_{\xi^j}(\gamma,\xi+t\eta)\eta^i\eta^j\Gamma_{ij}.$$

It follows that $\left|\nabla_t\frac{\partial u}{\partial t}\right|\le K|\eta|^2$. Similarly,

$$\left|\nabla_s\nabla_t\frac{\partial u}{\partial t}\right|\le K(|\eta|^2+(|\nabla\eta|+|\dot\gamma_0||\eta|)|\eta|),$$

where K is a constant bounding Γ and its derivatives and the derivatives of the exponential map on $TM(\varepsilon_0)$. Note that $\nabla_t \frac{\partial u}{\partial t}$ and $\nabla_s \nabla_t \frac{\partial u}{\partial t}$ vanish if $\xi = 0$. Putting all estimates together and assuming that $\varepsilon_0 < 1$, we get

$$\begin{aligned}|\Phi(1) - \Phi(0) - \Phi'(0)| &\leq K(|\eta|^2(|\dot\gamma_1| + |\nabla\eta| + |\dot\gamma_1||\eta|) + (1 + |\gamma_0|)|\eta|^2 + |\nabla\eta||\eta|) \\ &\leq K(|\eta|^2(|\dot\gamma_0| + 1) + |\nabla\eta||\eta|),\end{aligned}$$

which implies (2.8).

To deal with the first term of (2.7), let $v(t) = \tau^\eta(1,t)\mathcal{F}(u(1,1))$ and $w(r,t) = \tau^{\xi_t}(1,r)v(t)$. Thus we are to estimate the difference

$$w(0,1) - w(0,0) = \nabla_t w(0,0) + \int_0^1 (1-t)\tau^\eta(t,0)\nabla_t\nabla_t w(0,t)\,dt.$$

The second term is quadratic in η by the same argument as above applied to $u_1(\theta,t)$ such that $w(0,t) = \frac{\partial u_1}{\partial\theta}(0,t)$ and $u_1(0,t) = u(0,t)$. On the other hand, since $\nabla_r w = 0$ and $\nabla_t w(1,t) = 0$, it follows that

$$\begin{aligned}\nabla_t w(0,0) &= \nabla_t w(0,0) - \tau^\xi(1,0)\nabla_t w(1,0) = -\int_0^1 \tau^\xi(r,0)\nabla_r\nabla_t w(r,0)\,dr \\ &= -\int_0^1 \tau^\xi(r,0) R\left(\frac{\partial u}{\partial r}, \frac{\partial u}{\partial t}\right)\tau^\xi(1,r)v(0)\,dr \\ &= -\int_0^1 \tau^\xi(r,0)R(\operatorname{dexp}(r\xi)\xi, \operatorname{dexp}(r\xi)\eta)\tau^\xi(1,r)v(0)\,dr \\ &= A_\xi(\eta, \tau^\xi(1,0)v(0))\end{aligned}$$

and A is a bounded bilinear form on $TM(\varepsilon_0)$ (more precisely, if $\tau : TM \to M$ is the projection then A is a form on $\tau^*TM|_{TM(\varepsilon_0)}$). Summarizing, we have

$$\begin{aligned}&\left\|(\tau^{\xi_1}(1,0) - \tau^\xi(1,0)\tau^\eta(1,0))\mathcal{F}(u(1,1)) - A_\xi(\eta, \mathcal{F}_{\gamma_0}(\xi))\right\|_{L^2} \\ &\quad\leq \|A\|_{L^\infty}\|\eta\|_{L^\infty}\|\tau^\eta(1,0)\mathcal{F}(u(1,1)) - \mathcal{F}(u(1,0))\|_{L^2} \\ &\qquad + K\|\eta\|_{W^{1,2}}\|\eta\|_{L^\infty}(\|\dot\gamma_0\|_{L^2} + 1),\end{aligned} \tag{2.11}$$

which together with (2.8) implies that the linear operator

$$D_\xi : W^{1,2}(\gamma_0) \ni \eta \mapsto \tau^\xi(1,0)D_{\gamma_1} \circ \operatorname{dexp}(\xi)\eta + A_\xi(\eta, \mathcal{F}_{\gamma_0}(\xi)) \in L^2(\gamma_0)$$

is the differential of $\mathcal{F}_{\gamma_0}$ at the point ξ. Moreover, the quadratic estimates follow since in view of (2.8)

$$\|\tau^\eta(1,0)\mathcal{F}_{\gamma_0}(\xi+\eta) - \mathcal{F}_{\gamma_0}(\xi))\|_{L^2} \leq \|D_{\gamma_1}\frac{\partial u}{\partial t}(1,0)\|_{L^2} + K\|\eta\|^2_{W^{1,2}}(\|\dot\gamma_0\|_{L^2} + 1)$$

and the norm of $D_{\gamma_1} : W^{1,2}(\gamma_1) \to L^2(\gamma_1)$ is bounded by a universal constant. To estimate $D_\xi - D_0$, define $w_1(r) = D_{\gamma_r}\frac{\partial u}{\partial t}(r,0)$. Thus $w_1(0) = 0$ and $D_\xi\eta = \tau^\xi(1,0)w_1(1) + A_\xi(\eta, \mathcal{F}_{\gamma_0}(\xi))$, and

$$\nabla_r w_1 = \nabla_s\nabla_r\frac{\partial u}{\partial t} + R\left(\frac{\partial u}{\partial r}, \frac{\partial u}{\partial s}\right)\frac{\partial u}{\partial t} + \nabla_{\nabla_r\frac{\partial u}{\partial t}}X_\alpha + \nabla_r(\nabla X_\alpha)\frac{\partial u}{\partial t}.$$

On the other hand $\frac{\partial u}{\partial t}|_{r=0} = 0$ and $\nabla_r \frac{\partial u}{\partial t}|_{r=0} = \eta$, so $\nabla_r w_1(0) = D_0\eta$. We can also apply the same kind of argument as previously to obtain the estimate

$$|\nabla_r \nabla_r w_1| \leq K(|\xi| \cdot |\eta| + |\nabla\xi| \cdot |\eta| + |\xi| \cdot |\nabla\eta|).$$

Therefore

(2.12)

$$\begin{aligned}\|D_\xi \eta - D_0\eta\|_{L^2} &\leq \|\tau^\xi(1,0)w_1(1) - w_1(0) - \nabla_r w_1(0)\|_{L^2} + \|A_\xi(\eta, \mathcal{F}_{\gamma_0}(\xi))\|_{L^2} \\ &\leq K(\|\xi\|_{W^{1,2}} + \|\mathcal{F}_{\gamma_0}(\xi)\|_{L^2})\|\eta\|_{W^{1,2}}\end{aligned}$$

for $\xi, \eta \in B_{\varepsilon_0}(\gamma_0)$. A similar argument proves the continuity of the derivative in $B_{\varepsilon_0}(\gamma_0)$. □

If the flow satisfies the Morse–Smale condition, then Proposition 2.3.9 together with Theorem 2.3.2 implies that for every $\gamma \in \mathcal{M}(x^-, x^+)$ there is the implicit function

$$\phi_\gamma : \ker D_\gamma \to D^*(W^{2,2}(\gamma))$$

such that $\mathcal{F}(\exp_\gamma(\xi + \phi_\gamma(\xi))) = 0$ for ξ in a neighbourhood of 0 in $\ker D_\gamma$ which provides the manifold structure locally for $\mathcal{M}(x^-, x^+)$ near γ. The evaluation map

$$\mathrm{ev} : \mathcal{M}(x^-, x^+) \ni \gamma \mapsto \gamma(0) \in M$$

is an embedding with respect to this structure. Clearly, the evaluation $W^{1,2}(\gamma) \ni \xi \mapsto \xi(0) \in T_{\gamma(0)}M$ is a continuous linear map, and so $\psi_\gamma : \ker D_\gamma \to M$ defined by

$$\psi_\gamma(\xi) = \mathrm{ev} \circ \exp_\gamma(\xi + \phi_\gamma(\xi)) = \exp_{\gamma(0)} \xi(0) + \phi_\gamma(\xi)(0)$$

is differentiable. Moreover, $d\psi_\gamma(0)\xi = \xi(0)$ as $d\phi_\gamma(0) = 0$. Since $D_\gamma\xi = 0$ is a first order linear equation, $\xi(0) = 0$ implies $\xi = 0$ for $\xi \in \ker D_\gamma$.

PROPOSITION 2.3.9. *There is a constant K such that for any $c > 1$ the following holds. If γ_0 satisfies*

$$c\|D^*_{\gamma_0}\zeta\|_{L^2} \geq \|\zeta\|_{W^{1,2}} \qquad \textit{for all } \zeta \in W^{1,2}(\gamma_0),$$

then

(1) *the operator $D_{\gamma_0}D^*_{\gamma_0} : W^{2,2}(\gamma_0) \to L^2(\gamma_0)$ is invertible;*
(2) $\|D^*_{\gamma_0}\eta_0\|_{W^{1,2}} \leq KC_4(\|\dot\gamma_0\|^2_{L^2} + 1)\|D_{\gamma_0}D^*_{\gamma_0}\eta_0\|_{L^2}$ *for all $\eta_0 \in W^{2,2}(\gamma_0)$; and*
(3) *if $\xi_0 \in W^{1,2}(\gamma_0)$, $\|\xi_0\|_{W^{1,2}} \leq K^{-1}C_{-1}$ and $\gamma_1 = \exp_{\gamma_0}\xi_0$, then*

$$4c\|D^*_{\gamma_1}\zeta\|_{L^2} \geq \|\zeta\|_{W^{1,2}} \qquad \textit{for all } \zeta \in W^{1,2}(\gamma_1).$$

PROOF. We first establish some preliminary properties of D and D^*. Consider the norm of D:

$$\|\nabla\xi + \nabla_\xi X_\alpha\|_{L^2} \leq \|\xi\|_{W^{1,2}} + \|\nabla_\xi X_\alpha\|_{L^2}.$$

The covariant derivative ∇X_α is a smooth section of $T^*M \otimes TM$, and so $\|\nabla X_\alpha\|$ is bounded by some constant K. Hence

$$\|\nabla_\xi X_\alpha\|_{L^2} \leq K\|\xi\|_{L^2} \tag{2.13}$$

and $\|D\|$ is bounded by some universal constant K. The operators D and D^* are also bounded as operators from $W^{2,2}$ to $W^{1,2}$. We have

$$\|\nabla_{\dot\gamma}(D\xi)\|_{L^2} \leq \|\xi\|_{W^{2,2}} + \|\nabla_{\dot\gamma}\nabla_\xi\nabla f\|_{L^2}$$

and $\nabla(\nabla_\xi X_\alpha) = \nabla(\nabla X_\alpha)\xi + \nabla_{\nabla_\xi} X_\alpha$, so

$$\|\nabla_{\dot\gamma}(\nabla_\xi X_\alpha)\|_{L^2} \le \|\nabla(\nabla X_\alpha)\|_{L^\infty}\|\dot\gamma\|_{L^2}\|\xi\|_{L^\infty} + \|\nabla X_\alpha\|_{L^\infty}\|\nabla\xi\|_{L^2},$$

i.e.,

$$\|\nabla_\xi X_\alpha\|_{W^{1,2}} \le K(1+\|\dot\gamma\|_{L^2})\|\xi\|_{W^{1,2}}. \tag{2.14}$$

We can extend D and D^* to bounded operators from L^2 to $W^{-1,2} := (W^{1,2})^*$. Indeed, the mapping $\xi \mapsto -\nabla\xi$ extends to the adjoint of $\nabla : W^{1,2} \to L^2$, and from (2.13) it follows that $\nabla X_\alpha : L^2 \to L^2$ is bounded. Moreover ∇X_α is selfadjoint, and so the operator $D^* : L^2 \to W^{-1,2}$ is the adjoint of $D : W^{1,2} \to L^2$.

We can also obtain the following estimates for $\|\nabla_\xi X_\alpha\|_{W^{-1,2}}$ with $\xi \in L^2 \subset W^{-1,2}$. For any $\zeta \in W^{1,2}$ we have

$$\begin{aligned}|\nabla_\xi X_\alpha(\zeta)| &= |\langle\nabla_\xi X_\alpha, \zeta\rangle_{L^2}| = |\langle\xi, \nabla_\zeta X_\alpha\rangle_{L^2}| \le \|\xi\|_{W^{-1,2}}\|\nabla_\zeta X_\alpha\|_{W^{1,2}} \\ &\le \|\xi\|_{W^{-1,2}}K(1+\|\dot\gamma\|_{L^2})\|\xi\|_{W^{1,2}},\end{aligned} \tag{2.15}$$

and hence

$$\|\nabla_\xi X_\alpha\|_{W^{-1,2}} \le K(1+\|\dot\gamma\|_{L^2})\|\xi\|_{W^{-1,2}}. \tag{2.16}$$

Since all operators under consideration clearly have closed range, the condition $\|D^*_{\gamma_0}\eta\|_{L^2} \ge C_{-1}\|\eta\|_{W^{1,2}}$ implies that $DD^* : W^{-1,2} \to W^{-1,2}$ is invertible and

$$\|\eta\|^2_{W^{1,2}} \le c^2\langle D^*\eta, D^*\eta\rangle_{L^2} = c^2|DD^*\eta(\eta)| \le c^2\|\eta\|_{W^{1,2}}\|DD^*\eta\|_{W^{-1,2}},$$

i.e.,

$$\|(DD^*)^{-1}\| \le c^2. \tag{2.17}$$

Similarly, $DD^* : W^{2,2} \to L^2$ is invertible since $D^* : W^{2,2} \to W^{1,2}$ is injective. We shall investigate the norm of $(DD^*)^{-1} : L^2 \to W^{2,2}$. For $\eta \in L^2$ we have

$$\|\nabla\eta\|_{W^{-1,2}} \le \|\eta\|_{L^2} \qquad \text{and} \qquad \|\eta\|_{W^{-1,2}} \le \|\eta\|_{L^2}, \tag{2.18}$$

where $\nabla : L^2 \to W^{-1,2}$ is the adjoint of $\nabla : W^{1,2} \to L^2$. Hence, using (2.17), we get

$$\begin{aligned}\|\eta\|_{W^{2,2}} &\le \|\nabla\eta\|_{W^{1,2}} + \|\eta\|_{W^{1,2}} \le c^2(\|DD^*\eta\|_{W^{-1,2}} + \|DD^*\nabla\eta\|_{W^{-1,2}}) \\ &\le c^2(2\|DD^*\eta\|_{L^2} + \|\nabla DD^*\eta - DD^*\nabla\eta\|_{W^{-1,2}}).\end{aligned} \tag{2.19}$$

In the last expression the third order terms cancel out, namely

$$\begin{aligned}\nabla DD^*\eta - DD^*\nabla\eta = &-\nabla_{\nabla\nabla\eta}X_\alpha + \nabla_{\nabla_{\nabla_\eta X_\alpha}}X_\alpha + 2\nabla(\nabla_{\nabla\eta}X_\alpha) \\ &- \nabla\nabla\nabla_\eta X_\alpha + \nabla(\nabla_{\nabla_\eta X_\alpha}X_\alpha).\end{aligned}$$

Using (2.18), (2.13), (2.14), and (2.16) several times, we obtain

$$\|\nabla DD^*\eta - DD^*\nabla\eta\|_{W^{-1,2}} \le K(1+\|\dot\gamma\|^2_{L_2})\|\eta\|_{W^{1,2}}.$$

Combining this with (2.17) yields

$$\|\eta\|_{W^{2,2}} \le Kc^4(1+\|\dot\gamma\|^2_{L_2})\|DD^*\eta\|_{L_2}.$$

This proves part (2) of Proposition 2.3.9, as $\|D^*\eta\|_{W^{1,2}} \le K\|\eta\|_{W^{2,2}}$.

Proof of (3). Similarly to the proof of Proposition 2.3.8, let

$$u(s,t) = \exp_{\gamma_0(s)} t\xi(s),$$

so that $\gamma_i(s) = u(s,i)$ for $i = 0,1$. Let $\zeta_1 \in W^{1,2}(\gamma_1)$, and let $\zeta_t(s) = \zeta(s,t)$ be the continuous section of u^*TM such that $\zeta(s,\cdot)$ is a parallel transport of ζ_1 along $u(s,\cdot)$, i.e., $\nabla_t\zeta = 0$. Assume that $\|\xi\|_{W^{1,2}} \le \varepsilon$. We will find a constant K such that $\|\dot\gamma_0\|_{L^2} < c$ and $c\|D^*_{\gamma_0}\zeta_0\|_{L^2} \ge \|\zeta_0\|_{W^{1,2}}$ for $\zeta_0 = \zeta(\cdot,0)$ implies $4c\|D^*_{\gamma_1}\zeta_1\|_{L^2} \ge \|\zeta_1\|_{W^{1,2}}$ provided that $Kc\varepsilon < 1$. We first establish the relation between the $W^{1,2}$ norms of ζ_0 (resp. $D^*_{\gamma_0}\zeta_0$) and ζ_1 (resp. $D^*_{\gamma_1}\zeta_1$). Since $\nabla_t\zeta = 0$, the length $|\zeta|$ does not depend on t; in particular, $|\zeta_0| = |\zeta_1|$ and

$$|\nabla_t\nabla_s\zeta_t| = \left|R\left(\frac{\partial u}{\partial t}, \frac{\partial u}{\partial s}\right)\zeta_t\right| \le K(|\dot\gamma_0| + t|\nabla\xi|)|\xi||\zeta|$$

if only $\varepsilon \le \varepsilon_0$. Hence

$$|\nabla_s\zeta_1| = \left|\tau^\xi(1,0)\nabla_s\zeta_1\right| \le |\nabla_s\zeta_0| + \sup_{t\in I}|\nabla_t\nabla_s\zeta_t|,$$
$$\|\nabla_s\zeta_1\|_{L^2} \le \|\nabla_s\zeta_0\|_{L^2} + K(\|\dot\gamma_0\|_{L^2} + \|\nabla\xi\|_{L^2})\|\xi\|_{L^\infty}\|\zeta\|_{L^\infty}$$

and

$$(*)\qquad \|\zeta_1\|_{W^{1,2}} \le \|\nabla_s\zeta_0\|_{L^2} + 2K(c+\varepsilon)\varepsilon\|\zeta_0\|_{W^{1,2}} + \|\zeta_0\|_{L^2} \le 2\|\zeta_0\|_{W^{1,2}}$$

if $2K(c+\varepsilon)\varepsilon \le 1$. Analogously, using (2.7) and (2.8),

$$\nabla_t D^*_{\gamma_r}\zeta_r = R\left(\frac{\partial u}{\partial t}, \frac{\partial u}{\partial s}\right)\zeta_r + \nabla_t(X_\alpha(u))\,\zeta_r$$

and we obtain the estimate

$$(**)\qquad \|D^*_{\gamma_0}\zeta_0\|_{L^2} \le \|D^*_{\gamma_1}\zeta_1\|_{L^2} + K(c+\varepsilon+1)\varepsilon\|\zeta_1\|_{W^{1,2}}.$$

Combining $(*)$, $(**)$ and the assumed inequality $c\|D^*_{\gamma_0}\zeta_0\|_{L^2} \ge \|\zeta_0\|_{W^{1,2}}$ yields

$$\|\zeta_1\|_{W^{1,2}} \le 2c(\|D^*_{\gamma_1}\zeta_1\|_{L^2} + K(c+\varepsilon+1)\varepsilon\|\zeta_1\|_{W^{1,2}}),$$

and so $\|\zeta_1\|_{W^{1,2}} \le 4c\|D^*_{\gamma_1}\zeta_1\|_{L^2}$ if $Kc(c+\varepsilon+1)\varepsilon < \frac{1}{4}$.

PROOF OF THEOREM 2.3.3. We will use the notation of Proposition 2.3.8. We apply Newton's method of successive approximations in order to find a zero of the function $f := \mathcal{F}_\gamma \circ D^*_\gamma : W^{2,2}(\gamma) \to L^2(\gamma)$ in a neighbourhood of the origin in $W^{2,2}(\gamma)$. By Proposition 2.3.9 its derivative $Df(\eta) = D_{D^*_\gamma\eta}D^*_\gamma$ is invertible at 0. Thus we have to find $\lambda, \beta > 0$ such that if $\|f(0)\| < \frac{\beta}{\lambda}$ and $\|\eta\| < \beta$, then

$$(2.20)\qquad \|Df(\eta) - Df(0)\| < \lambda^{-1},$$

$$(2.21)\qquad Df(\eta)^{-1} \text{ exists and } \|Df(\eta)^{-1}\| < 2^{-1}\lambda.$$

Then the condition

$$\eta_{n+1} = \eta_n - Df(0)^{-1}f(\eta_n), \qquad \eta_0 = 0,$$

defines a Cauchy sequence, $\|\eta_n - \eta_{n-1}\| < 2^{-n}\beta$, and $\|f(\eta_n)\| < 2^{-n}\lambda^{-1}\beta$. This is proved inductively, since $\|\eta_{n+1} - \eta_n\| < 2^{-1}\lambda\|f(\eta_n)\|$ and

$$\|f(\eta_{n+1})\| = \|f(\eta_{n+1}) - f(\eta_n) - Df(0)(\eta_{n+1} - \eta_n)\| < \lambda^{-1}\|\eta_{n+1} - \eta_n\|$$

by the mean value theorem. Then the limit $\eta := \lim_{n\to\infty}\eta_n$ is a unique solution to $\mathcal{F}(\exp_\gamma D^*_\gamma\eta) = 0$ in the ball $B_\beta(0) \subset W^{2,2}$, and $\|D^*_\gamma\eta\| \le \|D^*_\gamma\|\|\eta\| \le K(1+c^2)\beta$.

By the assumptions and Proposition 2.3.9, $\|Df(0)^{-1}\| \le Kc^6 =: \lambda_1$. For any bounded operator $L : W^{2,2}(\gamma) \to L^2(\gamma)$ such that $\|L - Df(0)\| < \lambda_1^{-1}$, the inverse

L^{-1} exists and $\|L^{-1}\| \leq \frac{\lambda_1}{1-\lambda_1\|L-Df(0)\|}$. Hence if we set $\lambda = 3\lambda_1$, then the condition (2.20) implies (2.21). On the other hand, by Proposition 2.3.8 we have

$$\|f(\eta) - f(0)\| \leq \|D_\gamma D_\gamma^*\| \cdot \|\eta\| + K\|\eta\|^2(1+c^2)$$

and so

$$\begin{aligned}\|Df(\eta) - Df(0)\| &\leq K\|D_\gamma^*\|(\|D_\gamma^*\eta\| + \|f(\eta) - f(0)\| + \|f(0)\|)\\ &\leq K(1+c^2)\big((1+c^2)\|\eta\| + \|f(0)\|\big) < \frac{1}{\lambda},\end{aligned}$$

the last inequalities following if $\|\eta\| < \beta$, $\|f(0)\| < \frac{\beta}{\lambda}$ and

$$\beta < (K(1+c^2)(\lambda(1+c^2)+1))^{-1}.$$

This together with Proposition 2.3.9(3) completes the proof of Theorem 2.3.3(a).

In order to prove the remaining assertions of the theorem we have to look more closely at the size of the domain in which the implicit function is defined. Let $G = D_\gamma^*(D_\gamma D_\gamma^*)^{-1}$, i.e., $D_\gamma \circ G = \mathrm{id}_{L^2(\gamma)}$. If we go through the proof of the implicit function theorem (see e.g. [**Die69**]) we see that it is enough to find $\varepsilon_1 \geq \varepsilon_0 > 0$ such that

$$\|\mathcal{F}_\gamma(\xi_1) - \mathcal{F}_\gamma(\xi_2) - D_\gamma(\xi_1 - \xi_2)\|_{L^2} \leq (2\|G\|)^{-1}\|\xi_1 - \xi_2\|_{W^{1,2}}$$

for $\|\xi_i\|_{W^{1,2}} \leq \varepsilon_1$ and $\|\mathcal{F}_\gamma(\xi)\|_{L^2} \leq \frac{\varepsilon_1}{2}$ for $\|\xi\|_{W^{1,2}} \leq \varepsilon_0$.

1°. By Proposition 2.3.9, $\|G\| \leq Kc^6$.

2°. From Proposition 2.3.8 it follows that

$$\begin{aligned}&\|\mathcal{F}_\gamma(\xi_1) - \mathcal{F}_\gamma(\xi_2) - D_\gamma(\xi_1 - \xi_2)\|_{L^2}\\ &\quad\leq \|\mathcal{F}_\gamma(\xi_1) - \mathcal{F}_\gamma(\xi_2) - D_{\xi_2}(\xi_1 - \xi_2)\|_{L^2} + \|D_{\xi_2}(\xi_1 - \xi_2) - D_0(\xi_1 - \xi_2)\|_{L^2}\\ &\quad\leq Kc^2(\|\mathcal{F}_\gamma(\xi_2)\|_{L^2} + \|\xi_2\|_{W^{1,2}} + \|\xi_1 - \xi_2\|_{W^{1,2}})\,\|\xi_1 - \xi_2\|_{W^{1,2}}.\end{aligned}$$

3°. Again, by Proposition 2.3.8

$$\|\mathcal{F}_\gamma(\xi)\|_{L^2} \leq \|D_\gamma\xi\|_{L^2} + Kc^2\|\xi\|_{W^{1,2}}^2 \leq Kc^2\|\xi\|_{W^{1,2}}.$$

Putting these estimates together yields the result.

PROOF OF PROPOSITION 2.3.4. Since y is a hyperbolic fixed point of the flow, there are positive constants k and c such that

$$|\dot\gamma^-(s+R)| < c|\dot\gamma^-(s)|e^{-kR}, \qquad |\dot\gamma^+(s-R)| < c|\dot\gamma^+(s)|e^{-kR}$$

and

$$|\gamma^-(s+R)| < c|\gamma^-(s)|e^{-kR}, \qquad |\gamma^+(s-R)| < c|\gamma^+(s)|e^{-kR}$$

in normal coordinates at y, for R sufficiently large. It follows that $\tilde\gamma_R$ converges locally uniformly to y with $R \to +\infty$, and since for $s \in [-1,1]$ we have

$$\frac{d}{ds}\tilde\gamma_R(s) = \beta(s)\dot\gamma^-(s+R) + (1-\beta(s))\dot\gamma^+(s-R) + \dot\beta(s)\left(\gamma^-(s+R) - \gamma^+(s-R)\right)$$

and $|\dot\beta|$ is bounded, it follows that

$$|\dot{\tilde\gamma}_R(s)| < c_1e^{-k_1R} \qquad \text{for } s \in [-1,1].$$

Hence $\|\dot{\tilde{\gamma}}\|_{L^2}^2 \leq 2c_1 e^{-2k_1 R} + \ell^2(\gamma^-) + \ell^2(\gamma^+)$. Also, since X_α is locally Lipschitz,

$$|X_\alpha(\tilde{\gamma}_R(s))| < L|\tilde{\gamma}_R(s)| < c_2 e^{-k_2 R} \qquad \text{for } s \in [-1,1].$$

It follows that

$$\begin{aligned} \|\mathcal{F}(\tilde{\gamma}_R)\|_{L^2}^2 &= \int_{-\infty}^{+\infty} \left|\dot{\tilde{\gamma}}_R(s) + X_\alpha(\tilde{\gamma}_R(s))\right|^2 ds \\ &\leq \int_{-1}^{1} \left|\dot{\tilde{\gamma}}_R(s)\right|^2 + |X_\alpha(\tilde{\gamma}_R(s))|^2 \, ds \leq c_3 e^{-k_3 R}. \end{aligned} \tag{2.22}$$

In order to prove the second inequality, take a $\zeta \in W^{1,2}(\tilde{\gamma}_R)$. We first establish the estimate for $\alpha\zeta$ and $(1-\alpha)\zeta$, where $\alpha \in W^{1,2}(\mathbb{R},\mathbb{R})$ is given by

$$\alpha(s) = \begin{cases} 0 & \text{for } |s| \geq T+1, \\ 2 \pm \frac{s}{T} & \text{for } 1 \leq \pm s \leq T+1, \\ 1 & \text{for } |s| \leq 1 \end{cases}$$

with some $T > 0$. Since γ^-, γ^+ and y are solutions of (2.1), it follows that D_γ is onto and

$$c_0 \|D_\gamma^* \zeta\|_{L^2} \geq \|\zeta\|_{W^{1,2}} \quad \text{for} \quad \zeta \in W^{1,2}(\gamma),$$

where $\gamma = \gamma^-$, γ^+ or y. Consequently, if we write $(1-\alpha)\zeta = \zeta^- + \zeta^+$ with $\zeta^\mp \in W^{1,2}(\gamma^\mp)$, $\zeta^\mp(s) = 0$ for $\mp s \leq 0$, then

$$\begin{aligned} \|(1-\alpha)\zeta\|_{W^{1,2}} &= \|\zeta^-\|_{W^{1,2}} + \|\zeta^+\|_{W^{1,2}}, \\ \|D_{\tilde{\gamma}_R}^*(1-\alpha)\zeta\|_{L^2} &= \|D_{\gamma^-}^*\zeta^-\|_{L^2} + \|D_{\gamma^+}^*\zeta^+\|_{L^2}, \end{aligned}$$

and hence

$$c_0 \|D_{\tilde{\gamma}_R}^*(1-\alpha)\zeta\|_{L^2} \geq \|(1-\alpha)\zeta\|_{W^{1,2}}. \tag{$\dagger$}$$

Next, define $\xi_R \in W^{1,2}(y)$ by

$$\xi_R(s) = \begin{cases} e^{s+T+1}\tilde{\gamma}_R(-T-1) & \text{for } s \leq -T-1, \\ \tilde{\gamma}_R(s) & \text{for } s \in [-T-1, T+1], \\ e^{-s+T+1}\tilde{\gamma}_R(T+1) & \text{for } s \geq T+1. \end{cases}$$

Since ξ_R is a vector field along the constant path y, it follows that $\nabla_s \xi_R = \dot{\xi}_R$ and

$$\begin{aligned} \|\xi_R\|_{W^{1,2}}^2 &= \frac{3}{2}\left(|\tilde{\gamma}_R(-T-1)|^2 + |\tilde{\gamma}_R(T+1)|^2\right) \\ &\quad + \int_{-T-1}^{T+1} \left|\dot{\tilde{\gamma}}_R(s)\right|^2 + |\tilde{\gamma}_R(s)|^2 \, ds \leq c_4 e^{-k_4 R}. \end{aligned} \tag{2.23}$$

Clearly, $D_{\tilde{\gamma}_R}^*\alpha\zeta = D_{\xi_R}^*\alpha\zeta$, and if R is big enough it follows by Proposition 2.3.9(2) that

$$4c_0 \|D_{\tilde{\gamma}_R}^*\alpha\zeta\|_{L^2} \geq \|\alpha\zeta\|_{W^{1,2}}. \tag{$\dagger\dagger$}$$

Putting (†) and (††) together, we get

(2.24)
$$\begin{aligned}\|\zeta\|_{W^{1,2}} &\leq \|(1-\alpha)\zeta\|_{W^{1,2}} + \|\alpha\zeta\|_{W^{1,2}} \\ &\leq c_0\|D^*_{\tilde{\gamma}_R}(1-\alpha)\zeta\|_{L^2} + 4c_0\|D^*_{\tilde{\gamma}_R}\alpha\zeta\|_{L^2} \\ &\leq 5c_0(\|D^*_{\tilde{\gamma}_R}\zeta\|_{L^2} + \|\dot{\alpha}\|_{L^\infty}\|\zeta\|_{L^2}) \leq 5c_0\left(\|D^*_{\tilde{\gamma}_R}\zeta\|_{L^2} + \frac{1}{T}\|\zeta\|_{W^{1,2}}\right).\end{aligned}$$

Therefore Proposition 2.3.4 follows for $T > 5c_0$ and R sufficiently large. □

PROOF OF PROPOSITION 2.3.5. The family $\tilde{\gamma}_R$ converges to (γ^-, γ^+), and, since

$$\sup_{s\in\mathbb{R}} d(\tilde{\gamma}_R(s), \gamma_R(s)) < K_C e^{-kR},$$

γ_R converges to (γ^-, γ^+) as $R \to +\infty$ as well. Note that since

$$\lim_{R\to+\infty} \|\tilde{\gamma}_R\|^2_{L^2} = \ell^2(\gamma^-) + \ell^2(\gamma^+),$$

we have $\ell^2(\gamma_R) = \ell^2(\gamma^-) + \ell^2(\gamma^+)$. Define

$$U_\varepsilon = \{x \in M : \exists s \in \mathbb{R} \text{ such that } d(x, \gamma^-(s)) < \varepsilon \text{ or } d(x, y^+(s)) < \varepsilon\}.$$

It is easy to see that for any $\varepsilon > 0$ and a sequence $\{\gamma'_j\} \subset \mathcal{M}(x,z)$ converging to (γ^-, γ^+) there is a $j_0 \in \mathbb{N}$ such that $\gamma'_j(\mathbb{R}) \subset U_\varepsilon$ for $j \geq j_0$. Consequently, in order to complete the proof of Proposition 2.3.5 it is enough to show the following result.

LEMMA 2.3.10. *For any $\varepsilon_0 > \varepsilon > 0$ there is an $R_0 > 0$ such that the following holds. If $\gamma \in \mathcal{M}(x,z)$ and $\gamma(\mathbb{R}) \subset U_\varepsilon$, then for any $R > R_0$ there is an element $\xi \in W^{1,2}(\tilde{\gamma}_R)$ such that $\gamma = \exp_{\tilde{\gamma}_R}\xi$ (modulo time shift) and $\|\xi\|_{W^{1,2}} \leq K\varepsilon$, where K is independent of ε and γ.*

PROOF. We may assume that $d(\gamma(0), y) < \varepsilon$. There is an $R_0 > 0$ such that for $R > R_0$ we have $d(\gamma(0), \gamma^-(R)) < 2\varepsilon$ and $d(\gamma(0), \gamma^+(-R)) < 2\varepsilon$. Since the manifolds $W^u(x)$ and $W^s(z)$ are hyperbolic sets, there are positive constants C_1, k_1 such that

$$\begin{aligned}d(\gamma(-s), \gamma^-(-s+R)) &< 2\varepsilon C_1 e^{-k_1 s}, \\ d(\gamma(s), \gamma^+(s-R)) &< 2\varepsilon C_1 e^{-k_1 s}\end{aligned}$$

for $s \geq 0$. Therefore the required $\xi \in W^{1,2}(\tilde{\gamma}_R)$ satisfies $\|\xi\|^2_{L^2} \leq (2\varepsilon)^2(2T + \frac{C_1^2}{k_1})$ and $\|\xi\|_{L^\infty} < 2C_1\varepsilon$, and from Proposition 2.3.8 it follows that

$$\begin{aligned}\|\nabla\xi\|_{L^2} - \|\mathcal{F}(\tilde{\gamma}_R) + \nabla_\xi X_\alpha\|_{L^2} &\leq K\|\xi\|_{L^\infty}\|\xi\|_{W^{1,2}}(\|\dot{\tilde{\gamma}}_R\|^2_{L^2} + 1), \\ \frac{1}{2}\|\xi\|_{W^{1,2}} \leq \|\mathcal{F}(\tilde{\gamma}_R)\|_{L^2} + \|\nabla_\xi X_\alpha\|_{L^2} &\leq Ce^{-kR} + K_2\|\xi\|_{L^2} \leq K_3\varepsilon\end{aligned}$$

if $2\varepsilon < (K(\|\dot{\tilde{\gamma}}_R\|^2_{L^2} + 1))^{-1}$ and R is sufficiently large.

2.4. Orientation. To complete the proof of the identity $\partial^2 = 0$ we need the following result.

PROPOSITION 2.4.1. *Let x and z be two critical points of a, $\mathrm{ind}(x) = \mathrm{ind}(z) + 2 = k$. Assume that there is a connected component of $\widehat{\mathcal{M}}(x,z)$ whose ends converge to (γ_1^-, γ_1^+) and (γ_2^-, γ_2^+) with $\gamma_i^- \in \mathcal{M}(x, y_i)$, $\gamma_i^+ \in \mathcal{M}(y_i, z)$ and $\mathrm{ind}(y_1) = \mathrm{ind}(y_2) = k-1$. Then*

$$n_{\gamma_1^-} n_{\gamma_1^+} + n_{\gamma_2^-} n_{\gamma_2^+} = 0.$$

Since we deal with a connected component of $\widehat{\mathcal{M}}(x,z)$, we may assume that $\alpha = df$. For $f(x) > a > f(z)$ we have a diffeomorphism $\widehat{\mathcal{M}}(x,z) \cong \mathcal{M}_a(x,z) := W^u(x) \cap W^s(z) \cap f^{-1}(a)$. Indeed, let $\psi : \mathbb{R} \times M \to M$ denote the gradient flow of f. Since f decreases along nonconstant trajectories of the flow ψ for any non-critical point $x \in M$, there is at most one real number $\tau_a(x)$ such that $\psi(\tau_a(x), x) \in M_a$. It follows from the implicit function theorem applied to $f \circ \psi$ that τ_a is a smooth function defined on an open subset U_a of M. Then the diffeomorphism is induced by the map

$$\mathcal{M}(x,z) \ni \gamma \mapsto \gamma(\tau_a(\gamma(0))) \in f^{-1}(a).$$

If we take $a = f(\gamma_i^-(0))$, then $\gamma_i^-(0)$ is a boundary point of $\mathcal{M}_a(x,z)$. Let ν_i be the unit vector field tangent to $\mathcal{M}_a(x,z)$ pointing inward at $\gamma_i^-(0)$. Then $(\dot\gamma/\|\dot\gamma\|, \nu_i)$ form an orthonormal basis for $T\mathcal{M}(x,z)$. Thus the proposition follows immediately from the following result.

LEMMA 2.4.2.

$$n_{\gamma_i^-} n_{\gamma_i^+} = \begin{cases} +1 & \text{if } (\dot\gamma/\|\dot\gamma\|, \nu_i) \text{ matches the orientation of } \mathcal{M}(x,z), \\ -1 & \text{otherwise.} \end{cases}$$

PROOF. Recall that the orientation of $\mathcal{M}(x,y)$ is determined by the orientations $\langle x \rangle$ and $\langle y \rangle$ of $T_x W^u(x)$ and $T_y W^u(y)$ using the isomorphism

$$T_p W^u(x)/T_p\mathcal{M}(x,y) \cong T_p M/T_p W^s(y) \cong T_y M/T_y W^s(y) \cong T_y W^u(y), \tag{2.25}$$

and for $\mathrm{ind}(x) = \mathrm{ind}(y)+1$ the numbers n_γ are defined so that $n_\gamma \dot\gamma$ forms a positively oriented basis of $\mathcal{M}(x,y)$. We assume the convention that for oriented vector spaces V, W the natural isomorphism $V \cong (W \times V)/W$ induces the orientation of the quotient. Then we have the orientation preserving isomophisms ($\langle w \rangle$ denotes the vector space spanned by w)

$$\begin{aligned} T_x W^u(x) &\cong T_{\gamma_i^-} W^u(x)/\langle n_{\gamma_i^-} \dot\gamma_i^- \rangle \times \langle (-1)^k n_{\gamma_i^-} \dot\gamma_i^- \rangle \\ &\cong T_{y_i} W^u(y_i) \times \langle (-1)^k n_{\gamma_i^-} \dot\gamma_i^- \rangle, \\ T_{y_i} W^u(y_i) &\cong T_{\gamma_i^+} W^u(y_i)/\langle n_{\gamma_i^+} \dot\gamma_i^+ \rangle \times \langle (-1)^{k-1} n_{\gamma_i^+} \dot\gamma_i^+ \rangle \\ &\cong T_z W^u(z) \times \langle (-1)^{k-1} n_{\gamma_i^+} \dot\gamma_i^+ \rangle, \end{aligned}$$

and consequently

$$\begin{aligned} T_x W^u(x) &\cong T_z W^u(z) \times \langle n_{\gamma_i^-} \dot\gamma_i^- \rangle \times \langle n_{\gamma_i^+} \dot\gamma_i^+ \rangle, \\ T\mathcal{M}(x,z) &\cong \langle n_{\gamma_i^-} \dot\gamma_i^- \rangle \times \langle n_{\gamma_i^+} \dot\gamma_i^+ \rangle. \end{aligned} \tag{2.26}$$

Consider a solution of the equation

$$\nabla_{\dot\gamma_i^-} \xi = -\nabla_\xi X_\alpha(\gamma_i^-).$$

In local coordinates

$$\dot{\xi}(s) = A(s)\xi(s), \quad A(s) \to A_0, \tag{2.27}$$

where A_0 is the Hessian $-\nabla X_\alpha(y_i)$. We assume for simplicity that A_0 has the simple eigenvalues $\lambda_1, \ldots, \lambda_n$ (this holds for a generic metric). Let $p_1, \ldots, p_n$ denote the corresponding orthonormal basis of eigenvectors. Then the fundamental solution to (2.27) has the form

$$X(t) = \sum_{i=1}^{n} e^{\lambda_1 t}(P_i + R_i(t)),$$

where P_i denotes the projection onto p_i and the matrices R_i converge to zero as $t \to +\infty$ (see e.g. [**Hil65**]). If $\xi(t)$ is a solution with initial value

$$\xi(0) = \xi_0^{i_1} p_{i_1} + \cdots + \xi_0^{i_m} p_{i_m}, \qquad \xi_0^{i_j} \neq 0,$$

then it follows that

$$\lim_{t\to\infty} \frac{\xi(t)}{|\xi(t)|} = p_{i_{\max}}, \tag{2.28}$$

where $\lambda_{i_{\max}} = \max\{\lambda_{i_1}, \ldots, \lambda_{i_m}\}$. In particular, we have the limits

$$\lim_{s\to\infty} \frac{\dot{\gamma}_i^-}{\|\dot{\gamma}_i^-\|} = v_i^- \in T_{y_i} W^s(y_i),$$
$$\lim_{s\to-\infty} \frac{\dot{\gamma}_i^+}{\|\dot{\gamma}_i^+\|} = v_i^+ \in T_{y_i} W^u(y_i).$$

Let η_i^0 be a limit point of ν_i at γ_i^-, and $\eta_i(s)$ a solution to (2.27) with the initial value $\eta_i(0) = \eta_i^0$. Define

$$\eta_i^\infty := \lim_{s\to\infty} \frac{\eta_i}{\|\eta_i\|}.$$

Since γ is a solution to (2.1), η_i is an orbit of the tangent flow

$$\eta_i(s) = d\psi_s(\gamma_i^-(0))\eta_i^0.$$

Moreover, $\eta_i(s)$ is not collinear with $\dot{\gamma}_i^-$, and $d\psi_s \nu_i \in TW^u(x)$.

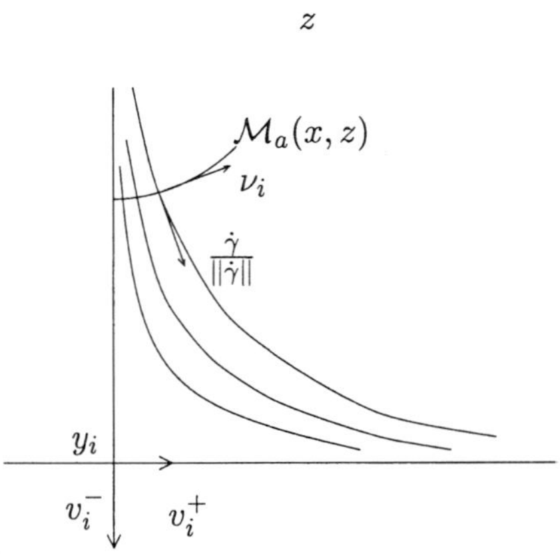

FIGURE 2

Therefore from the transversality of the intersection $TW^u(x) \cap TW^s(y_i)$ it follows that $\eta_i^\infty \in T_{y_i}W^u(y_i)$ (note that the limit (2.28) belongs either to $T_{y_i}W^s(y_i)$ or to $T_{y_i}W^u(y_i)$). On the other hand $d\psi_s\nu_i \in TW^s(z)$, and the Morse–Smale condition for y_i and z implies that $\eta_i^\infty = v_i^+$. Consequently, we have an orientation preserving isomorphism

$$\left\langle \frac{\dot\gamma}{\|\dot\gamma\|}, \nu_I \right\rangle \cong \langle v_i^-, v_i^+ \rangle \cong \langle \dot\gamma_i^-, \dot\gamma_i^+ \rangle,$$

which together with (2.26) proves the lemma. □

2.5. Transversality. In this section we show that the Morse–Smale condition can be achieved by an arbitrarily small perturbation of the metric g. This result is due to Smale ([**Sma61**]), and we will follow the argument presented in [**Flo88b**].

Let $\mathcal{G}$ denote the space of all smooth metrics on M. Then $\mathcal{G}$ is a subset of the space $\mathcal{C}^\infty(S^2TM)$, where $S^2TM \subset T^*M \otimes T^*M$ is the bundle of symmetric tensors. Thus with the identification $T^*M \otimes T^*M = \mathrm{Hom}(TM, T^*M)$ we may write $X_\alpha^g = g^{-1}\alpha$.

Let $\varepsilon = \{\varepsilon_k\}_{k\in N}$ be a sequence of positive numbers, and define the subspace $C_\varepsilon^\infty(S^2TM) \subset C^\infty(S^2TM)$ of all sections $A \in C^\infty(S^2TM)$ satisfying

$$\|A\|_\varepsilon := \sum_{k=0}^\infty \varepsilon_k \|A\|_{C^k(M)} < \infty,$$

where the $\mathcal{C}^k$ norms are taken with respect to a fixed metric h_0. This is a separable Banach space which is dense in $L^2(S^2TM)$ for a sufficiently rapidly decreasing sequence ε_k ([**Flo88b**]).

THEOREM 2.5.1. *Let α be a closed nondegenerate form on a compact manifold M. There is a dense set $\mathcal{G}_{\mathrm{reg}} \subset \mathcal{G}$ such that for every $g \in \mathcal{G}_{\mathrm{reg}}$ and $\gamma : \mathbb{R} \to M$ satisfying*

$$\mathcal{F}(\gamma, g) = \dot\gamma + g^{-1}\alpha(\gamma) = 0 \tag{2.29}$$

the operator $D_\gamma^g : W^{1,2}(\gamma) \to L^2(\gamma)$, $D_\gamma^g\xi = \nabla_{\dot\gamma}^g\xi + \nabla_\xi^g X_\alpha^g$ is onto.

Let x^-, x^+ be critical points of α. Given a metric $g_0 \in \mathcal{G}$, let N be an open subset of M such that $\gamma^{-1}(N) \neq \emptyset$ for any $\gamma \in \mathcal{M}_{g_0}(x^-, x^+)$. We define the closed subspace

$$G(N) \subset \mathcal{C}_\varepsilon^\infty(S^2TM)$$

of those sections which vanish outside N. Let U_{g_0} be a neighbourhood of zero in $G(N)$ such that $g_A = g_0 + A$ is a metric for $A \in U_{g_0}$. Moreover, if we take U_{g_0} small enough, then the g_A are uniformly equivalent to g_0, i.e.,

$$\sigma g_0(v,v) \le g_A(v,v) \le \sigma^{-1} g_0(v,v) \qquad \text{for all } v \in TM,$$

where σ is a constant independent of A. Then Theorem 2.5.1 is a consequence of the following result

THEOREM 2.5.2. *With the notation as above, the set of all $A \in U_{g_0}$ such that $D_\gamma^{g_A}$ is onto for every $\gamma \in \mathcal{M}_{g_A}(x^-, x^+)$ is residual in U_{g_0}.*

PROOF. We claim that the set

$$\mathcal{Z} = \{(\gamma, A) \in \mathcal{P} \times U_{g_0} : \mathcal{F}(\gamma, g_A) = 0\}$$

is a manifold. For $(\gamma, A_0) \in \mathcal{Z}$ consider the function

$$\bar{\mathcal{F}} = \bar{\mathcal{F}}_{\gamma, A_0} : B_{\varepsilon_0}(\gamma) \times U_{g_0} \to L^2(\gamma),$$
$$\bar{\mathcal{F}}(\xi, A) = \tau^{\xi}(1,0)\mathcal{F}(\exp_\gamma \xi, A),$$

where the exponential map and the parallel transport are taken with respect to g_A and the L^2-product with respect to g_{A_0}. As the g_A are uniformly equivalent to g_0, from the proof of Theorem 2.3.3(b) it follows that $\bar{\mathcal{F}}$ is continuously differentiable and

$$d\bar{\mathcal{F}}(0, A_0)(\xi, A) = \nabla^{g_{A_0}}\xi + \nabla_\xi^{g_{A_0}} X_\alpha^{g_{A_0}} - g_{A_0}^{-1} \circ A \circ g_{A_0}^{-1}\alpha = D_\gamma^{g_{A_0}}\xi + d_2\bar{\mathcal{F}}(A_0)A.$$

Since $D_\gamma^{g_{A_0}}$ is a Fredholm operator, the kernel of $d\bar{\mathcal{F}}(0, A_0)$ splits. Indeed, we can write

$$W^{1,2}(\gamma) = \ker D_\gamma^{g_{A_0}} \oplus V_1,$$

where $V_1 \subset W^{1,2}(\gamma)$ is a closed subspace. Clearly, $\ker d\bar{\mathcal{F}}(0, A_0) \cap V_1 = \{0\}$ and

$$\ker d\bar{\mathcal{F}}(0, A_0) \oplus V_1 = d\bar{\mathcal{F}}(0, A_0)^{-1}(D_\gamma^{g_{A_0}}(W^{1,2}(\gamma))).$$

Hence $\operatorname{codim} \ker d\bar{\mathcal{F}}(0, A_0) \oplus V_1 < \infty$, and so $\ker d\bar{\mathcal{F}}(0, A_0)$ has a closed complement in $W^{1,2}(\gamma) \times G(N)$. In order to show that $\mathcal{Z}$ is a submanifold of $\mathcal{P} \times U_{g_0}$ we prove that $d\bar{\mathcal{F}}(0, A_0)$ is onto. The range of $d\bar{\mathcal{F}}(0, A_0)$ is closed since D_γ is Fredholm. Hence it is enough to prove that for any $\eta \in L^2(\gamma)$ the condition

$$\left\langle d\bar{\mathcal{F}}(\xi, A), \eta \right\rangle_{L^2} = 0 \quad \text{for all } \xi \in W^{1,2}(\gamma) \text{ and } A \in G(N)$$

implies $\eta = 0$. The above condition means that

$$\langle D_\gamma \xi, \eta \rangle = 0$$

and

$$\int_{-\infty}^{+\infty} g_{A_0}(g_{A_0}^{-1} A X_\alpha^{g_{A_0}}, \eta) = 0.$$

The first equation yields

$$|\langle \nabla \xi, \eta \rangle| = |\langle \nabla_\xi X_\alpha, \eta \rangle| \leq K \|\eta\|_{L^2} \|\xi\|_{L^2},$$

i.e., $\eta \in W^{1,2}(\gamma)$, and it satisfies the first order linear equation $D^*\eta = 0$. Hence it is enough to show that the second equation implies that η vanishes at least at one point. We have

$$g_{A_0}(g_{A_0}^{-1} \circ A X_\alpha, \eta) = (A X_\alpha)\eta = (X_\alpha)^i A_j^i \eta^j,$$

where the last equality is understood with respect to some basis $\{e_i\}$ along γ. Suppose that $\eta^{j_0}(s_0) \neq 0$ for some index j_0 and $s_0 \in \gamma^{-1}(N)$. We find a symmetric matrix $\hat{A}$ such that

$$(X_\alpha(\gamma(s_0)))^T \hat{A} \eta(s_0) \neq 0.$$

If $(X_\alpha)^{j_0} \neq 0$ we take $\hat{A}_{j_0}^{j_0} = 1$, and $\hat{A}_j^i = 0$ otherwise. Else, there is an index $i_0 \neq j_0$ such that $(X_\alpha)^{i_0} \neq 0$, since X_α does not vanish along γ, and we may take

$\hat{A}^{i_0}_{j_0} = \hat{A}^{j_0}_{i_0} = 1$ and $\hat{A}^i_j = 0$ otherwise. Choosing an appropriate cutoff function, we can extend $\hat{A}$ to an element of $G(N)$ such that

$$\int_{-\infty}^{+\infty} X_\alpha(\gamma(s))^T \hat{A}(\gamma(s))\eta(s)\, ds \neq 0,$$

which leads to a contradiction.

To conclude the proof of Theorem 2.5.2, consider the projection

$$\pi : \mathcal{Z} \to U_{g_0}.$$

This is a differentiable map between Banach manifolds. The tangent space to $\mathcal{Z}$ at the point (γ, A_0) is the subspace of $W^{1,2}(\gamma) \times G(N)$ given by the equation

$$D_\gamma \xi + d_2\bar{\mathcal{F}}(A_0)A = 0.$$

Hence $\ker d\pi(\gamma, A_0) = \ker D_\gamma$ and $\operatorname{im} d\pi = d_2\bar{\mathcal{F}}^{-1}(\operatorname{im} D_\gamma)$. Consequently, we have $\dim \operatorname{coker} d\pi = \dim \operatorname{coker} D_\gamma$, as $d\bar{\mathcal{F}}$ is onto. It follows that $d\pi$ is Fredholm and A is a regular value of π if and only if $D^{g_A}_\gamma$ is onto for every γ satisfying $\dot{\gamma} + g_A^{-1}\alpha(\gamma) = 0$. By the Sard–Smale theorem ([**Sma73**]) the set of regular values of π is residual in U_{g_0}, which yields the result. □

2.6. Continuation. The homology groups we have constructed depend on the form α as well as on the metric g. We will denote the groups $H_*(C_*(M, \alpha, g); \Lambda_\alpha)$ introduced in the statement of Theorem 2.2.2 by $H_*(\alpha, g)$. The aim of this subsection is to prove

THEOREM 2.6.1. *Let α_0 and α_1 be closed nondegenerate forms on a compact manifold M such that the form $\alpha_1 - \alpha_0$ is exact, and let g_0 and g_1 be Riemannian metrics such that the corresponding flows generated by the X_{α_i}, $i = 0, 1$, satisfy the transversality condition for critical points. Then $\Lambda_{\alpha_0} \cong \Lambda_{\alpha_1}$ and*

$$H_*(\alpha_0, g_0) \cong H_*(\alpha_1, g_1).$$

PROOF. We begin by defining the Novikov complex in a slightly more general situation. An open set U in M is called an isolating neighbourhood for a flow ψ if the closure of the sum of all orbits contained in U is contained in U, i.e.,

$$S(U) := \operatorname{cl}\{x \in U : \psi^s(x) \in U \text{ for all } s \in \mathbb{R}\} \subset U. \tag{2.30}$$

If U is such a neighbourhood for the flow of X_α, we may define the Novikov complex relative to U by taking $C_*(U, \alpha, g)$ to be the submodule of elements $\xi \in C_*(M, \alpha, g)$ such that $\xi_{\tilde{x}} = 0$ whenever $\tilde{x} \notin \pi^{-1}(U)$, with the boundary operator $\partial^*(U, \alpha, g)$ defined as before except that we count only those connecting orbits which lie in $\pi^{-1}(U)$. That $\partial^2 = 0$ is clear because $\mathcal{M}(x, z) \cap U$ is open and closed in $\mathcal{M}(x, z)$.

In order to construct a chain homomorphism between $C_*(\alpha_0, g_0)$ and $C_*(\alpha_1, g_1)$, pick an $\varepsilon_1 > 0$ and smooth homotopies α_t, g_t which are constant near 0 and 1, i.e., $\alpha_t = \alpha_0$, $g_t = g_0$ if $t < \varepsilon_1$ and $\alpha_t = \alpha_1$, $g_t = g_1$ if $t > 1 - \varepsilon_1$, and such that $\frac{d}{dt}\alpha_t$ is exact. Thus the α_t induce the same homomorphism $\chi_{\alpha_t} = \chi_\alpha : \Gamma \to \mathbb{R}$, and the coefficient ring Λ_α. Let $\pi : \widetilde{M} \to M$ be a covering with $\pi_1(\widetilde{M}) = \ker[\alpha_t]$. Then there is a smooth function $f : [0, 1] \times \widetilde{M} \to \mathbb{R}$ with $df_t = \pi^*\alpha_t$ for $t \in [0, 1]$. The following function is well defined and smooth on $\widetilde{M} \times S^1$, where S^1 is parametrized by $t \mapsto e^{\pi i t}$:

$$F : \widetilde{M} \times S^1 \to \mathbb{R},$$
$$F(x,t) = f_{|t|} + \frac{K}{\pi} \cos \pi t \qquad \text{for } t \in [-1,1].$$

Moreover, $F(A\tilde{x},t) = F(\tilde{x},t) + \chi_\alpha(A)$ for any $A \in \Gamma$, and the differential dF descends to a closed form β on $M \times S^1$ with $\ker[\beta] = \ker[\alpha_t] \times \pi_1(S^1)$ and $\chi_\beta = \chi_\alpha : \Gamma \to \mathbb{R}$. Since M is compact, all critical points of β have the form (x,i), where x is a critical point of α_i, $i = 0,1$, if only K is sufficiently large. Moreover, $\beta(x,t) = \alpha_i(x) - K \sin \pi t dt$ for $|t-i| < \varepsilon_1$, and therefore these points are nondegenerate with $\operatorname{ind}(x,i) = \operatorname{ind} x + 1 - i$. Next we define the "product-like" metric on $M \times S^1$

$$G_{(x,t)} = g_t \oplus \mathbf{1},$$

where $\mathbf{1}$ denotes the metric on S^1 provided by the parametrization. For $|t-i| < \varepsilon_1$ the dual vector field X_β has the form

$$X^G_\beta(x,t) = (X^{g_i}\alpha_i(x), -K \sin \pi t \partial^t), \tag{2.31}$$

and so the submanifolds $M_i = M \times \{i\}$, $i = 0,1$, are invariant with respect to the flow and the restricted flow satisfies the Morse–Smale condition. If $t \neq 0,1$ and $\gamma = (\gamma_1, \gamma_2) : \mathbb{R} \to M \times S^1$ is a solution to $\dot{\gamma} = -X_\beta(\gamma)$ through a point (x,t), then from (7.1) it follows that $\lim_{s\to-\infty} \gamma_2(s) = 0$ and $\lim_{s\to+\infty} \gamma_2(s) = 1$. In particular, there are no orbits going from M_1 to M_0, and those connecting two critical points in M_0 (resp. M_1) are contained in M_0 (resp. M_1). Hence by Theorem 2.5.2 we can obtain the Morse–Smale condition by an arbitrarily small perturbation $\tilde{G}$ of G on the set $N_0 = \{(x,t) \in M \times S^1 : t \neq 0,1\}$. A homotopy α_t with such a metric will be called admissible.

The set $U = M \times (-\varepsilon_2, 1+\varepsilon_2) \subset M \times S^1$ with $0 < \varepsilon_2 < \frac{1}{2}$ is clearly an isolating neighbourhood, since X_β is transversal to ∂U. Thus we may consider the complex $(C^\beta_*, \Delta) := (C_*(U,\beta,\tilde{G}), \partial(U,\beta,\tilde{G}))$. Then $C^\beta_* = C^0_* \oplus C^1_*$ with C^i_* generated by $\langle(\tilde{x},i)\rangle$, where $\tilde{x}$ is a critical point of F, and we assume the orientations $\langle(\tilde{x},i)\rangle$ to be induced by the obvious isomorphisms

$$T_{(p,0)}W^u(x,0) \cong T_pW^u(x) \times T_0S^1,$$
$$T_{(p,1)}W^u(x,1) \cong T_pW^u(x)$$

for any $p \in M$. Using this identification and (2.25), we obtain the orientation preserving isomorphisms

$$\begin{aligned} T_pW^u(x) \times T_0S^1/T_{(p,0)}\mathcal{M}((x,0),(y,0)) &\cong T_{(y,0)}W^u((y,0)) \\ \cong T_pW^u(x)/T_p\mathcal{M}(x,y) \times T_0S^1 &\cong T_pW^u(x) \times T_0S^1/T_p\mathcal{M}(x,y), \\ T_{(p,1)}W^u(x,1)/T_{(p,1)}\mathcal{M}((x,1),(y,1)) & \\ \cong T_{(y,1)}W^u(y,1) &\cong T_pW^u(x)/T_p\mathcal{M}(x,y), \end{aligned}$$

and hence the orientation preserving inclusions $\mathcal{M}(x,y) \hookrightarrow \mathcal{M}((x,i),(y,i))$. (Recall that we adopt the convention that the natural isomorphism $V \cong (W \times V)/W$ is orientation preserving.) If we denote $\Delta = \sum_{i,j=0,1} \Delta_{ij}$, where $\Delta_{ij} : C_*^j \to C_*^i$, it follows that we have the identifications

$$\chi_i : (C_k(\alpha_i, g_i), \partial_{\alpha_i}) \to (C^i_{k+1-i}, \Delta_{ii}), \qquad \chi_i(\xi)\tilde{x} := \xi_{(\tilde{x},i)}.$$

Clearly the χ_i are well defined Λ_α homomorphisms, since the injections $\widetilde{M}_i \hookrightarrow \widetilde{M} \times S^1$ are Γ-equivariant. Since in view of the above remarks $\Delta_{01} = 0$, the equality $\Delta^2 = 0$ reduces to

$$\Delta_{ii}^2 = 0 \quad \text{for } i = 0, 1, \qquad \Delta_{10} \circ \Delta_{00} + \Delta_{11} \circ \Delta_{10} = 0.$$

In particular, the second equation means that

$$\psi_{10} = (-1)^k \chi_1^{-1} \circ \Delta_{10} \circ \chi_0 : C_*(\alpha_0, g_0) \to C_*(\alpha_1, g_1)$$

is a degree 0 chain homomorphism. If $\alpha_t = \alpha_0$, $g_t = g_0$ for all $t \in [0,1]$ and $\tilde{G} = G = g_0 \oplus \mathbf{1}$, then the set $\mathcal{M}((x,0),(y,1)) \subset U$ is nonempty and consists of a single orbit precisely when $x = y$. For any point (x,t) in $\mathcal{M}((x,0),(x,1))$ we have isomorphisms

$$\begin{aligned} T_{(x,t)} W^u(x,0)/T_{(x,t)} M((x,0),(x,1)) &\cong T_{(x,1)} W^u((x,1)) \\ &\cong T_x W^u(x) \cong T_x W^u(x) \times T_t S^1 / T_t S^1, \end{aligned} \tag{2.32}$$

and the last one preserves the orientation if and only if $\operatorname{ind}(x)$ is even. Hence in the case of constant homotopy ψ_{10} is an identity.

Note that in terminology of [**Spa66**] C_*^β is a chain cylinder of the homomorphism ψ_{01}.

We may repeat the whole construction for 2-parameter homotopies $\alpha_{t,r}, g_{t,r}$ joining four forms and metrics α_i, g_i for $i = 2t + r$, $t, r = 0, 1$, to obtain the Novikov complex $C_*^\beta = (C_*(\beta, \tilde{G}, U), \Delta)$ associated with the form β on the manifold $M \times S^1 \times S^1$ such that $\pi^*\beta + dF$ and

$$\begin{aligned} F &: \widetilde{M} \times S^1 \times S^1 \to \mathbb{R}, \\ F(x,t,r) &= f_{t,r}(x) + \frac{K}{\pi}(\cos \pi t + \cos \pi r), \end{aligned}$$

where $df_{t,r} = \pi^* \alpha_{t,r}$. Here $U = M \times (-\varepsilon_2, 1+\varepsilon_2)^2$. As previously, $C_*^\beta = \bigoplus_{0 \le i \le 3} C_*^i$,

$$\Delta = \begin{pmatrix} \Delta_{00} & 0 & 0 & 0 \\ \Delta_{10} & \Delta_{11} & 0 & 0 \\ \Delta_{20} & 0 & \Delta_{22} & 0 \\ \Delta_{30} & \Delta_{31} & \Delta_{32} & \Delta_{33} \end{pmatrix}$$

where $\Delta = \partial_*^\beta$ denotes the boundary operator, and we have natural identifications

$$\begin{aligned} &\chi_0 : (C_k^{\alpha_0}, \partial^{\alpha_0}) \to (C_0^{k+2}, \Delta_{00}), \\ &\chi_I : (C_k^{\alpha_i}, \partial^{\alpha_i}) \to (C_i^{k+1}, \Delta_{ii}) \quad \text{for } i = 1, 2, \\ &\chi_3 : (C_k^{\alpha_3}, \partial^{\alpha_3}) \to (C_3^k, \Delta_{33}). \end{aligned}$$

We also obtain chain homomorphisms

$$\begin{aligned} \psi_{10} = (-1)^{k+1} \chi_1^{-1} \Delta_{10} \chi_0, &\qquad \psi_{30} = -\chi_3^{-1} \Delta_{30} \chi_0, \\ \psi_{ij} = (-1)^k \chi_i^{-1} \Delta_{ij} \chi_j &\qquad \text{otherwise,} \end{aligned}$$

which fit into the diagram

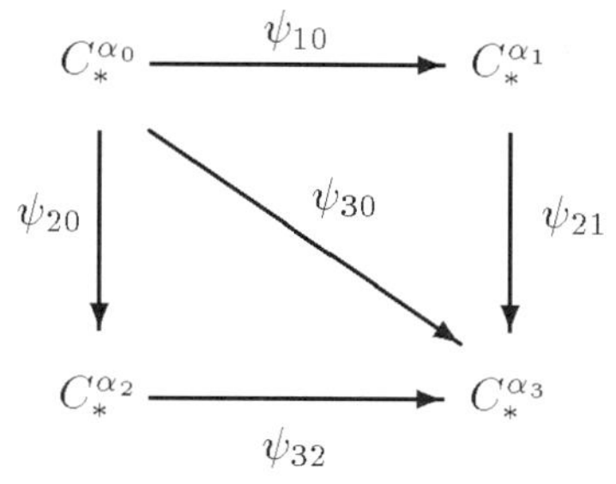

The condition $\Delta^2 = 0$ gives, among others, the equation

$$\Delta_{30} \circ \Delta_{00} + \Delta_{31} \circ \Delta_{10} + \Delta_{32} \circ \Delta_{20} + \Delta_{33} \circ \Delta_{30} = 0,$$

which means that the degree $+1$ homomorphism ψ_{30} provides the chain homotopy between $\psi_{31} \circ \psi_{10}$ and $\psi_{32} \circ \psi_{20}$.

Note that if all one-parameter homotopies (i.e., for t or $r = 0, 1$) are admissible, then we can obtain the transversality condition by perturbing the metric in the set $N_0 = \{(x, t, r) \in M \times S^1 \times S^1 : t \neq 0, 1 \text{ and } r \neq 0, 1\}$, thus without affecting the homomorphisms $\psi_{10}, \psi_{20}, \psi_{31}$ and ψ_{32}. If we take $\alpha_0 = \alpha_2$ and $\alpha_1 = \alpha_3$ and a homotopy which is constant with respect to t for $r = 0, 1$, then we can conclude that $h_{\alpha_0\alpha_1} = H_*(\psi_{10}) = H_*(\psi_{32})$ is independent of the particular choice of the homotopy joining α_0 and α_1. Similarly, with $\alpha_2 = \alpha_3$ we obtain

$$h_{\alpha_2\alpha_1} \circ h_{\alpha_1\alpha_0} = h_{\alpha_2\alpha_0},$$

and hence all $h_{\alpha_i\alpha_j}$ are isomorphisms as $h_{\alpha_i\alpha_i} = \mathrm{id}_{C_*^{\alpha_i}}$. This proves Theorem 2.6.1. □

2.7. Twisted coefficients. The reason for the Novikov construction is that for a nonexact form one cannot expect that the number of orbits connecting two critical points is always finite.

EXAMPLE 2.7.1. Consider a function on $\mathbb{R}^3$ given by

$$f(x, y, z) = y + (1 - \cos x)\cos y + \cos z.$$

Its differential descends to a 1-form α on the 3-torus T^3 (parametrized by $x \mapsto e^{ix}$ etc.),

$$\alpha = \sin x \cos y \, dx + (1 + (\cos x - 1)\sin y)\, dy - \sin z \, dz,$$

which is homologous to the constant form dy. Hence $\Gamma = \mathbb{Z}$, and the ring Λ_α consists of formal series $\sum_{i\in\mathbb{Z}} \lambda_i t^i$ in one variable t with a finite number of nonzero λ_i for negative i. The flow generated by the vector field dual to α with respect to the Euclidean metric on T^3 splits into the direct sum of flows in the (x, y)-direction and the z-direction. Its behaviour on the invariant 2-torus given by $\{z = 0\}$ or $\{z = \pi\}$

is sketched in Figure 3(a). We restrict our attention to an isolating neighbourhood $U_\varepsilon := \{(x,y,z) \in T^3 : -\varepsilon < z < \pi + \varepsilon\}$. There are eight critical points:

$$w = \left(\pi, \frac{\pi}{3}, 0\right) \qquad \text{of index 3,}$$
$$x_0 = \left(\pi, \frac{\pi}{3}, \pi\right);\ x_1 = \left(\frac{\pi}{2}, \frac{\pi}{2}, 0\right);\ x_2 = \left(\frac{3\pi}{2}, \frac{\pi}{2}, 0\right) \qquad \text{of index 2,}$$
$$y_0 = \left(\pi, \frac{2\pi}{3}, 0\right);\ y_1 = \left(\frac{\pi}{2}, \frac{\pi}{2}, \pi\right);\ y_2 = \left(\frac{3\pi}{2}, \frac{\pi}{2}, \pi\right) \qquad \text{of index 1,}$$
$$z = \left(\pi, \frac{2\pi}{3}, \pi\right) \qquad \text{of index 0,}$$

and two periodic orbits ($t \mapsto (0,-t,0)$ and $t \mapsto (0,-t,\pi)$). From the symmetry of the equation it follows that there are homoclinic orbits (i.e., connecting a point with itself) at the critical points x_i, y_i, $i = 1,2$. For the same reason the unstable manifold of x_1 intersects the stable manifold of y_1 along a 2-dimensional manifold. These are the only cases where the Morse–Smale condition fails. It is easy to see that there is one orbit connecting (in U_ε) each of the following pairs of points: w and x_i, $i = 0,1,2$, x_0 and y_i, x_i and y_0, y_i and z for $i = 1,2$, y_0 and z. We perturb the metric to destroy the homoclinic orbits and create one more orbit connecing w with x_i and y_i with z, $i = 1,2$, as follows. Choose $0 < \varepsilon < \frac{\pi}{2}$ and $0 < \alpha < 1$ and let $\psi : \mathbb{R} \to [0,\alpha]$ be a cutoff function satisfying

$$\psi(y) = \begin{cases} 0 & \text{if } |y - \frac{3\pi}{2}| \geq \varepsilon, \\ \alpha & \text{if } |y - \frac{3\pi}{2}| \leq \frac{\varepsilon}{2}. \end{cases}$$

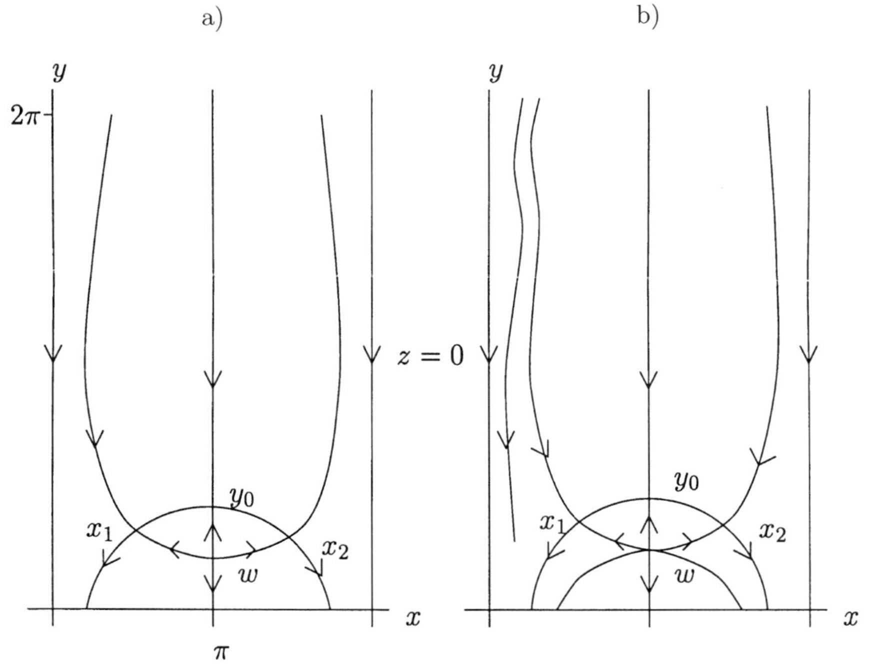

FIGURE 3

Define the new metric g by the formula

$$g = dz^2 + dx^2 + dy^2 - \psi(y) \cos z \sin x \, dxdy.$$

The perturbed system is shown in Figure 3(b). The stable manifold of x_1 is contained in the unstable manifold of w, and the part of the unstable manifold of x_1 which is contained in the 2-torus $\{z = 0\}$ consists of one orbit converging to y_0 and another one winding around the periodic orbit. A similar picture can be drawn for the torus $\{z = \pi\}$. Further perturbation of g in order to secure the transversality condition for x_i and y_i, $i = 1, 2$, can be made in the set

$$N = \{(x, y, z) \in T^3 : x \neq 0, |y - \tfrac{3\pi}{2}| \leq \varepsilon, 0 < z < \pi\},$$

thus without affecting any of the connecting orbits mentioned above. By considering the global section $S = \{y = \frac{3\pi}{2}\}$ of the flow and the behaviour of the Poincaré map of the two periodic orbits defined on an open subset of the section one can give a topological proof of the existence of infinitely many orbits connecting x_1 and y_1. There is, however, a simple algebraic argument proving this fact. We consider the Morse complex restricted to U_ε and write down the boundary operator (with the appropriate choice of the orientations)

$$\partial\langle w\rangle = \langle x_0\rangle + (1-t)\langle x_1\rangle - (1-t)\langle x_2\rangle,$$

$$\partial\langle x_0\rangle = \langle y_1\rangle - \langle y_2\rangle, \qquad \partial\langle x_1\rangle = \langle y_0\rangle + a_1\langle y_1\rangle, \qquad \partial\langle x_2\rangle = \langle y_0\rangle + a_2\langle y_2\rangle,$$

$$\partial\langle y_0\rangle = \langle z\rangle, \qquad \partial\langle y_1\rangle = (1-t)\langle z\rangle, \qquad \partial\langle y_2\rangle = (1-t)\langle z\rangle,$$

where a_1, a_2 are coefficients in Λ_α. The condition $\partial^2 = 0$ implies

$$1 + (1-t)a_1 = 1 + (1-t)a_2 = 0,$$

and so $a_1 = a_2 = -(1-t)^{-1} = -\sum_{k=0}^{\infty} t^k$. Thus there must be an infinite number of orbits connecting x_1 and y_1 (x_2 and y_2), each in a different homotopy class.

EXAMPLE 2.7.2. If M is a torus T^n, then $H^1_{DR}(M, \mathbb{R}) \cong \mathrm{Hom}(\mathbb{Z}^n, \mathbb{R})$, every homology class is represented by a constant form, and so the Novikov homology is trivial unless α is exact.

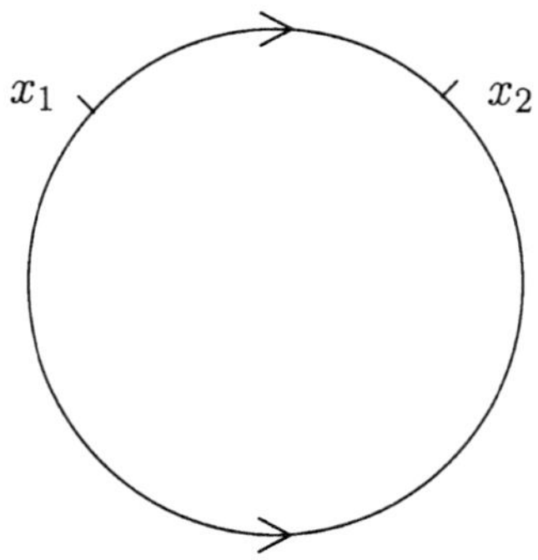

FIGURE 4

EXAMPLE 2.7.3. Consider the constant form $dx \in \mathcal{C}^\infty(T^*S^1)$ where the circle is parametrized by $\pi : \mathbb{R} \ni x \mapsto e^{ix} \in S^1$, and its exact perturbation $\alpha = (1 - 2\sin x)dx$. The associated flow has two critical points x_0 and x_1, of index 0 and 1, and two connecting orbits. Therefore it is a gradient flow for some Morse function $f : S^1 \to \mathbb{R}$ with respect to a different metric (see Figure 4). The ring Λ_α is as in Example 2.7.1, and if we consider the complex $C_*(S^1, \alpha)$ as a free Λ_α-module generated by x_0 and x_1, then the boundary operator ∂_*^N for Novikov homology has the form (with an appropriate choice of $\tilde{x}_i \in \pi^{-1}(x_i)$, $i = 0, 1$)

$$\partial_0^N \langle x_1 \rangle = (t-1)\langle x_0 \rangle.$$

This is an isomorphism, since $t - 1$ is invertible in Λ_α, and so $HN_*(S^1, [\alpha]) = 0$. On the other hand, we may consider the Morse complex for the function f tensored with Λ_α with the boundary

$$\mathbf{id}_{\Lambda_\alpha} \otimes \partial_0^N : \Lambda_\alpha \otimes_{\mathbb{Z}} C_1(S^1, f) \to \Lambda_\alpha \otimes_{\mathbb{Z}} C_0(S^1, f).$$

This operator is trivial, and in fact it is an evaluation of ∂_0^N at $t = 1$. Thus $H_i(C_*(S^1, f); \Lambda_\alpha) = \Lambda_\alpha$ for $i = 0, 1$, and we obtain the homology of S^1 with coefficients in Λ_α.

More generally, if we have a form α and a metric g such that the corresponding flow has no limit sets other than critical points, then it is a gradient flow for some Morse function f and a different metric ([**Sma61**]). In this case the number of connecting orbits in the definition of the boundary is finite, and the operator has the form

$$\partial_*^N \langle x \rangle = \sum_{\operatorname{ind}(y) = \operatorname{ind}(x) - 1} \Big(\sum_A n(y, x; A) t_1^{A_1} \cdots t_m^{A_m} \Big) \langle y \rangle,$$

where $n(y, x; A)$ counts the orbits in the homotopy class corresponding to A (there are several choices involved here) and A ranges over a finite set. Note that by taking $t = (1, \ldots, 1)$ we obtain the Morse complex for f with coefficients in Λ_α. In fact, by making a special choice of the gradient flow X_α we will construct an isomorphism between the Novikov complex and the Morse complex with twisted coefficients.

PROPOSITION 2.7.4. *Let M be an n-dimensional closed manifold and $a \in H^1(M, \mathbb{R})$. There exist a 1-form $\alpha \in a$, a Morse function f, and Riemannian metrics g_1, g_2 such that $g_1^{-1}\alpha = g_2^{-1}df$ and the flow generated by this vector field is of Morse–Smale type.*

PROOF. **Step 1**. Let $u_1, \ldots, u_k$ be a basis for the free part of $H_1(M, \mathbb{Z})$. We choose the corresponding dual manifolds $N_1, \ldots, N_k$ so that they intersect transversally.

Each u_i can be represented as a differential of a function $\tilde{u}_i : M \to S^1 = \mathbb{R}/\mathbb{Z}$ (unique up to homotopy) and an inverse image $N_i = \tilde{u}_i^{-1}(p_i)$ of any regular value p_i is a dual manifold of u_i, i.e., the value of u_i on any 1-cycle z is equal to the intersection number of $z \cdot N_i$ (see [**Mil68**]). Set $\tilde{u} : M \to \mathbb{T}^k$, $\tilde{u} = (\tilde{u}_1, \ldots, \tilde{u}_k)$, and let $\operatorname{pr}_{i_1 \cdots i_p} : \mathbb{T}^k \to \mathbb{T}^p$ denote the projection of the torus $\mathbb{T}^k$ onto $\mathbb{T}^p$, $\operatorname{pr}_{i_1 \cdots i_p}(x_1, \ldots, x_k) = (x_{i_1}, \ldots, x_{i_p})$. Choose a point $p = (p_1, \ldots, p_k) \in \mathbb{T}^k$ such that $(p_{i_1}, \ldots, p_{i_p})$ is a regular value of $\operatorname{pr}_{i_1 \cdots i_p} \circ \tilde{u}$ for any collection of subscripts $\{i_1, \ldots, i_p\} \subset \{1, \ldots, k\}$ (such a p exists by Sard's theorem, since the projections are open maps). Then

$$\bigcap_{j=1}^{p} N_{i_j} = \bigcap_{j=1}^{p} \tilde{u}_{i_j}^{-1}(p_{i_j}) = (\mathrm{pr}_{i_1\cdots i_p} \circ \tilde{u})^{-1}(p_{i_1}, \ldots, p_{i_p})$$

are manifolds of dimension $\dim\ker d(\mathrm{pr}_{i_1\ldots i_p} \circ \tilde{u}) = n - p$. If $x \in N_l \cap \bigcap_{j=1}^{p} N_{i_j}$ for $l \neq i_j$, $j = 1, \ldots, p$, then $T_x(N_l \cap \bigcap_{j=1}^{p} N_{i_j}) \not\subseteq T_x \bigcap_{j=1}^{p} N_{i_j}$. Therefore $N_l \pitchfork_x \bigcap_{j=1}^{p} N_{i_j}$, since $\operatorname{codim} N_l = 1$.

Step 2. We construct vector fields X_i defined in some neighbourhoods U_i of N_i such that

$$X_i \pitchfork N_i, \tag{2.33}$$
$$X_i|_{N_j} \subset TN_j \quad \text{for } i \neq j, \tag{2.34}$$
$$[X_i, X_j] = 0 \tag{2.35}$$

for $1 \leq i, j \leq k$.

First notice that for any $x_0 \in \bigcup_{i=1}^{k} N_i$ we have a neighbourhood V of x_0 and a chart $\psi_V : V \to \mathbb{R}^n$, $\psi_V(x_0) = 0$, such that, if $N_i \cap V \neq \emptyset$, then $x_0 \in N_i$ and $\psi_V(N_i \cap V)$ is contained in a hyperplane $\{(x_1, \ldots, x_n) \in \mathbb{R}^n : x_j = 0\}$ for some $1 \leq j \leq n$. In the neighbourhood V we define $X_i^V = \psi_V^* \frac{\partial}{\partial x_j}$, where $\frac{\partial}{\partial x_j} = (0, \ldots, \underset{j}{1}, \ldots, 0)$. We also choose ψ_V so that $d\tilde{u}_i(X_i^V)$ induces the positive orientation of S^1. Now we can use a partition of unity to obtain vector fields X_i' defined on some neighbourhood U_i' of N_i satisfying (2.33), (2.34), transversality following from the fact that $d\tilde{u}_i(X_i') \neq 0$ as a convex combination of $d\tilde{u}_i(X_i^V)$. Proceeding by induction, let us assume the existence of vector fields $X_1, \ldots, X_l$ satisfying (2.33), (2.34) for $i \leq l$, $j \leq k$ and (2.35) for $i, j \leq l$. Let ϕ_t^i be the local group of diffeomorphisms generated by the X_i, and let $\varepsilon_0 > 0$ be such that N_i is contained in the domain of ϕ_t^i for $|t| < \varepsilon_0$ and $i \leq l$. If we set $U_i^\varepsilon = \phi_{(-\varepsilon,\varepsilon)}^i(N_i)$ for $0 < \varepsilon < \varepsilon_0$, then $[X_i, X_j] = 0$ implies $\phi_t^i(U_j^\varepsilon) \subset U_j^\varepsilon$. If we denote

$$L_p = \bigcup_{\substack{\{i_1,\ldots,i_p\} \\ \subset\{1,\ldots,l\}}} \bigcap_{j=1}^{p} N_{i_j} \cap N_{l+1},$$

then $L_{l+1} = \emptyset$, $L_l = \bigcap_{j=1}^{l+1} N_j$, and $L_0 = N_{l+1}$. Starting with $p = l + 1$, we shall inductively construct vector fields Y_p in some neighbourhoods W_p of L_p and set $U_{l+1} = W_0$, $X_{l+1} = Y_0$.

Thus assume that Y_p satisfies (2.33), (2.34) and (2.35). Without loss of generality we may take

$$W_p = W_p^\varepsilon = \bigcup_{\substack{\{i_1,\ldots,i_p\} \\ \subset\{1,\ldots,l\}}} \bigcap_{j=1}^{p} \phi_{(-\varepsilon,\varepsilon)}^{i_j}(N_{i_j} \cap V),$$

where $V \subset U_{l+1}'$ is a neighbourhood of N_{l+1} and $0 < \varepsilon < \varepsilon_0$. Let $V' \subset U_{l+1}'$ be a neighbourhood of N_{l+1} such that $\phi_{(-\varepsilon/2,\varepsilon/2)}^i(V') \subset V$ for $i \leq l$. For any $\{i_1, \ldots, i_{p-1}\} \subset \{1, \ldots, l\}$ the set $K = K_{i_1\cdots i_{p-1}} := \bigcap_{j=1}^{p-1} N_{i_j} \cap V'$ is a submanifold of V' and $(K \setminus \overline{W}_p^{\varepsilon/2}, K \cap W_p^\varepsilon)$ is an open cover in $K_{i_1\cdots i_{p-1}}$. Using cutoff functions, we can glue together vector fields $X'_{l+1}|_{K\setminus \overline{W}_p^{\varepsilon/2}}$ and $Y_p|_{K\cap W_p^\varepsilon}$ to obtain a vector field $Y_{i_1\cdots i_{p-1}} : K_{i_1\cdots i_{p-1}} \to TM$ satisfying (2.33) and (2.34). Now notice that the map

$$\Phi = \Phi^{i_1\cdots i_{p-1}} : (-\varepsilon/2, \varepsilon/2)^{p-1} \times K_{i_1\cdots i_{p-1}} \to M,$$
$$(t_1, \dots, t_{p-1}, x) \mapsto \phi^{i_1}_{t_1} \circ \cdots \circ \phi^{i_{p-1}}_{t_{p-1}}(x)$$

is a diffeomorphism onto some neighbourhood $V_{i_1\cdots i_{p-1}}$ of $\bigcap_{j=1}^{p-1} N_{i_j} \cap N_{l+1}$. On $V_{i_1\cdots i_{p-1}}$ we set $Y_{p-1} = \Phi_* Y_{i_1\cdots i_{p-1}}$. Since $\phi_{i_j}(N_q) \subset N_q$ for all $q \leq k$, the vector field Y_{p-1} satisfies (2.33) and (2.34) and clearly $[Y_{p-1}, X_{i_j}] = 0$ for $j = 1, \dots, p-1$. On the set $\Phi((-\varepsilon/2, \varepsilon/2)^{p-1} \times (K \cap W_p^{\varepsilon/2}))$ we have $Y_{p-1} = Y_p$, since $[Y_p, X_{i_j}] = 0$. On the other hand, $\Phi((-\varepsilon/2, \varepsilon/2)^{p-1} \times (K \setminus W_p^{\varepsilon/2})) \cap U_q^{\varepsilon/2} = \emptyset$ if $q \neq i_j$, $j = 1, \dots, p-1$, and $q \leq l$. Hence $[Y_{p-1}, X_i] = 0$ on $V_{i_1\cdots i_{p-1}}$ for all $q \leq l$. Moreover our choice of V' guarantees that $V_{i_1\cdots i_{p-1}} \cap V_{i_1\cdots i_{p-2} i_p} \subset \Phi((-\varepsilon/2, \varepsilon/2)^{p-1} \times (K \cap W_p^{\varepsilon/2}))$, and therefore Y_{p-1} is well defined in

$$W_{p-1} = \bigcup_{\substack{\{i_1,\dots,i_{p-1}\}\\ \subset\{1,\dots,l\}}} V_{i_1\cdots i_{p-1}}.$$

This concludes the inductive step.

Step 3. We complete the proof with a simple rescaling argument. Define the function $\tau_i : U_i^\varepsilon \to \mathbb{R}$ by the condition $\tau_i(\phi_t^i(x)) = t$, where $x \in N_i$. If $\beta : \mathbb{R} \to \mathbb{R}$ is a smooth function with $\operatorname{supp} \beta' \subset (-\varepsilon, \varepsilon)$, then $d(\beta \circ \tau_i) = \beta' \circ \tau_i d\tau_i$ extends to a closed form on M. It is easy to see that $d(\beta \circ \tau_i)$ is homologous to $(\beta(\varepsilon) - \beta(-\varepsilon))\tilde{u}_i$. Indeed, without loss of generality assume that $N_i = \tilde{u}_I^{-1}(0)$ and $\beta(\pm\varepsilon) = \pm\frac{1}{2}$, i.e., $\beta \circ \tau_i : M \to S^1$, $\beta \circ \tau_i|_{M\setminus U_i^\varepsilon} = \frac{1}{2}$. We have to show that $\tilde{u}_i$ and $\beta \circ \tau_i$ are homotopic. If $\varepsilon_1 > 0$ is such that $\tilde{u}_i^{-1}([-\varepsilon_1, \varepsilon_1]) \subset U_i^\varepsilon$ and $\chi : S^1 \to S^1$ satisfies $\chi(S^1 \setminus (-\varepsilon_1, \varepsilon_1)) = \frac{1}{2}$ and $\chi^{-1}(0) = 0$, then $\tilde{v}_i = \chi \circ \tilde{u}_i$ is homotopic to $\tilde{u}_i$ and $\tilde{v}_i^{-1}(S^1 \setminus \{\frac{1}{2}\}) \subset U_i^\varepsilon$. Hence $\tilde{v}_i$ agrees with $\beta \circ \tau_i$ ouside U_i^ε, and the homotopy $t\beta \circ \tau_i + (1-t)\tilde{v}_i$ is well defined in U_i^ε.

Choose $0 < \varepsilon_1 < \varepsilon_2 < \varepsilon$ and smooth functions $\rho, h : \mathbb{R} \to \mathbb{R}$ such that $\operatorname{supp} h' \subset (\varepsilon_1, \varepsilon_2)$, $h(\varepsilon_1) = 0$, $h(\varepsilon_2) = 2$, $\rho > 0$, $\rho(1) = \rho(0) = 1$, $\operatorname{supp} \rho' \subset \operatorname{int}(\operatorname{supp} h')$ and $\int_0^1 \rho h' = 1$. Define

$$\beta(t) = \begin{cases} h(t) & \text{if } t \geq 0, \\ \frac{1}{2} h(-t) & \text{if } t \leq 0. \end{cases}$$

Thus $[d(\beta \circ \tau_i)] = u_i$, and $(\rho\beta') \circ \tau_i d\tau_i$ is exact since $\int_0^1 \rho\beta' = 0$. Therefore we may define $\alpha \in a$ by

$$\alpha = \sum_{i=1}^k \alpha_i d(\beta \circ \tau_i),$$

where $a = \sum \alpha_i u_i \in H^1_{DR}(M, \mathbb{R})$ and $\alpha_i \in \mathbb{R}$. Next we construct metrics g_1 and g_2 such that $g_1(X_i, X_j) = \delta_{ij}$, $g_2(X_i, X_j) = \delta_{ij} \rho \circ \tau_i$, and the X_i are perpendicular to $\ker d\tau_i$ with respect to both metrics. This can be done locally using coordinates induced by the X_i, since $[X_i, X_j] = 0$ and $X_j \in TN_i$ imply $d\tau_i(X_j) = 0$, and then we may use a partition of unity. Hence we have $g_1(X_i, \cdot) = d\tau_i$ and

$$g_1\left(\sum \alpha_i \beta' \circ \tau_i X_i, \cdot\right) = \alpha,$$
$$g_2\left(\sum \alpha_i \beta' \circ \tau_i X_i, \cdot\right) = \sum \alpha_i (\beta' \rho) \circ \tau_i d\tau_i =: df,$$

i.e., $\sum \alpha_i \beta' \circ \tau_i X_i = g_1^{-1}\alpha = g_2^{-1} df$. Notice that α does not vanish on the closure of the set $W = \{x \in M : g_1(x) \neq g_2(x)\}$. Therefore we may add a small Morse function with support away from W so that α becomes nondegenerate. Since no orbit connecting two critical points is contained entirely in $\overline{W}$, we may obtain the Morse–Smale condition by perturbing the metric in $M \setminus \overline{W}$, which does not affect the construction. □

In order to apply the above proposition we recall the concept of equivariant homology (see, e.g., [**Whi78**]). If $\pi : \tilde{M} \to M$ is a covering with the group of deck automorphisms Γ, then the action of Γ on the singular chain complex $\triangle_*(\tilde{M})$ of $\tilde{M}$ turns it into a $\mathbb{Z}\Gamma$-module, where $\mathbb{Z}\Gamma$ is the group ring of Γ. Let Λ be a ring on which Γ acts on the right, thus inducing on Λ a structure of $\mathbb{Z}\Gamma$-module. The (singular) equivariant homology $H_*^\Gamma(\tilde{M}; \Lambda)$ is defined as the homology of the $\mathbb{Z}\Gamma$-complex $\triangle_*$ with coefficients in Λ, or, equivalently, $H_*^\Gamma(\tilde{M}; \Lambda) = H_*(\Lambda \otimes \triangle_*/Q)$, where the submodule $Q = Q(C_*, \Lambda) \subset \Lambda \otimes_{\mathbb{Z}} \triangle_*$ is generated by the elements of the form $\lambda A \otimes c - \lambda \otimes Ac$ with $A \in \Gamma$, $\lambda \in \Lambda$. It is isomorphic to the singular homology of M with local coefficient system induced by the action of Γ on Λ ([**Whi78**]). Similarly we define the equivariant Morse complex. Although $\tilde{M}$ is not compact, a Morse complex $C_*(M, f)$ lifts to a complex $C_*(\tilde{M}, \tilde{f}, \tilde{g})$, $\tilde{f} = f \circ \pi$, $\tilde{g} = \pi^* g$, which is equivariant. We obtain the homology groups $H_*^\Gamma(\tilde{M}, \tilde{f}, \tilde{g}; \Lambda)$ defined analogously.

PROPOSITION 2.7.5. *The groups $H_*^\Gamma(\tilde{M}, \tilde{f}, \tilde{g}; \Lambda)$ are independent of the choice of the function f and the metric g. They are isomorphic to the singular equivariant homology $H_*^\Gamma(\tilde{M}; \Lambda)$.*

PROOF. The proof of independence of the choice of f and g is the same as for the ordinary Morse complex. The proof of the remaining part of the proposition is very much the same as in the ordinary case (see [**Flo89b**], [**Sal90**]). Thus we may choose a Morse function $f : M \to \mathbb{R}$ and a filtration $\emptyset = N_{-1} \subset \cdots \subset N_n = M$ so that (N_i, N_{i-1}) is an index pair for an isolated invariant set consisting of critical points of index i. We may also choose an index pair (N_x, L_x) homeomorphic to $(D^i \times D^{n-i}, S^{i-1} \times D^{n-1})$ for every critical point x of f ($\mathrm{ind}(x) = i$) so that $(N_x, L_x) \subset (N_i, N_{i-1})$ and this inclusion is a composition of an excision map and a homotopy equivalence, homotopy being induced by the gradient flow. If we set $\tilde{N}_i = \pi^{-1} N_i$, $\tilde{N}_x = \pi^{-1} N_x$, and $\tilde{L}_x = \pi^{-1} L_x$, it follows that the inclusion

$$\bigcup_{\mathrm{ind}(x)=i} (\tilde{N}_x, \tilde{L}_x) \subset (\tilde{N}_i, \tilde{N}_{i-1}),$$

being a composition of an equivariant excision map and an equivariant homotopy equivalence, induces isomorphism in the equivariant homology. Thus $H_k^\Gamma(\tilde{N}_i, \tilde{N}_{i-1})$ is nonzero only if $k = i$ and the natural transformation from the universal coefficients theorem

$$\chi_X : \Lambda \otimes_{\mathbb{Z}\Gamma} H_*(\tilde{N}_i, \tilde{N}_{i-1}) \to H_*^\Gamma(\tilde{N}_i, \tilde{N}_{i-1}; \Lambda)$$

is an isomorphism. Thus there is a commutative diagram

$$\begin{array}{ccc}
\bigoplus_{\mathrm{ind}(x)=i} \Lambda \otimes_{\mathbb{Z}\Gamma} H_i(\tilde{N}_x, \tilde{L}_x) & \xrightarrow{\cong} & \bigoplus_{\mathrm{ind}(x)=i} H_i^{\Gamma}(\tilde{N}_x, \tilde{L}_x; \Lambda) \\
\cong \downarrow & & \downarrow \cong \\
\Lambda \otimes_{\mathbb{Z}\Gamma} H_i(\tilde{N}_i, \tilde{N}_{i-1}) & \xrightarrow{\cong} & H_i^{\Gamma}(\tilde{N}_i, \tilde{N}_{i-1}) \\
1\otimes\partial_i \downarrow & & \downarrow \partial_i^{\Gamma} \\
\Lambda \otimes_{\mathbb{Z}\Gamma} H_{i-1}(\tilde{N}_{i-1}, \tilde{N}_{i-2}) & \xrightarrow{\cong} & H_{i-1}^{\Gamma}(\tilde{N}_{i-1}, \tilde{N}_{i-2}) \\
\cong \uparrow & & \uparrow \cong \\
\bigoplus_{\mathrm{ind}(x)=i-1} \Lambda \otimes_{\mathbb{Z}\Gamma} H_i(\tilde{N}_x, \tilde{L}_x) & \xrightarrow{\cong} & \bigoplus_{\mathrm{ind}(x)=i-1} H_i^{\Gamma}(\tilde{N}_x, \tilde{L}_x; \Lambda),
\end{array}$$

where $\partial, \partial^{\Gamma}$ are the boundary homomorphisms in the exact sequence of the triple $(\tilde{N}_i, \tilde{N}_{i-1}, \tilde{N}_{i-2})$. Now notice that $(\tilde{N}_x, \tilde{L}_x)$ is a disjoint union of pairs $(N_{\tilde{x}}, L_{\tilde{x}})$ over all $\tilde{x} \in \pi^{-1}(x)$. Therefore we may consider the composition $\Lambda \otimes_{\Gamma} \triangle$ of the left vertical arrows separately for each pair $(\tilde{x}, \tilde{y})$ of critical points of $\tilde{f}$, $\mathrm{ind}(\tilde{x}) = \mathrm{ind}(\tilde{y}) + 1$. But these are contained in a compact invariant set, and therefore $\triangle$ is isomorphic to the boundary operator in the Morse complex ([**Flo89b**]). On the other hand, a standard algebraic argument shows that the homology of the complex $(H_*^{\Gamma}(\tilde{N}_i, \tilde{N}_{i-1}), \partial^{\Gamma})$ is isomorphic to the equivariant homology of $\tilde{M}$. □

COROLLARY 2.7.6. *If $a \in H^1(M, \mathbb{R})$ and $\Gamma = \pi_1/\ker a$, then, with the same notation as in Theorem 2.2.2,*

$$HN_*(M, a; \mathbb{Z}) \cong_{\Lambda_a} H_*^{\Gamma}(\tilde{M}; \Lambda_a(\mathbb{Z})),$$

with the action of Γ on Λ_a given by $(\lambda A)(B) = \lambda(BA^{-1})$.

PROOF. We describe the chain isomorphism between the Novikov complex of α and the equivariant Morse complex of f with $\alpha \in a$ and f chosen as in Proposition 2.7.4. Thus the corresponding flows are identical, and we define

$$\Phi : \Lambda_a \otimes C_*(\tilde{M}, \tilde{f}) \to C_*(M, \alpha),$$
$$\Phi\left(\sum_x \lambda_x \otimes x\right)(x_0) = \sum_A \lambda_{Ax_0}(A^{-1}).$$

It is easy to see that Φ is an epimorphism and $Q \subset \ker \Phi$. If $\sum_x \lambda_x \otimes x \in \ker \Phi$, then

$$0 = \Phi\Big(\sum_x \lambda_x \otimes x\Big)(Ax_0) = \sum_B \lambda_{BAx_0}(B^{-1})$$
$$= \sum_B \lambda_{Bx_0}(AB^{-1}) = \Big(\sum_B \lambda_{Bx_0} B\Big)(A),$$

i.e., $\sum_B \lambda_{Bx_0} B = 0$. Hence

$$\sum_x \lambda_x \otimes x = \sum_B \sum_{x_0} \lambda_{Bx_0} \otimes Bx_0 - \sum_{x_0} \sum_B \lambda_{Bx_0} B \otimes x_0 \in Q,$$

where we pick one x_0 in every orbit of Γ. Thus Φ is an isomorphism. Furthermore, we have

$$\begin{aligned}
\partial\Phi\left(\sum_x \lambda_x \otimes x\right)(y) &= \sum_{x'} n(x', y_0)\Phi\left(\sum_x \lambda_x \otimes x\right)(x') \\
&= \sum_{x'}\sum_A n(x', y)\lambda_{Ax'}(A^{-1}) = \sum_A\sum_x n(A^{-1}x, y)\lambda_x(A^{-1}), \\
\Phi\partial^\Gamma\left(\sum_x \lambda_x \otimes x\right)(y) &= \Phi\left(\sum_x\sum_{y'} \lambda_x n(x, y') \otimes y'\right)(y) \\
&= \sum_A\sum_x n(x, Ay)\lambda_x(A^{-1}),
\end{aligned}$$

and so Φ is a chain map since $n(Ax, Ay) = n(x, y)$. This proves the last assertion of Theorem 2.2.2 □

3. Floer theory

3.1. Introduction. In this section we prove a technical result which, we hope, can be used in computing Floer homology for a Lagrangian intersection whenever it has a simple structure. The two extreme cases are, on one hand, when two Lagrangian submanifolds L_0, L_1 intersect transversally at a single point, and on the other hand, when they are equal. If we are interested in the "local" behaviour of Floer's complex near the intersection, then the first case is trivial: there is one generator of the complex and no connecting orbits. In the case $L_0 = L_1$ it can be seen locally as the zero section of the cotangent bundle T^*L_0. This has been investigated in [**Flo88a**], and it turns out that the Floer complex can be identified with the Morse complex of a function on L. If we want to draw an analogy with the finite dimensional Morse theory, then the first case corresponds to a single critical point of a Morse function, and the second to a constant function, i.e., the whole manifold consists of critical points. An "intermediate" stage occurs when the critical set of a function consists only of nondegenerate critical manifolds. Then there is a spectral sequence whose entries are the homology of critical manifolds and which converges to the homology of the manifold ([**Bot82**]). We think that the clean intersection should play the role of the nondegenerate critical manifold in Floer theory. The main theorem of this section (Theorem 3.4.11) states that the Floer homology in a neighbourhood of a connected clean intersection is isomorphic to the singular homology of the intersection. The difficulty one encounters trying to obtain a global result is the possibility of the existence of "large" gradient lines, i.e., connecting orbits which cannot be deformed into a small neighbourhood of $L_0 \cap L_1$. We can avoid this by imposing quite restrictive topological assumptions, or sometimes these additional solutions do not exist for "dimensional" reasons.

The set-up is described in the next two subsections. Floer presented his theory in a series of papers ([**Flo88b**], [**Flo89b**], [**Flo88a**]) under the assumption that the symplectic form vanishes over $\pi^2(M, L_i)$. Details in the monotone case have been carried out in [**Oh93a**], and we shall base our set-up on these papers. The analysis necessary to construct Floer homology is a vast subject, and we do not make an attempt to reproduce all details here. We state the facts necessary for understanding and proving the results of 3.4. Subsection 3.5 contains some straightforward consequences of Theorem 3.4.11.

3.2. Preliminaries and notation. A 2-form ω on a manifold M is said to be *symplectic* if and only if it is closed and nondegenerate. This means that M is necessarily of even dimension $2n$, and that the exterior power $\omega^n = \omega \wedge \cdots \wedge \omega$ is a volume form on M. An example of a symplectic manifold is the total space of the cotangent bundle $\pi_N : T^*N \to N$ of any manifold N, equipped with the symplectic form $\omega_{T^*N} = -d\lambda_{T^*N}$, where λ_{T^*N} is the canonical 1-form on T^*N:

$$\lambda_{T^*N}(\alpha) = \alpha \circ d\pi_N, \qquad \alpha \in T^*N.$$

A diffeomorphism $\phi : (M_1, \omega_1) \to (M_2, \omega_2)$ between two symplectic manifolds is called a *symplectomorphism* if $\phi^*\omega_2 = \omega_1$. The Darboux theorem asserts that any symplectic manifold is locally symplectomorphic to an open subset of $\mathbb{R}^{2n}$ with the standard symplectic form

$$\omega_{\mathbb{R}^{2n}} = \sum_{i=1}^{n} dq_i \wedge dp_i.$$

An n-dimensional submanifold L of (M, ω) is said to be *Lagrangian* if ω vanishes on TL. An ingenious argument of Moser, used in the proof of the Darboux theorem, also yields the Lagrangian Neighbourhood Theorem: *A neighbourhood of a Lagrangian submanifold L in M is symplectomorphic to a neighbourhood of the zero section in T^*L.*

Any symplectic manifold (M, ω) admits an almost complex structure, that is, an endomorphism J of the tangent bundle such that $J^2 = -I$. This structure can be chosen to be compatible with ω, meaning that

$$\omega(Jv, w) + \omega(v, Jw) = 0 \quad \text{for } v, w \in TM$$

and $g_J := \omega \circ (\mathrm{id} \times J)$ is a Riemannian metric on M. This gives the reduction of the structure grop of TM to $U(n)$. Moreover, the space of ω-compatible almost complex structures is contractible.

3.3. Floer homology. Let L_0, L_1 be two compact Lagrangian submanifolds of a symplectic manifold (M, ω). Define the space of paths

$$\Omega(L_0, L_1) = \{\gamma \in C^\infty([0,1], M) \text{ such that } \gamma(i) \in L_i,\ i = 0, 1\} \tag{3.1}$$

and let $\Omega(L_0, L_1, \gamma_0)$ denote the path component of $\gamma_0 \in \Omega(L_0, L_1)$. Since the L_i are Lagrangian, the formula

$$\hat{\omega}(\gamma)\xi = \int_0^1 \omega(\xi, \dot{\gamma})\, dt, \tag{3.2}$$

where $\gamma \in \Omega(L_0, L_1)$ and ξ is a vector field along γ, defines a closed 1-form on $\Omega(L_0, L_1)$ in the sense that if a map $u : S^1 \times I \to M$, $u(s, \cdot) \in \Omega(L_0, L_1)$, represents a contractible loop in $\Omega(L_0, L_1)$, then

$$[\hat{\omega}](u) = \int u^*\omega = 0.$$

This way ω induces a homomorphism

$$I_\omega : \pi_1(\Omega(L_0, L_1), \gamma_0) \to \mathbb{R}.$$

If I_ω vanishes, then $da_\omega = \hat{\omega}$, where a_ω is a real valued function. In general, the symplectic action functional is defined only on the universal cover $\tilde{\Omega}$ of Ω,

$$\begin{array}{ccc} \tilde{\Omega}(L_0, L_1, \gamma_0) & \xrightarrow{a_\omega} & \mathbb{R} \\ \downarrow & & \downarrow \\ \Omega(L_0, L_1, \gamma_0) & \longrightarrow & \mathbb{R}/\mathrm{im}\, I_\omega, \end{array}$$

where $a_\omega(\tilde{\gamma}) = \int u^*\omega$. Here $u : I \times I \to M$ represents an element $\tilde{\gamma} \in \tilde{\Omega}$; that is, $u(0,t) = \gamma_0(t)$, $u(1,t) = \gamma(t)$ and $u(s,i) \in L_i$.

3.3.1. *Description of* $\pi_1(\Omega(L_0, L_1))$ *and* $\pi_0(\Omega(L_0, L_1))$. The path components, the fundamental group of $\Omega(L_0, L_1)$, and the homomorphism induced on the group $\pi_1(\Omega(L_0, L_1))$ by the symplectic form ω can be easily visualized. On the other hand, it may be convenient to describe them more directly in terms of $\pi_k(M)$ and $\pi_k(L_i)$. In what follows we may assume that $\Omega(L_0, L_1, \gamma_0)$ consists of all continuous paths connecting L_0 and L_1, since this space has the same weak homotopy type.

Fix a base point $\gamma_0 \in \Omega(L_0, L_1)$ and set $x_i = \gamma_0(i)$ for $i = 0, 1$. We introduce the notation

$$\begin{aligned} &l_i : L_i \hookrightarrow M, \\ &k_i : (M, x_i) \hookrightarrow (M, L_i), \\ &p_i : \Omega(L_0, L_1) \ni \gamma \mapsto \gamma(i) \in L_I, \end{aligned}$$

where k_i, l_i are the inclusions.

LEMMA 3.3.1. *If the L_i are connected, then there is a bijection*

$$\pi_0(\Omega(L_0, L_1)) \cong \mathrm{im}\, \pi_1(l_0) \backslash \pi_1(M) / \mathrm{im}\, \pi_1(l_1), \tag{3.4,}$$

where on the right hand side two elements $a, b \in \pi_1(M)$ are in the same equivalence class if and only if there are $c_i \in \pi_1(L_i)$ such that $a = \pi_1(l_0)(c_0) b \pi_1(l_1)(c_1)$.

Moreover, there are short exact sequences

$$0 \to \pi_2(M)/\mathrm{im}(\pi_2(l_1) - \pi_2(l_0)) \to \pi_1(\Omega(L_0, L_1)) \to \ker \pi_1(l_0)^{-1}\pi_1(l_1) \to 0 \tag{3.5}$$

and

$$\begin{aligned} &0 \to \pi_2(M, L_0) \oplus \pi_2(M, L_1) / (\pi_2(k_0), \pi_2(k_1))\pi_2(M) \\ &\quad \to \pi_1(\Omega(L_0, L_1)) \to \mathrm{im}\pi_1(l_0) \cap \mathrm{im}\pi_1(l_1) \to 0. \end{aligned} \tag{3.6}$$

PROOF. The evaluation map

$$p := (p_0, p_1) : \Omega(L_0, L_1) \ni \gamma \mapsto (\gamma(0), \gamma(1)) \in L_0 \times L_1$$

is a Serre fibration, and the fiber $p^{-1}(\gamma_0) \overset{j}{\hookrightarrow} \Omega(L_0, L_1)$ is homotopy equivalent to the space of loops

$$\begin{aligned} \Omega(M, x_0) = \{\gamma : [0,1] \to M : \gamma(0) = x_0 = \gamma(1) \in L_1\}, \\ p_0^{-1}(\gamma_0) \ni \gamma \mapsto \gamma * \gamma_0^{-1} \in \Omega(M, x_0). \end{aligned}$$

Therefore $\pi_k(p_0^{-1}(\gamma_0)) \cong \pi_{k+1}(M, x_0)$, and there is the exact homotopy sequence of homotopy groups

$$\begin{aligned}(3.7)\qquad \cdots \longrightarrow \pi_{k+1}(M) \xrightarrow{(\pi_k(j))} \pi_k(\Omega(L_0, L_1)) \xrightarrow{(\pi_k(p_0), \pi_k(p_1))} \pi_k(L_0) \oplus \pi_k(L_1) \\ \xrightarrow{\pi_k(l_1) - \pi_k(l_0)} \pi_k(M) \rightarrow \cdots,\end{aligned}$$

from which (3.4) and (3.5) follow. Note that although the map

$$\pi_1(l_0)^{-1}\pi_1(l_1) : \pi_1(L_0) \oplus \pi_1(L_1) \ni (a, b) \mapsto \pi_1(l_0)a^{-1}\pi_1(l_1)b \in \pi_1(M)$$

is not a homomorphism, its kernel is a subgroup of $\pi_1(L_0) \oplus \pi_1(L_1)$. Then the map

$$(\pi_2(k_0), 0) : \pi_2(M) \rightarrow \pi_2(M, L_0) \oplus \pi_2(M, L_1)$$

induces the isomorphism

$$\pi_2(M)/(\operatorname{im}\pi_2(l_0) + \operatorname{im}\pi_2(l_1)) \cong \pi_2(M, L_0) \oplus \pi_2(M, L_1)/(\pi_2(k_0), \pi_2(k_1))\pi_2(M).$$

We also have the isomorphism

$$\ker \pi_1(l_0)^{-1}\pi_1(l_1) \ni (a_0, a_1) \mapsto \pi_1(l_0)a_0 = \pi_1(l_1)a_1 \in \operatorname{im}\pi_1(l_0) \cap \operatorname{im}\pi_1(l_1),$$

and thus (3.6). □

3.3.2. *Monotonicity.* There is another homomorphism defined on π_1 of the path space

$$I_\mu = I_{\mu, L_0, L_1} : \pi_1(\Omega(L_0, L_1), \gamma_0) \rightarrow \mathbb{Z}.$$

Namely, $I_\mu(u)$ is the Maslov class of the bundle $u^*TM \rightarrow S^1 \times I$. This may be described as follows. The bundle u^*TM admits a symplectic trivialization $\Psi : u^*TM \rightarrow S^1 \times I \times \mathbb{C}^n$ which, restricted to $S^1 \times \{i\}$, gives a loop

$$\lambda_i : S^1 \ni t \mapsto \Psi(u^*TL_i(t, i)) \subset \mathbb{C}^n$$

of Lagrangian planes in $\mathbb{C}^n$. We set

$$I_\mu^\Psi(u) = \mu(\lambda_1) - \mu(\lambda_0),$$

where μ is the Maslov index. If Ψ' is another symplectic trivialization, then

$$\Psi' \circ \Psi^{-1}(t, s, z) = (t, s, \psi_s(t)z)$$

and $\psi_s(t) \in \mathrm{Symp}(2n)$. We then have the formula for the Maslov index of $\lambda_i' : t \mapsto \Psi'(u^*TL_i(t, i))$, namely

$$\mu(\lambda_i') = \mu(\psi_i \circ \lambda_i) = \mu(\lambda_i) + \mu(\psi_i),$$

where in the last term μ denotes the isomorphism between $\pi_1(\mathrm{Symp})$ and $\mathbb{Z}$. But the loops ψ_i are homotopic in $\mathrm{Symp}(2n)$, and so $I_\mu^{\psi'}(u) = I_\mu^\Psi(u)$. We say that a triple (M, L_0, L_1, γ_0) is *monotone* if

$$I_\omega = \lambda I_\mu$$

for some $\lambda \geq 0$.

REMARK 3.3.2. In view of the decomposition of (3.6), the monotonicity means that I_ω and I_μ are proportional over $\pi_2(M, L_i)$ and that the induced homomorphism

$$(I_\omega - \lambda I_\mu)_\# : \operatorname{im} \pi_1(l_0) \cap \operatorname{im} \pi_1(l_1) \to \mathbb{R}$$

vanishes. This happens, for example, if one of the groups $\operatorname{im} \pi_1(l_i)$ is torsion or if $L_0 = L_1$, since in the latter case any element in $\operatorname{im} \pi_1(l_i)$ is represented in Ω by a loop u of constant paths, and so $I_\omega[u] = I_\mu[u] = 0$. Note that these are the topological assumptions in [**Oh93a**].

If $A \in \pi_2(M)$ and $\pi_1(j) : \pi_2(M) \to \pi_1(\Omega)$ is as in (3.7), then

$$I_\mu \circ \pi_1(j)(A) = 2c_1(A),$$

where c_1 is the first Chern class of TM (see [**Flo88a**], [**Oh93a**]). In particular, if N_{c_1} is the minimal Chern number and N_μ the minimal Maslov number, i.e., the positive generator of $\operatorname{im} I_\mu$, then N_μ necessarily divides $2N_{c_1}$. Similarly, if Σ_i denote the generators of $\operatorname{im} I_\mu|_{\pi_2(M,L_i)}$, then $N_\mu | \Sigma_i | 2N_{c_1}$. Also, if a triple (M, L_0, L_1) is monotone, then M is itself monotone, i.e.,

$$[\omega]_{\pi_2(M)} = 2\lambda c_{1\,\pi_2(M)}.$$

Conversely, if M is monotone and the groups $\pi_1(L_i)$ are torsion, then it follows from (3.5) that (M, L_0, L_1) is monotone as well.

3.3.3. *The gradient flow.* In order to develop the Morse theory we have to introduce a Hamiltonian perturbation of (3.2), since the form $\tilde{\omega}$ may be degenerate. Let X_t be a Hamiltonian vector field on M with the Hamiltonian $H : [0,1] \times M \to \mathbb{R}$, that is,

$$i_{X_t}\omega = dH_t, \qquad t \in [0,1],$$

and let ϕ_t be the family of symplectomorphisms generated by X_t. Consider the perturbed form

$$\hat{\omega}_H(\gamma)\xi = \hat{\omega}(\gamma)\xi + \int_0^1 dH_t(\xi(t))\,dt = \int_0^1 \omega(\xi, \dot{\gamma} - X_t(\gamma))\,dt.$$

It is exact if $\hat{\omega}$ is exact and $\hat{\omega}_H = da_H$ with

$$a_H(\gamma) = \int u^*\omega + \int_0^1 H_t(\gamma(t)\,dt,$$

where u is as in (3.3). The critical points of a_H are precisely the paths $x \in \Omega(L_0, L_1)$ satisfying

$$\dot{x}(t) = X_t(x(t)), \tag{3.8}$$

which in turn are in one-to-one correspondence with the intersection points of L_0 and $\phi_1^{-1}(L_1)$. A solution to (3.8) is a *nondegenerate* critical point of a_H if

$$d\phi^{-1}(T_{x(1)}L_1) \cap T_{x(0)}L_0 = \{0\},$$

that is, if L_0 intersects $\phi_1^{-1}(L_1)$ transversally. We shall say that a Hamiltonian H is *regular* if all critical points of a_H are nondegenerate. This holds for a generic function in $\mathcal{H} := C^\infty(I \times M, \mathbb{R})$. Let $\mathcal{J}_\omega$ denote the space of all smooth t-dependent families $J : [0,1] \times M \to \mathrm{End}(TM)$ of ω-compatible almost complex structures on M. For a given $J \in \mathcal{J}_\omega$ the L^2-gradient of a_H is given by

$$\operatorname{grad} a_H(\gamma) = J(\dot{\gamma} - X_t(\gamma)),$$

and we may write the gradient equation

$$\overline{\partial}_{J,H}(u) = \frac{\partial u}{\partial s} + J_t(u)\frac{\partial u}{\partial t} + \nabla H_t(u) = 0, \tag{3.9}$$

$$u : \mathbb{R} \times I \to M, \qquad u(s,i) \in L_I, \tag{3.10}$$

where $\nabla H_t = -J_t X_t$ is the gradient of H_t with respect to the metric g_{J_t}.

This is an elliptic equation, and so the Cauchy problem is not well posed. On the other hand, by the elliptic regularity, any $W^{1,p}_{\mathrm{loc}}$ solution to (3.9), (3.10) with $p > 2$ is smooth ([**Flo88b**], Lemma 2.1). Instead of trying to define a gradient flow of a_H globally, Floer considered the solutions with bounded energy ℓ^2:

$$\ell^2(u) := \int_{-\infty}^{+\infty} \int_0^1 \left\| \frac{\partial u}{\partial s} \right\|_{L^2}^2 dt\, ds,$$
$$\mathcal{M}_{J,H}(L_0, L_1) := \{u : \overline{\partial}_{J,H}(u) = 0 \text{ and } \ell^2(u) < \infty\}.$$

It turns out that these are precisely the solutions which satisfy the limit conditions

$$u(s,\cdot) \to x^\pm, \qquad s \to \pm\infty,$$

for some $x^-, x^+ \in \operatorname{Crit} a_H$. We then have

$$\ell^2(u) = \int u^*\omega + \int_0^1 H_t(x^+(t)) - H_t(x^-(t))\, dt.$$

We denote the set of these solutions by $\mathcal{M}_{J,H}(x^-, x^+)$. We fix a $p > 2$, and similarly as in the previous section we introduce the space of paths $\mathcal{P}(x^-, x^+)$ connecting x^- and x^+ in Ω:

$$\begin{aligned}\mathcal{P}(x^-, x^+) = \Big\{ u : \mathbb{R} \times I \to M \text{ such that } & u \in W^{1,p}_{\mathrm{loc}}(\mathbb{R} \times I, M), \\ & \exists T > 0, \exists \xi_- \in W^{1,p}((-\infty, -T] \times I; (x^-)^*TM), \\ & \exists \xi_+ \in W^{1,p}([T, \infty) \times I; (x^+)^*TM) \text{ such that} \\ & u(s,t) = \exp_{x^\pm(t)} \xi_\pm(s,t) \text{ for } \pm s \geq T \text{ and } t \in I \Big\}.\end{aligned}$$

It is clear that the definition does not depend on the choice of a metric on M. If $x^\pm$ are nondegenerate, then $\mathcal{P}(x^-, x^+)$ is a Banach manifold ([**Flo88b**], Theorem 3), whose tangent space at $u \in \mathcal{P}(x^-, x^+) \cap C^\infty(\mathbb{R} \times I, M)$ consists of $W^{1,p}$-sections of the pullback bundle u^*TM satisfying $\xi(s,i) \in T_{u(s,i)}L_i$, which we denote by $W^{1,p}_{L_0,L_1}(u^*TM)$. Then $\overline{\partial}_{J,H}$ can be seen as a section of the Banach vector bundle $\mathcal{E} \to \mathcal{P}$ with fibres $\mathcal{E}_u = L^p(u^*TM)$. For any $u \in \mathcal{P}$ the linearization of $\overline{\partial}_{J,H}$ along $u \in \mathcal{P}(x^-, x^+)$,

$$D_{J,H}(u)\xi = \nabla_s \xi + J_t(u)\nabla_t \xi + \nabla_\xi J_t(u)\frac{\partial u}{\partial t} + \nabla_\xi \nabla H(u), \tag{3.11}$$

defines a bounded operator $D_{J,H}(u) : W^{1,p}_{L_0,L_1}(u^*TM) \to L^p(u^*TM)$.

THEOREM 3.3.3. *If x^-, x^+ are nondegenerate critical points of a_H, then the operator $D_{J,H}(u)$ is Fredholm. There is a generic set $(\mathcal{J} \times \mathcal{H})_{\text{reg}}(L_0, L_1) \subset \mathcal{J} \times \mathcal{H}$ such that $D_{J,H}(u)$ is onto for any $(J, H) \in (\mathcal{J} \times \mathcal{H}_{\text{reg}})$ and $u \in \mathcal{M}_{J,H}(L_0, L_1)$, and then $\mathcal{M}_{J,H}(x^-, x^+)$ is a collection of finite dimensional manifolds with*

$$\dim_u \mathcal{M}_{J,H}(x^-, x^+) = \operatorname{ind} D_{J,H}(u).$$

The fact that $D_{J,H}(u)$ is Fredholm was proved in [**Flo88b**]. The proof of the transversality result in that paper contains some gaps, which were filled in [**FHS94**]. The main idea is essentially as in 2.5. We may assume without loss of generality that $H = 0$ (see Remark 3.13 below). We take $\mathcal{J}$ as the parameter space and consider the Fredholm section

$$\mathcal{F} : \mathcal{P} \times \mathcal{J} \ni (u, J) \mapsto \overline{\partial}_J(u) \in \mathcal{E}.$$

We have to prove that $\mathcal{F}$ is transversal to the zero section. Then the set

$$\mathcal{Z} = \{(u, J) \in \mathcal{P} \times \mathcal{J} : \overline{\partial}_J(u) = 0\}$$

is an infinite dimensional Banach manifold and the projection

$$j : \mathcal{Z} \to \mathcal{J}$$

is a Fredholm map. The set $\mathcal{J}_{\text{reg}}$ consists precisely of the regular values of j, and thus is residual in $\mathcal{J}$ by the Sard–Smale theorem. It remains to prove that the vertical part of $d\mathcal{F}$,

$$D\mathcal{F}(u, J)(\xi, Y) = D_J(u)\xi + Y(u)\frac{\partial u}{\partial t},$$

is onto for any $(u, J) \in \mathcal{P} \times \mathcal{J}$. As in 2.5 it is enough to show that $D\mathcal{F}(u, J)$ has a dense range. Suppose then, for a contradiction, that a non-zero vector field $\eta \in L^q(u)$, where $\frac{1}{q} + \frac{1}{p} = 1$, satisfies the conditions

$$D_J^*(u)\eta = 0,$$

$$\int_{-\infty}^{+\infty} \int_0^1 \left\langle \eta, Y_t(u)\frac{\partial u}{\partial t} \right\rangle dt\, ds = 0. \tag{3.12}$$

Since $D_J^*(u)$ is an elliptic operator with smooth coefficients, the vanishing of η in a non-empty open set yields, by the Aronszajn unique continuation theorem, $\eta \equiv 0$. Now, define

$$R(u) := \{(s,t) \in \mathbb{R} \times (0,1) : \frac{\partial u}{\partial s} \neq 0,\ u(s,t) \neq x^{\pm} \text{ and } u(s,t) \notin u(\mathbb{R} - \{s\}, t)\}.$$

A calculation in [**SZ94**] shows that if $\eta(s,t) \neq 0$ for some $(s,t) \in R(u)$, then there exists $Y \in T_J\mathcal{J}$ such that the integral (3.12) is positive. The subtlety of the argument lies in proving that $R(u)$ is non-empty. The proof, based on the Carleman similarity principle, is formulated in [**FHS94**], although in a slightly different set-up, for pseudoholomorphic maps defind on the whole of $\mathbb{R}^2$. Theorem 4.3 in that paper states that $R(u)$ is open and dense in $\mathbb{R}^2$. However, the argument has a local character, and so we may simply disregard the boundary points in our case.

REMARK 3.3.4. We assumed above that $H = 0$. It is easy to see that this does not affect generality of the proof: Let ϕ be an exact symplectomorphism and ϕ_t a Hamiltonian isotopy joining id and ϕ. There is a bijection

$$\Phi^* : \Omega(L_0, L_1, \gamma_0) \to \Omega(L_0, \phi^{-1}(L_1), \phi_\cdot^{-1} \circ \gamma_0) \tag{3.13}$$

given by $\Phi(\gamma)(t) = \phi_t^{-1}(\gamma(t))$, and this gives the one-to-one correspondence between the trajectory spaces

$$\mathcal{M}_{J,H}(L_0, L_1) \cong \mathcal{M}_{\tilde{J},0}(L_0, \phi^{-1}(L_1)),$$

where $\tilde{J}_t = \phi_t^* J_t$. Indeed, if $u : \mathbb{R} \times I \to M$ satisfies (3.9) and (3.10), then $\tilde{u}(s,t) := \phi_t^{-1}(u(s,t))$ defines a solution to

$$\frac{\partial \tilde{u}}{\partial s} + \tilde{J}_t(u) \frac{\partial \tilde{u}}{\partial t} = 0$$

with $\tilde{u}(s,0) \in L_0$ and $\tilde{u}(s,1) \in \phi^{-1}(L_1)$. Similarly, the correspondence $\xi \mapsto \tilde{\xi}$, $\tilde{\xi}(s,t) := d\phi_t^{-1}\xi(s,t)$, defines the isomorphisms $\Phi_L^* : L^p(u^*TM) \to L^p(\tilde{u}^*TM)$, $(\Phi_W^* : W^{1,p}_{L_0,L_1}(u^*TM) \to W^{1,p}_{L_0,\psi^{-1}L_1}(\tilde{u}^*TM))$ and

$$D_{\tilde{J},0}(\tilde{u}) \circ \Phi_W^* = \Phi_L^* \circ D_{J,H}(u).$$

This is because the Levi–Civita connection $\nabla^{\tilde{J}_t}$ associated with $g_{\tilde{J}_t}$ is equal to the pullback $\phi_t^* \nabla^{J_t}$, and also because

$$\begin{aligned}
\bar{\partial}_{\tilde{J},0}(\tilde{u}) &= \frac{\partial}{\partial s}\left(\phi_t^{-1} \circ u\right) + d\phi_t^{-1} \circ J \circ d\phi_t \left(\frac{\partial}{\partial t}\left(\phi_t^{-1} \circ u\right)\right) \\
&= d\phi_t^{-1}\left(\frac{\partial u}{\partial s} + J_t(u)\left(\frac{\partial u}{\partial t} - X_t(u)\right)\right) = \bar{\partial}_{J,H}(u).
\end{aligned}$$

Hence the operators $D_{\tilde{J},0}(\tilde{u})$ and $D_{J,H}(u)$ have the same index, and $(\tilde{J},0) \in (\mathcal{J} \times \mathcal{H})_{\text{reg}}(L_0, \phi^{-1}(L_1))$ if and only if $(J,H) \in (\mathcal{J} \times \mathcal{H})_{\text{reg}}(L_0, L_1)$ (more generally, a pair (J, H_1) corresponds to $(\tilde{J}, \tilde{H}_1)$ with $\tilde{H}_1 = (H - H_1) \circ \phi$).

Clearly, the map (3.13) preserves also the homomorphisms I_ω and I_μ,

$$I_{\omega, L_0, \phi^{-1}(L_1)} \circ \Phi^* = I_{\omega, L_0, L_1} \qquad \text{and} \qquad I_{\mu, L_0, \phi^{-1}(L_1)} \circ \Phi^* = I_{\mu, L_0, L_1}.$$

The index $\mu_u := \operatorname{ind} D_{J,H}(u)$ depends on the homotopy class of u and is equal to the Viterbo index. This is defined as follows ([**Vit87**], [**Flo88a**]). In view of the remark above we may assume without loss of generality that $H = 0$ and let $u : I^2 \to M$ be a path in $\Omega(L_0, L_1)$ joining two nondegenerate intersection points, $u(s,i) \in L_i$, $u(0,t) = x^-$, $u(1,t) = x^+$. We may choose a symplectic trivialisation $\Psi : u^*TM \to I^2 \times \mathbb{C}^n$ which is constant at the end points $u(i,t)$ and such that $\Psi(T_{x^\pm} L_1) = i\Psi(T_{x^\pm} L_0)$. Then the index m_u is defined as the Maslov class of the loop

$$\lambda(\tau) = \begin{cases} \Psi(T_{u(4\tau,0)} L_0) & \text{for } 0 \le \tau \le \frac{1}{4}, \\ -ie^{2\pi i \tau}\Psi(T_{x^+} L_0) & \text{for } \frac{1}{4} \le \tau \le \frac{1}{2}, \\ \Psi(T_{u(3-4\tau,1)} L_1) & \text{for } \frac{1}{2} \le \tau \le \frac{3}{4}, \\ ie^{2\pi i \tau}\Psi(T_{x^-} L_1) & \text{for } \frac{3}{4} \le \tau \le 1. \end{cases}$$

In [**Flo88a**] Floer proved that these indices coincide. More precisely, let $\beta : \mathbb{R} \to I$ be a cutoff function such that $\beta((-\infty, 0]) = 0$ and $\beta([1, +\infty)) = 1$, and let

$u \in \mathcal{M}_{J,0}(L_0, L_1)$. Then $m_{u'} = \mu_u$, where $u' : I^2 \to M$ is chosen so that u and $u'(\beta(s), t)$ are homotopic in $\Omega(L_0, L_1)$ with the end points fixed. There is a straightforward relation between the Viterbo index and I_μ.

LEMMA 3.3.5. *If $u_1, u_2 : I^2 \to M$ connect two intersection points x^-, x^+, then*

$$I_\mu(u_1 * (-u_2)) = m_{u_1} - m_{u_2}.$$

Here we use the notation

$$(u_1 * (-u_2))(s,t) = \begin{cases} u_1(2s, t) & \text{if } s \leq \frac{1}{2}, \\ u_2(1-2s, t) & \text{if } s \geq \frac{1}{2}. \end{cases}$$

This result together with Remark 3.3.4 implies

COROLLARY 3.3.6. *If the triple (M, L_0, L_1) is monotone and H is regular, then for any $u_1, u_2 \in \mathcal{M}_{J,H}(L_0, L_1)$*

$$\ell^2(u_1) = \ell^2(u_2) \Leftrightarrow \mu_{u_1} = \mu_{u_2}.$$

In order to construct the boundary operator we need a compactness property for 1- and 2-dimensional parts of $\mathcal{M}_{J,H}(L_0, L_1)$ similar to Proposition 2.2.1. We say that a sequence $\{u_\nu\}_{\nu=1}^\infty \subset \mathcal{C}^\infty(\mathbb{R}\times I, M)$ converges to a split trajectory $(u^1, \ldots, u^m)$ if there are collections of critical points $x^+ = x_0, x_1, \ldots, x_m = x^-$ of $\hat{\omega}_H$, trajectories $u^k \in \mathcal{M}_{J,H}(x_k, x_{k-1})$, and sequences of time shifts $\{s_\nu^k\}_{\nu=1}^\infty$, $k = 1, \ldots, m$, such that $u_\nu(\cdot + s_\nu^k, \cdot)$ converges with its all derivatives on compact sets to u^k.

The following propositions summarize the compactness results. For the proof we refer to [**Flo88b**], [**Flo89a**], [**Oh93a**]. Let $u_\nu \in \mathcal{M}_{J_\nu, H_\nu}(L_0, L_1)$ be a sequence of trajectories with (J_ν, H_ν) converging to some (J, H) in C^1.

PROPOSITION 3.3.7. *Assume that $\omega|_{\pi_2(M,L_i)} = 0$, $i = 0, 1$, and the energy of u_ν is bounded,*

$$\ell^2(u_\nu) < C.$$

Then a subsequence of u_ν converges to a split trajectory $\underline{u} \subset \mathcal{M}_{J,H}(L_0, L_1)$, and

$$\sum \ell^2(u^i) \leq \limsup \ell^2(u_\nu).$$

PROPOSITION 3.3.8. *Assume that the triple (M, L_0, L_1) is monotone, $L_0 \pitchfork L_1$, and $\Sigma_i \geq 3$ for $i = 0, 1$. Let $u_\nu \in \mathcal{M}_{J_\nu, H_\nu}(L_0, L_1)$ be a sequence of trajectories of index $\eta_{u_\nu} = 1$ or 2. Then a subsequence of u_ν converges to a split trajectory $\underline{u} \in \mathcal{M}_{J,H}(L_0, L_1)$.*

LEMMA 3.3.9. *If $\ell^2(u_\nu) \to 0$, then u_ν converges uniformly to a constant map.*

In the statements above we assume the energy bound for u_ν, that is, a bound in the $W^{1,2}$ topology. If the derivatives of the u_ν are bounded in L^∞, then they are bounded in L^p for $p > 2$, and the elliptic bootstrapping argument yields a subsequence convergent in C^∞_{loc}. Otherwise, a non-constant pseudoholomorphic disc or sphere would bubble off. Such a disc must have positive energy, which is impossible under the assumptions of Proposition 3.3.7 or Lemma 3.3.9. If the intersection is transverse, then the index of the disc is well defined and cannot exceed the index of u_ν. This guarantees convergence in Proposition 3.3.8 (recall that Σ_i denote the generators of $\operatorname{im} I_\mu|_{\pi_2(M,L_i)}$).

The Floer complex $C_*(L_0, L_1, \gamma_0, J, H)$ is defined for a regular pair (J, H). It is generated as a $\mathbb{Z}_2$ module by the critical points of a_H, and the boundary operator $\partial : C_* \to C_*$ is given by

$$\partial x = \sum \langle y, \partial x \rangle y, \tag{3.14}$$

where $\langle y, \partial x \rangle$ is the number modulo 2 of one-dimensional components of $\mathcal{M}_{J,H}(x, y)$. Propositions 3.3.7 and 3.3.8 imply that the number of these components is finite, and thus (3.14) is well defined. Then compactness of the 2-dimensional parts together with the glueing argument yields $\partial \circ \partial = 0$ as in the finite dimensional case. The following theorem is the central result of Floer's papers [**Flo88a**], [**Flo88b**], [**Flo89b**]. The details in the monotone case have been carried out in [**Oh93a**].

THEOREM 3.3.10. *Assume that $I_\omega = 0$ or that the triple (M, L_0, L_1, γ_0) is monotone with $\Sigma_i \geq 3$. Then the boundary operator is well defined, and $\partial \circ \partial = 0$. For any two regular pairs (J^0, H^0) and (J^1, H^1) there is a natural isomorphism between the resulting homology groups*

$$HF_*(L_0, L_1, \gamma_0, J^0, H^0) \cong HF_*(L_0, L_1, \gamma_0, J^1, H^1). \tag{3.15}$$

We denote these groups by $HF_*(L_0, L_1, \gamma_0)$.

REMARK 3.3.11. a) Obviously the Floer complex is symplectically invariant, i.e., for any symplectomorphism ϕ,

$$C_*(L_0, L_1, \gamma_0, J, H) \cong C_*(\phi(L_0), \phi(L_1), \phi \circ \gamma_0, \phi_* J, H \circ \phi^{-1}).$$

b) The map (3.13) yields the isomorphism between the Floer complexes

$$C_*(L_0, L_1, \gamma_0, J, H) \cong C_*(L_0, \phi^{-1}(L_1), \phi_\cdot^{-1} \circ \gamma_0, \phi_t^* J_t, 0)$$

for any regular pair (J, H). Consequently, we obtain an isomorphism

$$HF_*(L_0, L_1) \cong HF_*(L_0, \phi^{-1}(L_1)).$$

However, unlike the isomorphism in (3.15), this is not naturally defined, because it depends a priori on the homotopy class of the Hamiltonian isotopy ϕ_t.

3.3.4. *Floer homology of an isolated invariant set.* We can define Floer homology in a slightly more general setting. Let $\mathcal{U} \subset \Omega(L_0, L_1)$ be a closed subset and assume that $\mathcal{U}$ is *bounded*, i.e., the evaluation map $\mathcal{U} \times I \ni (\gamma, t) \mapsto \gamma(t) \in M$ has precompact image. Let $\mathcal{M}_{J,H}(L_0, L_1, \mathcal{U})$ denote the set of $u \in \mathcal{M}_{J,H}(L_0, L_1)$ such that $u(s, \cdot) \in \mathcal{U}$ for all $s \in \mathbb{R}$. Then the maximal invariant subset $\mathcal{S}_{J,H}(\mathcal{U})$ of $\mathcal{U}$ is defined to be the image of $\mathcal{M}_{J,H}(L_0, L_1, \mathcal{U})$ under the evaluation map

$$\mathrm{ev} : \mathbb{R} \times \mathcal{M}_{J,H}(L_0, L_1) \to \Omega(L_0, L_1), \qquad \mathrm{ev}(s, u)(t) = u(s, t).$$

We say that $\mathcal{S}_{J,H}(\mathcal{U})$ is *isolated* if its closure is contained in the interior of $\mathcal{U}$ (cf. (2.30)).

If $\mathcal{S}_{J,H}(\mathcal{U})$ is isolated and $u_\nu \in \mathcal{M}_{J,H}(L_0, L_1, \mathcal{U})$ converges to an element $u \in \mathcal{M}_{J,H}(L_0, L_1)$, then $u \in \mathcal{M}_{J,H}(L_0, L_1, \mathcal{U})$. Therefore we can build the local Floer homology $H_*(C_*(L_0, L_1, \mathcal{U}, J, H))$ out of all critical points of $\hat{\omega}_H$ in $\mathcal{U}$ and the connecting orbits in $\mathcal{M}_{J,H}(L_0, L_1, \mathcal{U})$. The following proposition is the analogue of Proposition 1.2 in [**Flo89a**] in the Lagrangian intersection setting.

PROPOSITION 3.3.12. *Assume that $\mathcal{U}$ is bounded, $\mathcal{S}_{J,H}(\mathcal{U})$ is isolated and the symplectic action a_H is defined on $\mathcal{U}$. There is an $\varepsilon > 0$ such that if $\|J' - J\|_{C^1} < \varepsilon$ and $\|H' - H\|_{C^1} < \varepsilon$, then $\mathcal{S}_{J',H'}(\mathcal{U})$ is isolated, and if moreover the pairs (J, H), (J', H') are regular, then*

$$H_*(C_*(L_0, L_1, \mathcal{U}, J', H')) \cong H_*(C_*(L_0, L_1, \mathcal{U}, J, H)).$$

In this context we say that $\mathcal{S}_{J',H'}(\mathcal{U})$ is a *continuation* of $\mathcal{S}_{J,H}(\mathcal{U})$.

3.4. Clean intersection. The aim of this section is to prove that the local Floer homology in a neighbourhood of a Lagrangian intersection is isomorphic to its singular homology provided that the intersection is appropriately regular. We say that two submanifolds L_0, L_1 of M have a *clean intersection along a manifold* N if and only if $N \subset L_0 \cap L_1$ and $T_xN = T_xL_0 \cap T_xL_1$ for $x \in N$. The special case $L_0 = L_1 = N$ was considered by Floer, who proved the existence of the isomorphism in [**Flo88a**]. The results of this section should be seen as a generalization of Floer's theorem.

We sketch the main points of the proof. If two Lagrangian submanifolds L_0, L_1 of M intersect cleanly along a compact manifold N, then, seen as a subset of the pathspace, $N \subset \Omega(L_0, L_1)$ is an isolated invariant set for the unperturbed symplectic action. This is because N is a strong deformation retract of its neighbourhood $\mathcal{U}$ in $\Omega(L_0, L_1)$, and therefore $\hat{\omega}$ is exact in $\mathcal{U}$. Moreover, the critical points of the action functional a in $\mathcal{U}$ are the constant paths at the points of N, and if N is connected they lie in the same level set of a. Since a decreases along its gradient lines there are no connecting orbits in $\mathcal{U}$ except for the constant ones, and $\mathcal{U}$ isolates N. The critical points in N are degenerate; in fact, the kernel of the Hessian of a at x is exactly T_xN. Then perturb the functional in the direction of N only by adding a Hamiltonian term. In other words, the Hamiltonian H may be constant in the direction orthogonal to N. But then the gradient of H is tangent to N, and the critical points and the gradient lines of $H|_N$ coincide with those of a_H which are t-independent. The crucial point of the proof is that they are the only ones in $\mathcal{U}$ and therefore the Floer complex coincides with the Morse complex of H_N.

The section begins with the description of a normal form neighbourhood of a clean Lagrangian intersection. In this form we then construct the appropriate Hamiltonian H, which, however, is not compactly supported. Finally, putting these results together, we formulate and prove the main theorem of this section—Theorem 3.4.11.

3.4.1. *The normal form of a clean Lagrangian intersection.* Let L be a manifold and N a submanifold of L. Let $TN^{\text{annih}} \subset T^*L$ be the subbundle of $T^*L|_N$ annihilating TN:

$$TN^{\text{annih}} := \{\alpha \in T^*L|_N : \alpha|_{TN} = 0\}.$$

This is a Lagrangian submanifold of (T^*L, ω_{T^*L}), since $\dim TN^{\text{annih}} = \dim L$ and the canonical 1-form λ_{T^*L} vanishes on TN^{annih}. Clearly $N = L \cap TN^{\text{annih}}$ is a clean intersection, since $T_x(TN^{\text{annih}}) = T_xN \oplus T_xN^{\text{annih}}$ for $x \in N$. We prove that every clean Lagrangian intersection can be put in this form. Thus the following proposition may be seen as a generalization of the Lagrangian neighbourhood theorem.

PROPOSITION 3.4.1. *Let (M, ω) be a symplectic manifold and L_0, L_1 two Lagrangian submanifolds of M which intersect cleanly along a compact manifold N.*

There exist a vector bundle $\tau : L \to N$, a neighbourhood V_0 of N in T^*L, a neighbourhood U_0 of N in M and a symplectomorphism $\phi : U_0 \to V_0$ such that

$$\phi(L_0 \cap U_0) = L \cap V_0 \qquad \textit{and} \qquad \phi(L_1 \cap U_0) = TN^{\text{annih}} \cap V_0. \tag{3.16}$$

PROOF. By the Lagrangian neighbourhood theorem there is a symplectomorphism $\chi_0 : (W, \omega_{T^*L_0}) \to (M, \omega)$ defined on a neighbourhood W of the zero section in T^*L_0 such that $\chi_0|_{L_0} = \text{id}$. Fix any metric on L_0, and take $L = TN^\perp \subset TL_0$ to be the orthogonal complement of TN in TL_0. Then the exponential map is a diffeomorphism between neighbourhoods of N in L_0 and in L, which induces a natural symplectomorphism of the cotangent bundles (restricted to these neighbourhoods). Therefore we may assume without loss of generality that $L_0 = L$ and $M = T^*L$.

Denote $L_2 = TN^{\text{annih}} \subset T^*L$. The proof of the proposition is based on the observation that both L_1 and L_2 are transversal to the "complement" of N in T^*L, and therefore L_1 can be seen as a graph over L_2. More precisely, we shall prove the following result.

LEMMA 3.4.2. *There exists a symplectomorphism $\chi_1 : U_1 \to V_1$, where $U_1 \subset T^*L$ and $V_1 \subset T^*L_2$ are neighbourhoods of N, such that*

$$\chi|_{L_2 \cap U_1} = \text{id}, \qquad \chi_1(L \cap U_1) \subset T^*L_2|_N, \qquad \chi_1(L_1 \cap U_1) = \text{graph}\,\alpha,$$

where α is a 1-form on L_2.

Then α is closed as L_1 is Lagrangian, and moreover $N \subset \text{graph}\,\alpha$, or in other words $\alpha|_N = 0$. Hence the map

$$\psi_\alpha : T^*_x L_2 \ni \beta \mapsto \beta - \alpha(x) \in T^*_x L_2, \qquad x \in L_2 \cap U_1,$$

is a symplectomorphism. We have $\psi_\alpha(\text{graph}\,\alpha) \subset L_2$, and ψ_α restricted to $T^*L_2|_N$ is the identity. In particular, $\psi_\alpha(V_2) \subset \chi_1(U_1)$ for a sufficiently small neighbourhood V_2 of N in T^*L_2. Now, a symplectomorphism

$$\phi(x) = \chi_1^{-1}\psi_\alpha\chi_1(x)$$

is well defined and clearly satisfies (3.16) if we take $U_0 = \chi_1^{-1}(V_2)$. □

PROOF OF LEMMA 3.4.2. A submanifold of T^*L_2 is (locally) a graph of a 1-form if it has the dimension of L_2 and is transversal to the fibres $T^*_x L_2$, $x \in L_2$. Since transversality is an open condition, it is enough to show that $\chi_1(L_1) \pitchfork_x T^*_x L_2$ for $x \in N$.

Let $E = \ker d\tau$ be the vertical subbundle of TL and E^{annih} the subbundle of T^*L annihilating E. For $x \in N$ consider the following decomposition:

$$T_x(T^*L) = T_xL \oplus T^*_xL = E_x \oplus T_xN \oplus TN^{\text{annih}}_x \oplus E^{\text{annih}}_x.$$

The clean intersection $L_1 \cap L$ implies that $T_xL_1 \cap E_x = \{0\}$. The first three summands of the above decomposition form the symplectic complement $T_xN^\perp$ of T_xN. On the other hand, $T_xL_1 \subset T_xN^\perp$, and so $T_xL_1 \cap E^{\text{annih}}_x = \{0\}$ as well. Now, $E_x \oplus E^{\text{annih}}_x$ is the tangent space at x to the restricted bundle $E^{\text{annih}}|_{L_x}$ (seen as a submanifold of T^*L). Thus L_1 is transversal to $E^{\text{annih}}|_{L_x}$ at x. Hence it is enough to construct a symplectomorphism χ_1 in such a way that

$$\chi_1(E^{\text{annih}}|_{L_x} \cap U_1) \subset T^*_x L_2.$$

Note that the last condition implies also that $\chi_1(L) \subset T^*L_2|_N$. Now the existence of the required symplectomorphism χ_1 follows immediately from the two lemmas below. $\square$

LEMMA 3.4.3. *There is a vector bundle $\sigma : T^*L \to TN^{\mathrm{annih}}$ whose fibres are Lagrangian submanifolds of T^*L. In particular, $\sigma^{-1}(x) = E^{\mathrm{annih}}|_{L_x}$ for $x \in N$.*

PROOF. First, we have the restriction map $T^*L \to E^*$. Second, E is isomorphic to the pullback bundle τ^*L, which gives the isomorphism of the dual bundles, $E^* \cong \tau^*L^*$. Third, $TL|_N = L \oplus TN$, and the restriction

$$TN^{\mathrm{annih}} \ni \alpha \to \alpha|_L \in L^*$$

is the bundle isomorphism. Putting all these facts together, we obtain a vector bundle map σ

$$\sigma : T^*L \to E^* \cong \tau^*L^* \to L^* \cong TN^{\mathrm{annih}}.$$

Clearly, $\sigma^{-1}(x) = E^{\mathrm{annih}}|_{L_x}$ if $x \in N$. To see that σ carries a vector bundle structure, consider the local coordinates

$$\psi : W \times \mathbb{R}^k \ni (q, q') = (q_1, \dots, q_n, q'_1, \dots, q'_k) \mapsto \psi(q, q') \in L, \quad W \subset \mathbb{R}^n,$$

induced by a local trivialization of $\tau : L \to N$; that is, $\psi(q, 0) \in N$, $L_{\psi(q,0)} = \psi(\{q\} \times \mathbb{R}^k)$, and if $\bar{\psi} : \overline{W} \times \mathbb{R}^k \to L$ is another chart of this type, then the change of the coordinates takes form

$$\bar{\psi}^{-1}\psi(q, q') = (A(q), B(q)q') = (A_1(q), \dots, A_n(q), B_i^1(q)q'_i, \dots, B_i^k(q)q'_i),$$

where $A : W \cap \psi^{-1}\bar{\psi}(\overline{W}) \to \mathbb{R}^n$ is a diffeomorphism and $B : W \cap \psi^{-1}\bar{\psi}(\overline{W}) \to GL(\mathbb{R}^k)$. Let

$$\Psi : W \times \mathbb{R}^k \times \mathbb{R}^{n+k} \ni (q, q', p, p') \mapsto \Psi(q, q', p, p') \in T^*L \tag{3.17}$$

denote the natural coordinates on T^*L induced by ψ. In these coordinates the map σ becomes the projection

$$\sigma_\Psi = \Psi^{-1}\sigma\Psi : (q, q', p, p') \mapsto (q, 0, 0, p'),$$

and this induces the vector structure in the fibre

$$\sigma^{-1}(\Psi(q, 0, 0, p')) \cong \{q\} \times \mathbb{R}^k \times \mathbb{R}^n \times \{p'\}.$$

This is because the change of coordinates $\Theta = \bar{\Psi}^{-1} \circ \Psi$ is given by

$$\begin{aligned} \Theta(q, q', p, p') &= \left(\bar{\psi}^{-1}\psi(q, q'), (p, p')d_{(q,q')}(\bar{\psi}^{-1}\psi)^{-1}\right) \\ &= \left(A(q), B(q)q', pC(q)^{-1} - D(q, p')q', p'B(q)^{-1}\right), \end{aligned}$$

where $C : W \to GL(\mathbb{R}^n)$, $D : W \times \mathbb{R}^k \to L(\mathbb{R}^k, \mathbb{R}^n)$,

$$C(q)^i_j = \frac{\partial A_i}{\partial q_j} \quad \text{and} \quad D(q, p')^i_j = p'_k \left(B(q)^{-1}\right)^k_l \frac{\partial B(q)^i_l}{\partial q_m} \left(C(q)^{-1}\right)^m_j.$$

Thus Θ restricted to $\{q\} \times \mathbb{R}^k \times \mathbb{R}^n \times \{p'\}$ is linear, and so the atlas consisting of the charts as above yields the structure of a vector bundle on $\sigma : T^*L \to L_2$. Moreover, from the local description of σ follows clearly that a fibre $\sigma^{-1}(x)$ is a Lagrangian submanifold of T^*L. $\square$

LEMMA 3.4.4. *Let $\sigma : V \to L$ be a vector bundle such that (V, ω_0) is a symplectic manifold and the zero section $L \subset V$ and the fibres V_x are Lagrangian. Then for every compact subset $K \subset L$ there is a fibre preserving symplectomorphism χ defined in a neighbourhood U of K in V,*

$$\chi : (U, \omega_0) \to (T^*L, \omega_{T^*L}), \qquad \pi_L \circ \chi = \sigma|_U, \qquad \chi|_L = \mathrm{id}.$$

PROOF. This is a special case of [**Wei71**], Theorem 7.1. Define a vector bundle isomorphism

$$\eta : V \to V \oplus TL = TV|_L \xrightarrow{-i\omega} TV|_L^* = (V \oplus TL)^* \to T^*L,$$

that is, $\eta(v) = -i_v\omega|_{TL}$. Now $\omega_0|_L = \eta^*\omega_{T^*L}|_L$, and so ω_0 and $\omega_1 = \eta^*\omega_{T^*L}$ are diffeomorphic in some neighbourhood of K. Moreover, a close look at Moser's proof reveals that this diffeomorphism is fibre preserving. For the sake of completeness we recall the argument. Let $\theta = \omega_1 - \omega_0$. Then $d\theta = 0$, and θ vanishes on $TV|_L$. Moreover, $\theta|_{V_x} = 0$, as the fibres V_x are Lagrangian with respect to both ω_0 and ω_1. Then θ is exact in a neighbourhood of K, $\theta = d\rho$, and one defines the vector field X_t by

$$i_{X_t}\omega_t = -\rho, \text{ where } \omega_t = \omega_0 + t\theta.$$

If ϕ_t is the family of diffeomorphisms generated by X_t, then

$$\frac{d}{dt}\phi_t^*\omega_t = \phi_t^*(L_{X_t}\omega_t + \theta) = \phi_t^*(di_{X_t}\omega_t + d\rho) = 0$$

and so $\phi_1^*\omega_1 = \omega_0$. The diffeomorphism ϕ_1 is fibre preserving if only X_t is tangent to the fibres. As fibres are Lagrangian with respect to ω_t, this is equivalent to $\rho|_{V_x} = 0$. Define a smooth map $\psi : \mathbb{R} \times V \to V$, $\psi(t, w) = tw$, and the family of 1-forms $\rho_t : TV \to \mathbb{R}$ by the formula

$$\rho_t(y)w = \theta(\psi(t, w))\left(\frac{\partial \psi}{\partial t}(t, y), d\psi_t(y)w\right).$$

Then $\rho = \int_0^1 \rho_t$. But $\theta|_{V_x} = 0$ and $\frac{\partial \psi}{\partial t}$ is tangent to the fibres, and so $\rho_t|_{V_x} = 0$, which completes the proof of the lemma and of the proposition. □

3.4.2. *The t-independence of the gradient lines of a_H in the normal form.* Let N be a compact, connected manifold and $\tau : L \to N$ a vector bundle over N. Recall that $\pi : T^*L \to L$ denotes the projection.

For any function $f : N \to \mathbb{R}$ we define its extension to T^*L by composing it with the projections

$$f_L := f \circ \tau : L \to \mathbb{R}, \qquad H_f := f_L \circ \pi : T^*L \to \mathbb{R}.$$

Let g_N be a metric on N. Our next step is to extend g_N to T^*L. Consider a metric g on the manifold L such that

- $g|_N = g_N$,
- the fibres $L_x = \tau^{-1}(x)$ are orthogonal to N,
- N is a totally geodesic submanifold of L.

Note that the second condition implies that

$$\nabla^g f_L(x) = \nabla^{g_N} f(x)$$

for $x \in N$. The Levi–Civita connection associated with g provides the splitting of $T_\xi(T^*L)$ into horizontal and vertical subspaces $T_\xi(T^*L) = H_\xi \oplus V_\xi$, and $V_\xi = T_\xi(T^*_{\pi(\xi)}L)$ is canonically isomorphic to $T^*_{\pi(\xi)}L$ while $d\pi$ gives the isomorphism $d_\xi\pi|_{H_\xi} : H_\xi \to T_{\pi(\xi)}L$. These give the "diagonal" lift of g to a metric g^D on T^*L (the Kaluza–Klein metric) such that $H_\xi \perp V_\xi$ and the above isomorphisms become isometries. In the canonical coordinates (q,p) we have

$$g^D(\partial_{q_i}, \partial_{q_j}) = g_{ij} + g^{kl}\Gamma^k_{ir}\Gamma^l_{js}p_r p_s,$$
$$g^D(\partial_{p_i}, \partial_{q_j}) = g^{ik}\Gamma^k_{jl}p_l,$$
$$g^D(\partial_{p_i}, \partial_{p_j}) = g^{ij}.$$

Here Γ are the Christoffel symbols in the cotangent bundle. In particular, if $x \in L$ then $T(T^*L) = TL \oplus t^*L$ and $g^D(x) = g(x) \oplus g^*(x)$. The metric g^D is compatible with the canonical symplectic structure, and the induced almost complex structure $J = J_g$ maps a horizontal vector $w \in H_\xi$ to the vertical lift of $i_{d_\xi \pi w} g \in T^*_{\pi(\xi)}L$. Thus in the canonical coordinates

$$J = \begin{pmatrix} -A & -I \\ I + A^2 & A \end{pmatrix},$$

where $A^k_i = \Gamma^k_{ij}P_j$ is the connection matrix.

LEMMA 3.4.5. *The action functional is well defined in* $\Omega(T^*L, L, TN^{\text{annih}})$. *Moreover,*

$$|a_\omega(\gamma) - a_\omega(x)| \leq \|\dot\gamma\|^2_{L^2}$$

for any $x \in N$ *and* $\gamma \in \Omega$.

PROOF. The symplectic form ω is exact, $\omega = d\lambda$, and $\lambda|_L = \lambda_{TN^{\text{annih}}} = 0$. Therefore, if $u : I \times S^1 \to T^*L$, $u_0(s) = u(0,s) \in L$, $u_1(s) \in TN^{\text{annih}}$, represents a loop in ω, then

$$[\hat\omega][u] = \int u^*\omega = \int u_1^*\lambda - \int u_0^*\lambda = 0$$

and $\hat\omega$ is exact. We can also assume that $a_\omega(\gamma) = 0$ if $\gamma \subset L$. Let $\gamma \in \Omega$. In the canonical coordinates, $\gamma = (q,p)$ with $p(0) = 0$ and $p(1) \in T_{q(1)}N^{\text{annih}}$. Define $u : I \times I \to T^*L$, $u(s,t) = (q(t), sp(t))$. Then

$$a_\omega(\gamma) = \int_{I^2} u^*\omega = \int_{\partial I^2} u^*\lambda = \int_0^1 p\dot q\, dt = \langle p, \dot q\rangle_{L^2}.$$

Now $|\dot\gamma^H| = |\dot q|$ and $|\dot\gamma^V| = |\nabla p|$, and $|p|_{L^2} \leq 2|\nabla p|_{L^2}$ since $p(0) = 0$, and so

$$|a_\omega(\gamma)| \leq |p|_{L^2}|\dot q|_{L^2} \leq 2|\nabla p|_{L^2}|\dot q|_{L^2} \leq |\dot\gamma^V|^2_{L^2} + |\dot\gamma^H|^2_{L^2} = |\dot\gamma|^2_{L^2}. \qquad \square$$

The main result of this subsection is the following proposition, which is a generalization of Floer's Theorem 2 in [**Flo88a**].

PROPOSITION 3.4.6. *Let* (N, g_N) *be a compact Riemannian manifold,* $\tau : L \to N$ *a vector bundle over* N *and* $f : N \to \mathbb{R}$ *a* C^2 *function on* N. *Let* $J = J_g$ *and* $H = H_f$ *be constructed as above and suppose that there is a neighbourhood* U *of* N *in* L *such that*

$$\|\nabla^g df_L(x)\| \leq 1 \tag{3.18}$$

for $x \in U$. Then the following hold:

(a) *All critical points and gradient lines (with respect to J) of the action functional a_H in $\Omega(\pi^{-1}(U), U, TN^{\text{annih}})$ are t-independent, and so they are in one-to-one correspondence with the critical points and the gradient lines of f (with respect to g_N).*

(b) *The critical points of a_H are nondegenerate if f is a Morse function. In this case, if $x^-, x^+ \in \operatorname{Crit} f$ and $u : \mathbb{R} \to N$ is a t-independent element of $\mathcal{P}(x^-, x^+)$, then the linearized operator $D_{J,H}(u)$ is onto if and only if the operator $D_f(u) : W^{1,p}(u^*TN) \to L^p(u^*TN)$ given by*

$$D_f(u)\xi = \nabla_s \xi + \nabla_\xi \nabla f(u)$$

is onto and the assignment $\xi \mapsto \xi'(s,t) = \xi(s)$ gives the isomorphism $\ker D_f(u) \cong \ker D_{J,H}(u)$.

PROOF. Consider the local coordinates (q, q', p, p') as in (3.17). Then

$$\frac{\partial H}{\partial p} = \frac{\partial H}{\partial p'} = \frac{\partial H}{\partial q'} = 0 \quad \text{and} \quad X_H = (0, 0, df(q), 0),$$

and the Hamiltonian flow associated with H takes the form

$$\phi_t(q, q', p, p') = (q, q', p + t\,df(q), p').$$

If $x = (q, q', p, p') \in L$ then $p = p' = 0$. If $\phi_1(x) \in TN^{\text{annih}}$ then $q' = 0$ and $p + df(q) = 0$. Thus the only solutions to (3.8) with the boundary conditions $x(0) \in L$, $x(1) \in TN^{\text{annih}}$ are the constant paths $x(t) = (q, 0, 0, 0)$, where q is a critical point of f. Furthermore, in these coordinates we have

$$D\phi_1(x) = \begin{pmatrix} I & 0 & 0 & 0 \\ 0 & I & 0 & 0 \\ D^2 f(q) & 0 & I & 0 \\ 0 & 0 & 0 & I \end{pmatrix},$$

where the Hessian $D^2 f(q)$ is nondegenerate. Let $v = (Q, Q', 0, 0)$ be a vector tangent to L. The vector $D\phi_1(x)v = (Q, Q', D^2 f(q)Q, 0)$ is tangent to TN^{annih} if and only if $Q' = D^2 f(q)Q = 0$. But then also $Q = 0$, which shows that $D\phi_1(x)(T_xL) \cap T_xTN^{\text{annih}} = \{0\}$, i.e., x is nondegenerate. □

REMARK 3.4.7. Obviously, the established correspondence implies that the $\mathbb{Z}_2$ intersection number $\#_2(L_0, L_1)$ is equal to the Euler characteristic $\chi(N, \mathbb{Z}_2)$ of N. It is not difficult to see that this is true generally, in the differential topology category and also, up to an appropriate sign, over $\mathbb{Z}$.

Since $dH = df_L \circ d\pi$ vanishes on the vertical subbundle $V \subset T(T^*L)$, the gradient of H with respect to the metric g^D belongs to the horizontal subspace, $\nabla H(\xi) \in H_\xi$. Furthermore $d_\xi \pi|_{H_\xi}$ is an isometry and hence $d_\xi\pi(\nabla H(\xi)) = \nabla H(\pi(\xi)) = \nabla^g f_L(\pi(\xi))$; that is, ∇H is the horizontal lift of $\nabla^g f_L$. In particular, if $x \in N$ then

$$\nabla H(x) = \nabla^{g_N} f(x) \in TN.$$

Therefore if $u : \mathbb{R} \to N$ is a gradient line of f, then $u(s,t) := u(s)$ is a solution to (3.9) satisfying the appropriate boundary conditions.

Conversely, let $u : \mathbb{R} \times I \to U$ be a solution to (3.9), (3.10). We want to show that $\frac{\partial u}{\partial t} \equiv 0$. Let $x(s,t)$ and $y(s,t)$ be the horizontal and the vertical components of $u(s,t)$. Strictly speaking, $x = \pi \circ u$ and y is a section of T^*L along x, i.e., $y \in \Gamma(x^*T^*L)$. Then $\frac{\partial u}{\partial s}$ (and similarly $\frac{\partial u}{\partial t}$) decomposes into the horizontal and vertical parts, where $(\frac{\partial u}{\partial s})^H$ is the horizontal lift of $\frac{\partial x}{\partial s}$ and $(\frac{\partial u}{\partial s})^V$ is the vertical lift of $\nabla_s y \in T_x^*L$. Thus we may write the equation (3.9) in the form

$$\frac{\partial x^*}{\partial s} - \nabla_t y + dh_1(x) = 0,$$
$$\nabla_s y + \frac{\partial x^*}{\partial t} = 0,$$

where $v^* = g(v,\cdot) = J(v)$ for $v \in TL$. The boundary conditions are

$$y(s,0) = 0, \qquad x(s,1) \in N, \qquad y(s,1) \in T_{x(s,1)}N^{\text{annih}},$$
$$\lim_{s\to\pm\infty} y(s,t) = 0 \qquad \text{and} \qquad \lim_{s\to\pm\infty} x(s,t) = x^{\pm}$$

for some critical points x^+, x^- of f. We have to show that $y \equiv 0$. Then $\frac{\partial x}{\partial t} = 0$ and $x : \mathbb{R} \to N$ is a gradient line of f. This is in principle an application of Green's identity, since y is a solution of the elliptic equation

$$\nabla y - \langle df_L, \nabla_s y\rangle = 0.$$

Define

$$\gamma(s) = \frac{1}{2}\int_0^1 |y(s,t)|^2\,dt = \frac{1}{2}\|y(s)\|_{L^2}^2.$$

Since $\lim_{s\to\pm\infty}\gamma(s) = 0$ and γ is nonnegative, it attains a maximum in $\mathbb{R}$. On the other hand, the following lemma shows that γ is convex, and so $\gamma \equiv 0$.

LEMMA 3.4.8. *If $\|\nabla^g f_L\|_{L^\infty(U)} < 1$, then $\gamma''(s) \geq 0$ for all $s \in \mathbb{R}$.*

PROOF ([**Flo88a**], Lemma 5.2). Since the Levi–Civita connection is torsion free, it follows that

$$\nabla_s \frac{\partial x}{\partial t} = \nabla_t \frac{\partial x}{\partial s},$$

and we may compute

$$\frac{\partial^2}{\partial s^2}\frac{1}{2}|y(s,t)|^2 = |\nabla_s y|^2 + \langle \nabla_s^2 y, y\rangle = |\nabla_s y|^2 - \left\langle \nabla_t \frac{\partial x}{\partial s}, y\right\rangle$$
$$= |\nabla_s y|^2 - \langle \nabla_t^2 y, y\rangle + \langle \nabla_{\partial x/\partial t} df_L, y\rangle.$$

Integrating by parts, for the second term we get

$$\int_0^1 \langle \nabla_t^2 y, y\rangle\,dt = \langle \nabla_t y(s,1), y(s,1)\rangle - \langle \nabla_t y(s,0), y(s,0)\rangle - \int_0^1 |\nabla_t y|^2\,dt.$$

Now, $y(s,0) = 0$ and $y(s,1) \in T_{x(s,1)}N^{\text{annih}}$. On the other hand, $\frac{\partial x}{\partial s}(s,1) \in T_xN$ and $\nabla f_L(x(s,1)) \in T_xN$. Then we have

$$\langle \nabla_t y(s,1), y(s,1)\rangle = \left\langle y(s,1), \frac{\partial x^*}{\partial s}(s,1) + df_L(x)\right\rangle$$
$$= y(s,1)\left(\frac{\partial x}{\partial s}(s,1) + \nabla f_L(x)\right) = 0$$

for all $s \in \mathbb{R}$. Consequently

$$\begin{aligned}\gamma''(s) &= \|\nabla_s y(s)\|_{L^2}^2 + \|\nabla_t y(s)\|_{L^2}^2 - \int_0^1 \langle \nabla_{\nabla_s y} df_L(x), y\rangle\, dt \\ &\geq \|\nabla_s y(s)\|_{L^2}^2 + \|\nabla_t y(s)\|_{L^2}^2 - \|\nabla df_L\|_{L^\infty}\|\nabla_s y\|_{L^2}\|y\|_{L^2}.\end{aligned}$$

Since $y(s,0)=0$, the Poincaré inequality gives $\|y(s)\|_{L^2} \leq 2\|\nabla_t y(s)\|_{L^2}$, and therefore $\gamma'' \geq 0$ if only $\|\nabla df_L\|_{L^\infty} = \sup_{x\in U}|\nabla df_L(x)| < 1$. □

The isomorphism $\ker D_f(u) \cong \ker D_{J,H}(u)$ can be proved in a similar manner. A vector field $\xi \in \Gamma(u^*T(T^*L))$ decomposes into

$$\xi = (\zeta, \eta) \in T_u L \oplus T_u^* L.$$

Since $\xi(s,t) \in T_{u(s)}(T^*L)$ for all $t\in[0,1]$, we have $\nabla_t \xi = \frac{\partial \xi}{\partial t}$. Furthermore, direct computation of the Christoffel symbols of the Levi–Civita connection associated with g^D yields

$$\nabla_{\partial_{q_i}}\partial_{q_j}(q,0) = \nabla^g_{\partial_{q_i}}\partial_{q_j}(q) \in t_q L \quad\text{and}\quad \nabla_{\partial_{q_i}}\partial_{p_j}(q,0) \in T_L^*.$$

Therefore

$$\nabla_s\zeta = \nabla_s^g \zeta \in T_u L \quad\text{and}\quad \nabla_s \eta \in T_u^* L.$$

Since $dH(q,p) = \partial_{q_i} f_L dq_i$, we have also

$$\nabla_\zeta dH(u) = \nabla_\zeta^g df_L(u) \quad\text{and}\quad \nabla_\eta dH(u) = 0.$$

Thus the linearized equation (3.14) takes the form

$$\begin{aligned}\nabla_s^g \zeta^* - \frac{\partial \eta}{\partial t} + \nabla^g_\zeta df_L(u) &= 0, \\ \nabla_s \eta + \frac{\partial \zeta^*}{\partial t} &= 0\end{aligned} \tag{3.19}$$

with the boundary conditions

$$\begin{aligned}&\eta(s,0)=0, \qquad \zeta(s,1)\in T_{u(s)}N, \qquad \eta(s,1)\in T_{u(s)}N^{\text{annih}}, \\ &\lim_{s\to\pm\infty}\eta(s,t)=0 \quad\text{and}\quad \lim_{s\to\pm\infty}\zeta(s,t)=0.\end{aligned} \tag{3.20}$$

If $\zeta \in T_u N$, then, since N is totally geodesic in L, we have $\nabla_s^g\zeta = \nabla_s^{g_N}\zeta$ and $\nabla^g_\zeta df_L = \nabla_\zeta^{g_N} f$, and so any vector field $\zeta \in \ker D_f(u)$ is a solution to (3.19), (3.20). If we set

$$\gamma_1(s) = \frac{1}{2}\int_0^1 \|\eta(s,t)\|^2\, dt,$$

then the converse follows from the lemma similar to Lemma 3.4.8.

LEMMA 3.4.9. *If $\|\nabla^g df_L\|_{L^\infty(U)} < 1$, then $\gamma_1''(s) \geq 0$ for all $s\in\mathbb{R}$.* □

Finally, the same argument provides the isomorphism between the kernels of the adjoint operators

$$\ker D_f^*(u) \cong \ker D_{J,H}^*(u),$$

and so $D_f(u)$ is onto if and only if $D_{J,H}(u)$ is onto, which completes the proof of Proposition 3.4.1. □

REMARK 3.4.10. Consider the case $L = N = TN^{\mathrm{annih}}$. It is easy to see that the exactness was not used in the proof; that is, the result holds if we replace $df = df_L$ by a closed nondegenerate form α on L such that

$$\left\|\nabla^g \alpha\right\|_g \leq 1$$

with respect to some regular metric g on L. But this condition can always be satisfied if we rescale the metric

$$g_\varepsilon := \varepsilon^2 g.$$

This is because the Levi–Civita connection is invariant under rescaling, $\nabla^{g_\varepsilon} = \nabla^g$, and $\|\cdot\|_{g_\varepsilon} = \varepsilon\|\cdot\|_g$. Clearly, g_ε is regular if g is regular. We shall draw some conclusions from this fact in 3.5.3.

3.4.3. *The main result.*

THEOREM 3.4.11. *Let (M, ω) be a symplectic manifold and L_0, L_1 two Lagrangian submanifolds of M. Let N be a compact connected component of $L_0 \cap L_1$ such that L_0 and L_1 intersect cleanly along N. Fix a base point $x_0 \in N$. Let U be a relatively compact neighbourhood of N such that*

(a) *there are no critical points of a other than those in N in the connected component $\Omega(U, L_0, L_1, x_0)$ of x_0 in the path space $\Omega(U, L_0, L_1)$, and*

(b) *the action functional of ω is well defined in $\Omega(U, L_C, L_1, x_0)$.*

Then $\mathcal{U} = \Omega(U, L_0, L_1, x_0)$ is an isolating neighbourhood and $S_{J,0}(\mathcal{U}) = N$ for any almost complex structure J. There exists an almost complex strucure J_0 and a Hamiltonian $H_0 : M \to \mathbb{R}$ such that

1. *$S_{J_0,H_0}(\mathcal{U})$ is a continuation of N,*
2. *(J_0, H_0) is a regular pair, and if $g_N = g_J|_N$ and $f = H_0|_N$, then (g_N, f) is a regular pair in the sense of (2.5.1),*
3. *the Floer complex of $CF_*(\mathcal{U}, J_0, H_0)$ coincides with the Morse complex $C_*^{\mathrm{Morse}}(N, g_N, f)$ of f, and thus*

$$HF_*(\mathcal{U}, \mathbb{Z}_2) \cong H_*^{\mathrm{sing}}(N, \mathbb{Z}_2).$$

PROOF. Let $\phi : U_0 \to V_0$ be as in Proposition 3.4.1. Then clearly U_0 satisfies (a) and (b). Let U_1 be another neighbourhood of N such that $U_1 \subset \bar{U}_1 \subset U_0$. Fix a Morse function $f : N \to \mathbb{R}$ and a regular metric g_N on N such that

$$\|\nabla df\|_{L^\infty} < 1,$$

and take an almost complex structure J_0 and a Hamiltonian H_0 on M such that

$$J_0\big|_{U_1} = \phi^* J_g\big|_{U_1} \quad \text{and} \quad H_0\big|_{U_1} = H_f \circ \phi\big|_{U_1},$$

where J_g and H_f are constructed as in Proposition 3.4.6. Thus if u is a critical point or a gradient line of a_{H_0} contained in U_1, then $\phi \circ u$ is a critical point or a gradient line of a_{H_f}. If we choose U_1 small enough, then the assumption (3.18) of Proposition 3.4.6 is satisfied and so u is t-independent. Now the following lemma completes the proof of the theorem.

LEMMA 3.4.12. *Let U be a neighbourhood of N satisfying conditions* (a) *and* (b) *of Theorem* 3.4.11, *and let $U_1 \subset U$. Then there exists an $\varepsilon > 0$ such that if $\|H\|_{C^1} < \varepsilon$, then any solution to* (3.8) *and* (3.9) *contained in U lies in fact in U_1.*

PROOF. Let $x : [0,1] \to U$ satisfy $\dot{x} = X_H(x)$ and $x(0) \in L_0$. Then we have

$$\operatorname{dist}(x(0), x(t)) \leq \int_0^t |\dot{x}(s)|\, ds \leq \|H\|_{C^1}.$$

Choose a neighbourhood U_2 of N such that $\overline{U}_2 \subset U_1$. Then

$$d = \min(\operatorname{dist}(U_2, M \setminus U_1), \operatorname{dist}(L_0 \cap (U \setminus U_2), L_1))$$

is positive. Assume $\|H\|_{C^1} < d$. It follows that $x(0) \in U_2$ implies $x(t) \in U_1$ for all $t \in [0,1]$, and if $x(0) \notin U_2$, then $x(1) \notin L_1$.

Suppose now that there is a sequence $\{\varepsilon_n\}_{n \in N} \subset \mathbb{R}_+$ converging to 0 for every $n \in \mathbb{N}$ and a solution $u_n' : I \times \mathbb{R} \to U$ to

$$\frac{\partial u_n'}{\partial s} + J_n(u_n)\frac{\partial u_n'}{\partial t} + \nabla H_n(u) = 0,$$

where $\|H_n\|_{C^1} < \varepsilon_n$ and J_n is a regular almost complex structure, with the limit conditions $u_n'(s,i) \in L_i$, $i = 0,1$,

$$\lim_{s \to \pm\infty} u_n'(s) = x_n^{\pm} \subset U_1,$$

and such that $u_n'(I \times \mathbb{R}) \not\subset U_1$. As $\overline{U}$ is compact, there exist a subsequence $\{u_n\} \subset \{u_n'\}$ and a sequence (s_n, t_n) such that $\lim_{n\to\infty} u_n(s_n, t_n) = y_0 \in U \setminus U_1$. We may assume that $\lim_{n\to\infty} t_n = t_0$ and, using time shift, that $s_n = 0$ for all $n \in \mathbb{N}$. Now, since $\hat{\omega}$ is exact in $\mathcal{U}$, we have

$$\ell^2(u_n) = a_\omega(x^-) - a_\omega(x^+) + \int_0^1 (H_n \circ x_n^+ - H_n \circ x_n^-)\, dt \to 0$$

by Lemma 3.4.5, and it follows from Lemma 3.3.9 that a subsequence of $\{u_n\}$ converges locally uniformly to a constant map u_0. Because of the limit conditions, $u_0 \in N$. On the other hand $u_0(0, t_0) = \lim u_n(0, t_n) = y_0$, a contradiction. □

COROLLARY 3.4.13. *Assume that (M, L_0, L_1, γ_0) is a monotone triple such that $\Sigma_i \geq 3$, $i = 0, 1$. Assume that L_0 intersects L_1 cleanly along a connected manifold $N = L_0 \cap L_1$ such that* $\dim N + 1 < N_\mu$. *Then*

$$HF_*(M, L_0, L_1, \mathbb{Z}_2) \cong H_*^{\text{sing}}(N, \mathbb{Z}_2).$$

PROOF. Clearly, it is enough to show that all trajectories of index 1 must lie in a small neighbourhood U_1 of N in M if only the Hamiltonian perturbation is sufficiently small. We proceed as in the proof of Lemma 3.4.12. Thus suppose that there are sequences $J_n \to J$, $H_n \to 0$ and $u_n \in \mathcal{M}_{J_n,H_n}(x_n^-, x_n^+)$ such that $\mu_{u_n} = 1$ and $u_n(\mathbb{R} \times I) \not\subset U$. We may assume that H is constructed as in Proposition 3.4.6, so that $x^\pm$ are t-independent. For each n choose a path $v_n : \mathbb{R} \to N$ connecting x_n^- and x_n^+; that is, $v_n \in \mathcal{P}(x_N^-, x_N^+)$. Then Proposition 3.4.6 together with Theorem 2.3.2 implies

$$\mu_{v_n} = \operatorname{ind} D_{J_n,H_n}(v_n) = \operatorname{ind} D_f(v_n) = \operatorname{ind} x_n^- - \operatorname{ind} x_n^+.$$

Consequently, $|I_\mu(u_n*(-v))| = |\mu_{u_n} - \mu_v| \le \dim N + 1 < N_\mu$. But this means that $I_\mu(u_n*(-v)) = 0$. Since (M, L_0, L_1) is monotone and obviously $\int v_n^*\omega = 0$, this yields

$$\int u_n^*\omega = \int u_n^*\omega - \int v_n^*\omega = I_\omega(u_n*(-v)) = 0.$$

The conclusion follows as before. □

3.5. Applications.

3.5.1. *Example: Intersection of linear Lagrangian tori in* T^{2k}. Consider the torus T^{2k} with the standard symplectic structure $\omega_0 = \sum dx_i \wedge dy_i$. For $i = 0, 1$, let $\tilde{A}_i : T^k \to T^{2k}$ be an inclusion which is induced by a linear map of the universal covers $A_i : \mathbb{R}^k \to \mathbb{R}^{2k}$. We have $\operatorname{rk} A_i = k$, and $\tilde{A}_i$ is an injection if and only if $A_i x \in \mathbb{Z}^{2k}$ implies that $x \in \mathbb{Z}^k$— that is, if and only if

$$A_i\mathbb{Q}^k \cap \mathbb{Z}^{2k}/A_i\mathbb{Z}^k = \operatorname{Tor}(\mathbb{Z}^{2k}/A_i\mathbb{Z}) = 0,$$

where Tor denotes the torsion part of the group. The latter condition is equivalent to the greatest common divisor of the k-minors of A_i being equal to 1. Assume that $L_i := \tilde{A}_i(T^k)$ is a Lagrangian submanifold of T^{2k} (that is, $\left(A_i'^T A_i''\right)^T = A_i'^T A_i''$, where $A_i = (A_i', A_i'')$). In order to describe $\pi_0(\Omega(L_0, L_1))$ and $\pi_1(\Omega(L_0, L_1))$ we use (3.4) and (3.6). We have $\pi_1(L_i) = \mathbb{Z}^k$, $\pi_1(M) = \mathbb{Z}^{2k}$, $\pi_2(M) = 0$ and

$$\pi_1(i_0)^{-1}\pi_1(i_1) = \overline{A} = -A_0 \oplus A_1 : \mathbb{Z}^{2k} \to \mathbb{Z}^{2k}.$$

Consequently

$$\begin{aligned} \pi_0(\Omega(L_0, L_1)) &\cong \mathbb{Z}^{2k}/\overline{A}\mathbb{Z}^{2k} = \operatorname{coker}\overline{A}, \\ \pi_1(\Omega(L_0, L_1)) &\cong \ker \overline{A}. \end{aligned} \tag{3.21}$$

The intersection of $L_0 \cap L_1$ is a closed subgroup of T^{2k}, so $L_0 \cap L_1 = T^r \oplus G$, where G is discrete and $r = \dim \operatorname{im} A_0 \cap \operatorname{im} A_1 = \dim \ker\overline{A}$. The injection

$$\sigma : L_0 \cap L_1 \to \pi_0(\Omega(L_0, L_1))$$

can be described as follows. Let $p \in L_0 \cap L_1$ and $[x_p] = \tilde{A}_0^{-1}(p)\,[y_p] = \tilde{A}_1^{-1}(p)$. Then there is an $n_p \in \mathbb{Z}^{2k}$ such that $A_1 y_p - A_0 x_p + n_p = 0$. Obviously, $n_p \in \overline{A}\mathbb{Q}^{2k}$. Moreover, if $x' \in [x_p]$ and $y' \in [y_p]$, then $x_p = x' + m_0$, $y_p = y' + m_1$ and

$$A_1 y' - A_0 x' + n_p + A_1 m_1 - A_0 m_0 = 0,$$

and so n_p is well defined modulo $\overline{A}\mathbb{Z}^{2k}$. It is easy to see that $\sigma(p) = [n_p] \in \operatorname{Tor}(\mathbb{Z}^{2k}/\overline{A}\mathbb{Z}^{2k})$. Indeed, the homotopy $\gamma_s(t) = [(1-s)A_0 x_p + stn_p] \in T^{2k}$ gives a path in $\Omega(L_0, L_1)$ connecting the constant path at p and the loop $[tn_p]$ representing the class $[n_p]$ under the isomorphism (3.21). The fibres of σ are connected: if $A_1 y - A_0 x + n = 0$, $A_1 y' - A_0 x' + n' = 0$ and $n - n' = A_1 m_1 - A_0 m_0$, then

$$\gamma(t) = [A_0(t(x + m_0) + (1-t)x')] = [A_1(t(t + m_0) + (1-t)y')]$$

connects $p = [A_0 x]$ and $p' = [A_0 x']$. Summarizing, $L_0 \cap L_1$ consists of the r-dimensional tori T_α, $\alpha \in \operatorname{Tor}(\mathbb{Z}^{2k}/\overline{A}\mathbb{Z}^{2k})$, each in the different component of $\Omega(L_0, L_1)$ represented by the nontorsion elements of $\mathbb{Z}^{2k}/\overline{A}\mathbb{Z}^{2k}$.

Let $[n] \in \mathbb{Z}^{2k}/\overline{A}\mathbb{Z}^{2k}$. Then the isomorphism $\pi_1(\Omega(L_0, L_1, [tn])) \cong \ker \overline{A}$ can be written explicitly:

$$\ker \overline{A} \ni m = (m_0, m_1) \mapsto [F_m] \in \pi_1(\Omega(L_0, L_1, [tn])),$$
$$F_m(s,t) = [sA_0 m_0 + tn] = [sA_1 m_1 + tn] \in T^{2k}.$$

Thus

$$[\hat{\omega}] : \ker \overline{A} \ni m \mapsto \int_{I\times I} F_m^* \omega = n^T J_0 A_0 m_0 \in \mathbb{R},$$

and $[\hat{\omega}]$ vanishes if and only if n annihilates $A_0\mathbb{R}^k \cap A_1\mathbb{R}^k$, that is, $n \in \overline{A}\mathbb{R}^{2k}$, or, equivalently, $[n] \in \mathrm{Tor}(\mathbb{Z}^{2k}/\overline{A}\mathbb{Z}^{2k})$. We may then apply Theorem 3.4.11 to each component of $L_0 \cap L_1$ with $U = T^{2k}$. We summarize the above remarks in the following corollary.

COROLLARY 3.5.1. *For $i = 0, 1$ let $A_i = (A_i', A_i'') : \mathbb{Z}^k \to \mathbb{Z}^{2k}$ satisfy the following conditions:*

- $\mathrm{rk}\, A_i = k$,
- *the* gcd *of the k-minors of A_i is equal to* 1,
- $\left(A_i'^T A_i''\right)^T = A_i'^T A_i''$.

Then $\tilde{A}_0(T^k)$ and $\tilde{A}_1(T^k)$ are Lagrangian submanifolds, and

$$HF_*(\tilde{A}_0(T^k), \tilde{A}_1(T^k)) \cong H_*^{\mathrm{sing}}(T^r) \times \mathrm{Tor}(\mathbb{Z}^{2k}/\overline{A}\mathbb{Z}^{2k}),$$

where $\overline{A} = -A_0 \oplus A_1$ and $2k - r = \mathrm{rk}\,\overline{A}$. Consequently, if $L_i = \psi(\tilde{A}_i(T^k))$, where ψ_i is an exact symplectomorphism and L_0 intersects L_1 transversally, then

$$\#L_0 \cap L_1 \geq 2^r \cdot \mathrm{gcd} \text{ of the } k\text{-minors of } \overline{A}.$$

3.5.2. *Floer homology of a symplectomorphism.* Let (M, ω) be a compact symplectic manifold and $\phi : M \to M$ a symplectomorphism. There are two ways of constructing Floer homology detecting its fixed points, $\mathrm{Fix}\,\phi$.

First, the graph of ϕ is a Lagrangian submanifold of $(M \times M, (-\omega) \times \omega)$ and its fixed points correspond to the intersection points of graph ϕ with the diagonal $\triangle = \{(x,x) \in M \times M\}$. Thus we have the Floer homology of the Lagrangian intersection $HF_*(M \times M, \triangle, \mathrm{graph}\,\phi)$. This intersection is transversal if the fixed points of ϕ are nondegenerate, i.e., if 1 is not an eigenvalue of $d\phi(x)$, for $x \in \mathrm{Fix}\,\phi$.

The second approach was mentioned in [**Flo88a**] and presented with details in [**DS93b**], [**DS93a**], [**FHS94**]. There the action

$$\hat{\omega}_H(\gamma)\xi = \int_0^1 \omega(\xi, \dot{\gamma} - X_t(\gamma))\, dt \tag{3.22}$$

is defined on the space of paths

$$\Omega_\phi := \{\gamma : \mathbb{R} \to M : \gamma(t+1) = \phi(\gamma(t))\}.$$

The tangent vector field $\xi \in \Gamma(\gamma^* TM)$ in (3.22) satisfies $\xi(t+1) = d\phi\xi(t)$, and X_t is a Hamiltonian vector field with the Hamiltonian $H : \mathbb{R} \times M \to \mathbb{R}$ satisfying the periodicity conditions $H_{t+1} \circ \phi = H_t$. The critical points of (3.22) correspond to the fixed points of $\psi_H^{-1} \circ \phi$, where ψ_H is the time one map of the Hamiltonian system $\dot{x} = X_t(x)$. Similarly, we choose an ω-compatible almost complex structure

$J : \mathbb{R}\times M \to \operatorname{End}(TM)$ such that $\phi^* J_{t+1} = J_t$ and write down the Cauchy–Riemann equations

$$\begin{gathered}\bar\partial_{\phi,J,H}(u) = \frac{\partial u}{\partial s} + J_t(u)\frac{\partial u}{\partial t} + \nabla H_t(u) = 0, \\ u : \mathbb{R}\times\mathbb{R}\to M, \qquad u(s,t+1) = \phi(u(s,t)).\end{gathered} \tag{3.23}$$

Finally, we set

$$\mathcal{M}_{J,H}(M,\phi) := \{u : \bar\partial_{\phi,J,H}(u) = 0 \text{ and } \ell^2(u) < \infty\}.$$

As before, for a generic J and H, this is a collection of finite dimensional manifolds, and one can construct a $\mathbb{Z}_2$-complex $CF_*(M,\phi)$ generated by the fixed points of ϕ. It is worth mentioning that this construction provides a link between symplectic and instanton Floer homology (see [**DS93a**], [**DS93b**]).

LEMMA 3.5.2. *For a symplectomorphism $\phi : M \to M$ and J_t, H_t as above there exist an almost complex structure $\tilde J_t$ and a Hamiltonian $\tilde H_t$ on $(M\times M, (-\omega)\times\omega)$ and a bijection*

$$\mathcal{M}_{J,H}(M,\phi) \ni u \mapsto \tilde u \in \mathcal{M}_{\tilde J,\tilde H}(M\times M, \triangle, \operatorname{graph}\phi).$$

Moreover, $\tilde u$ is regular if and only if u is regular.

PROOF. The lemma is a simple consequence of the observation that there is a bijection

$$\begin{gathered}\Omega_\phi \ni \gamma \mapsto \tilde\gamma \in \Omega(\triangle, \operatorname{graph}\phi), \\ \tilde\gamma(t) = (\tilde\gamma_1(t), \tilde\gamma_2(t)) = (\gamma(\tfrac{1-t}{2}), \gamma(\tfrac{1+t}{2})),\end{gathered}$$

between the path spaces. It is easy to see that this correspondence translates all necessary concepts of Floer's construction from one setup to the other. Indeed, let $\gamma \in \Omega_\phi$ satisfy (3.8). Then

$$\begin{aligned}\dot{\tilde\gamma}_1(t) &= -\tfrac12 X_{\frac{1-t}{2}}(\tilde\gamma_1(t)), \\ \dot{\tilde\gamma}_2(t) &= \tfrac12 X_{\frac{1+t}{2}}(\tilde\gamma_2(t)).\end{aligned}$$

The vector field $\tilde X_t := \frac12(X_{\frac{1-t}{2}}, X_{\frac{1+t}{2}})$ has the Hamiltonian

$$\tilde H_t := \tfrac12(H_{\frac{1-t}{2}}\circ \mathrm{pr}_1 + H_{\frac{1+t}{2}}\circ\mathrm{pr}_2),$$

where $\mathrm{pr}_i : M\times M\to M$ denote the projections. If we set

$$\tilde J_t := (-J_{\frac{1-t}{2}})\times J_{\frac{1+t}{2}},$$

then the map $\tilde u : \mathbb{R}\times I \to M\times M$,

$$\tilde u(s,t) = (\tilde u_1(s,t), \tilde u_2(s,t)) = (u(\tfrac{s}{2}, \tfrac{1-t}{2}), u(\tfrac{s}{2}, \tfrac{1+t}{2})),$$

satisfies

$$\frac{\partial\tilde u}{\partial s} + \tilde J_t(\tilde u)(\frac{\partial \tilde u}{\partial t} - \tilde X_t(\tilde u)) = 0 \tag{3.24}$$

for any $u \in \mathcal{M}_{J,H}(M,\phi)$. Conversely, any solution to (3.24) defines a $W^{1,p}_{\mathrm{loc}}$, and thus smooth, solution $\hat{u}$ to (3.23),

$$\hat{u}(s,t) = \begin{cases} \phi^{[t]} \circ \tilde{u}_1(2s, 1-2t') & \text{for } t' \leq \frac{1}{2}, \\ \phi^{[t]} \circ \tilde{u}_2(2s, 2t'-1) & \text{for } t' \geq \frac{1}{2}, \end{cases}$$

where $t = [t] + t'$ and $[t]$ is the integer part of t. Similarly, there is an isomorphism between the kernels of the linearized operators

$$\ker D^{\phi}(u)_{J,H} \cong \ker D_{\tilde{J},\tilde{H}}(\tilde{u})$$

and their adjoints, which shows that $\operatorname{ind} D^{\phi}(u)_{J,H} = \operatorname{ind} D_{\tilde{J},\tilde{H}}(\tilde{u})$ and $\tilde{u}$ is regular if and only if u is regular. □

COROLLARY 3.5.3. *If J, H is a regular pair, then there is an isomorphism*

$$HF_*(M,\phi,J,H) \cong HF_*(M \times M, \triangle, \operatorname{graph}\phi, \tilde{J}, \tilde{H})$$

induced by an isomorphism of the chain complexes.

The concept of clean intersection translates to a nondegenerate manifold of fixed points, i.e., a submanifold $N \subset M$ such that $N \subset \operatorname{Fix}\phi$ and $\operatorname{Fix} d\phi(x) = T_x N$ for $x \in N$.

COROLLARY 3.5.4. *Let (M,ω) be a symplectic manifold and $\phi : M \to M$ a symplectomorphism with a connected nondegenerate manifold of fixed points, $\operatorname{Fix}\phi = N$. Assume that $I_{\omega,\phi} = 0$. Then*

$$HF_*(M,\phi) \cong H_*^{\mathrm{sing}}(N,\mathbb{Z}_2).$$

3.5.3. *The Novikov homology in the cotangent bundle.* We end this section with some remarks on a "non-exact" intersection, based rather on Floer's original result ([**Flo88a**]) than on the generalization of the previous section. Let L_0, L_1 be two Lagrangian submanifolds of (M,ω). Assume that

$$\omega|_{\pi_2(M,L_i)} = 0 \qquad \text{for } i = 0,1.$$

Then, according to the decomposition (3.6) of $\pi_1(\Omega(L_0,L_1))$ the homomorphism I_ω descends to

$$\omega_\# : \operatorname{im}\pi_1(l_0) \cap \operatorname{im}\pi_1(l_1) \to \mathbb{R}.$$

This can be described directly as the integral of ω over a cylinder which represents a homotopy in M between two loops $\alpha_i \subset L_i$ such that $\pi_1(l_0)[\alpha_0] = \pi_1(l_1)[\alpha_1]$.

Proposition 3.3.7 provides the compactness property for the subsets of $\mathcal{M}(L_0,L_1)$ with bounded energy. Therefore we may construct the Floer complex over an appropriate Novikov ring similarly as in 2.2. Let

$$\Gamma = \operatorname{im}\pi_1(l_0) \cap \operatorname{im}\pi_1(l_1) / \ker\omega_\#,$$

and let $\phi_\omega : \Gamma \to \mathbb{R}$ be the induced homomorphism. Fix a regular pair (J,H) and, for any solution x to (3.8), choose a path v_x connecting x with the base point in $\Omega(L_0,L_1)$, i.e.

$$v_x : I \times I \to M, \quad v_x(s,i) \in L_I, \quad v_x(0,\cdot) = \gamma_0, \quad v_x(1,\cdot) = x.$$

For any class $A \in \Gamma$ and $x^{\pm} \in \operatorname{Crit} a_H$, denote

$$\mathcal{M}_1(x^-, x^+, A) := \{u \in \mathcal{M}_1(x^-, x^+) : \mu_u = 1,\ v_{x^-} * u * (-v_{x^+}) \in A\}.$$

If $u \in \mathcal{M}_1(x^-, x^+, A)$, then its energy is bounded:

$$\ell^2(u) = \int u^*\omega + \int_0^1 H_t(x^-(t)) - H_t(x^+(t))\, dt$$

and

$$\int u^*\omega = \phi_\omega(A) - \int_0^1 v_{x^-}^*\omega + \int_0^1 v_{x^-}^*\omega.$$

We define the Floer complex as a free module generated by the zeros of ω_H over the Novikov ring

$$\Lambda_\omega = \Lambda(\Gamma, \phi_\omega, \mathbb{Z}_2).$$

The boundary operator $\partial = \partial_{\Lambda_\omega}$ is given by (3.14), where $\langle y, \partial x\rangle \in \Lambda_\omega$ and

$$\langle y, \partial x\rangle_A = n(x, y, A) = \#\mathcal{M}_1(x^-, x^+, A)/\mathbb{R}.$$

Proposition 3.3.7 assures that ∂ is well defined.

PROPOSITION 3.5.5. $\partial_{\Lambda_\omega}^2 = 0$, *and the resulting homology is naturally isomorphic for different choices of a regular pair* (J, H).

REMARK 3.5.6. Let X_t, $t \in [0,1]$ be a symplectic vector field, α the flux form of X_t,

$$\alpha = \operatorname{Flux}[X_t] = \int_0^1 i_{X_t}\omega\, dt,$$

and ϕ_t the family of symplectomorphisms generated by X_t. Then, as in Remark 3.13, there is a bijection between $\Omega(L_0, L_1)$ and $\Omega(L_0, \phi_1^{-1}(L_1))$, and it is easy to see that

$$\omega_{\#}^{\alpha} := \omega_{\#}(L_0, \phi_1^{-1}(L_1)) = \omega_{\#}(L_0, L_1) + [\alpha]|_{\operatorname{im}\pi_1(l_0)\cap\operatorname{im}\pi_1(l_1)}.$$

Moreover, the Floer complex $CF_*(L_0, \phi_1^{-1}(L_1))$ can be identified with the complex $CF_*(L_0, L_1, X_t, \Lambda_{\omega^\alpha})$ which is built from the solutions to

$$\dot{x}(t) = X_t(x(t)), \tag{3.25}$$

$$\frac{\partial u}{\partial s} + J_t(u)\left(\frac{\partial u}{\partial t} - X_t(u)\right) = 0. \tag{3.26}$$

If Y_t is another symplectic vector field with flux homologous to α, generating the family of symplectomorphisms ψ_t, then ϕ_1 and ψ_1 can be connected by a Hamiltonian isotopy (Banyaga, [**Ban78**]), and thus

$$CF_*(L_0, L_1, X_t, \Lambda_{\omega^\alpha}) \cong CF_*(L_0, L_1, Y_t, \Lambda_{\omega^\alpha}).$$

We denote the homology by $HF_*(L_0, L_1, [\alpha])$, where $[\alpha] \in H^1_{DR}(M, \mathbb{R})$, although this is an abuse, since the above isomorphism is not natural (Remark 3.3.11).

Now let L be a compact manifold and α a closed nondegenerate form on L. If we define the family of symplectomorphisms of T^*L by

$$\phi_t : T^*L \ni \beta \mapsto \beta + t\alpha(\pi(\beta)) \in T^*L,$$

then the flux of the associated vector field is just $\alpha \circ \pi$, and

$$\omega^\alpha = [\alpha] : \pi_1(L) \to \mathbb{R}.$$

Therefore Proposition 3.4.6 together with Remark 3.4.10 implies that all solutions to (3.25), (3.26) are t-independent. This yields

COROLLARY 3.5.7. *The Floer homology $HF_*(T^*L, L, L, [\alpha \circ \pi])$ is isomorphic to the Novikov homology $HN_*(L, [\alpha], \mathbb{Z}_2)$.*

Note that this result is not so strong because, unlike in the exact case, it cannot be immediately generalized to an arbitrary Lagrangian submanifold of M, even under the assumption that $\omega|_{\pi_2(M,L)} = 0$. The argument of Lemma 3.4.12 fails simply because we do not control the energy of u_n.

References

[Ban78] A. Banyaga, *Sur la structure du groupe des difféomorphismes qui préservent une forme symplectique*, Comment. Math. Helv. **53** (1978), 174–227.

[Bot82] R. Bott, *Lectures on Morse theory, old and new*, Bull. Amer. Math. Soc. (N.S.) **7** (1982), 331–358.

[Die69] J. Dieudonné, *Foundations of modern analysis*, Academic Press, New York, 1969.

[DS93a] S. Dostoglou and D. Salamon, *Instanton homology and symplectic fixed points*, London Math. Soc. Lecture Notes, vol. 192, Cambridge Univ. Press, Cambridge, 1993, pp. 57-93.

[DS93b] ———, *Self-dual instantons and holomorphic curves*, Ann. of Math. **139** (1994), 581–640.

[Flo88a] A. Floer, *Morse theory for lagrangian intersections*, J. Diff. Geom. **28** (1988), 513–547.

[Flo88b] ———, *Unregularized gradient flow for the symplectic action*, Comm. Pure Appl. Math. **41** (1988), 775–813.

[Flo89a] ———, *Symplectic fixed points and holomorphic spheres*, Comm. Math. Phys. **120** (1989), 575–611.

[Flo89b] ———, *Witten's complex and infinite dimensional Morse theory*, J. Diff. Geom. **30** (1989), 207–221.

[FHS94] A. Floer, H. Hofer, and D. Salamon, *Transversality in elliptic Morse theory for symplectic action*, Duke Math. J. **80** (1995), 251–292.

[Gro85] M. Gromov, *Pseudo holomorphic curves in symplectic manifolds*, Invent. Math. **82** (1985), 307–347.

[Hil65] Einar Hille, *Lectures on ordinary differential equations*, Addison-Wesley, Reading, MA, 1965.

[Hir70] M. Hirsch, *Differential topology*, Springer-Verlag, Heidelberg, 1970.

[HS92] H. Hofer and D. Salamon, *Floer cohomology and Novikov rings*, Floer Memorial Volume, Progress in Math., vol. 133, Birkhäuser, Basel, pp. 483–524.

[Kli78] W. Klingenberg, *Lectures on closed geodesics*, Springer–Verlag, Berlin, New York, 1978.

[LO93] Lê Hông Vân and K. Ono, *Symplectic fixed points, Calabi invariant and Novikov homology*, Topology **34** (1995), 155–176.

[Mil63] J. Milnor, *Lectures on the h-cobordism theorem*, Princeton Univ. Press, Princeton, NJ, 1963.

[Mil68] ———, *Infinite cyclic coverings*, Conference on the Topology of Manifolds (Michigan State University, E. Lansing, Mich., 1968), Prindle, Weber and Schmidt, Boston, 1968, pp. 115–133.

[Nov81] S. P. Novikov, *Multivalued functions and functionals. An analogue of Morse theory*, Dokl. Akad. Nauk SSSR **260** (1981), 33–37; English transl., Soviet Math. Dokl. **24** (1981), 222–226.

[Nov82] ———, *The Hamiltonian formalism and a many-valued analogue of Morse theory*, Uspekhi Mat. Nauk **37** (1982), no. 5, 3–49; English transl., Russian Math. Surveys **37** (1982), no. 5, 1–56.

[Oh93a] G.-Y. Oh, *Floer homology of Lagrangian intersections and pseudo-holomorphic discs*, I, Comm. Pure Appl. Math. **46** (1993), 949–993.

[Ono93] K. Ono, *The Arnold conjecture for weakly monotone symplectic manifolds*, Invent. Math. **119** (1995), 519–537.

[Sal90] D. Salamon, *Morse theory, the Conley index and Floer homology*, Bull. London Math. Soc. **22** (1990), 113–140.

[SZ94] D. Salamon and E. Zehnder, *Morse theory for periodic solutions of Hamiltonian systems and the Maslov index*, Comm. Pure Appl. Math. **45** (1994), 1303–1360.

[Sch93] M. Schwarz, *Morse homology*, Progress in Mathematics, vol. 111, Birkhäuser, Basel, 1993.

[Sma60] S. Smale, *Morse inequalities for a dynamical system*, Bull. Amer. Math. Soc. **66** (1960), 43–49.

[Sma61] ———, *On gradient dynamical systems*, Ann. of Math. **74** (1961), 199–206.

[Sma73] ———, *An infinite dimensional version of Sard's theorem*, Amer. J. Math. **87** (1973), 213–221.

[Spa66] E. Spanier, *Algebraic topology*, MacGraw-Hill, New York, Toronto, London, 1966.

[Vit87] C. Viterbo, *Intersection de sous-variétés lagrangiennes, fonctionelles d'action et indice des systèmes hamiltoniens*, Bull. Soc. Math. France **115** (1987), 361–390.

[Wei71] A. Weinstein, *Symplectic manifolds and their submanifolds*, Adv. in Math. **6** (1971), 329–346.

[Whi78] G. W. Whitehead, *Elements of homotopy theory*, Springer-Verlag, Heidelberg, 1978.

[Wit82] E. Witten, *Supersymmetry and Morse theory*, J. Diff. Geom. **17** (1982), 661–692.

UNIVERSITY OF WARWICK

Amer. Math. Soc. Transl.
(2) Vol. **196**, 1999

Surgery, Quantum Cohomology and Birational Geometry

Yongbin Ruan

1. Introduction

Recently, some amazing relations between quantum cohomology and birational geometry have been discovered. Surgeries play a fundamental role in these recent discoveries. This expository article surveys these new relations.

In the early days of quantum cohomology theory, the author [**R3**] showed that Mori's extremal ray theory can be generalized from projective manifolds to symplectic manifolds. This gave the first evidence that there exists a link between quantum cohomology and birational geometry. Subsequently, attention was diverted to establishing a mathematical foundation for quantum cohomology. Here the progress was slow. There are several papers ([**Wi2**], [**Wi3**] and others) extending results of [**R3**] to Calabi–Yau 3-folds and crepant resolutions. It is time to think about these mysterious connections again. The breakthrough appeared in the paper of Li and Ruan [**LR**], where the change of quantum cohomology under flops and small extremal transitions for 3-folds was calculated. Their results showed a surprising naturality property of quantum cohomology under these surgeries. At present, this naturality property is best understood in complex dimension three, whereas its general scope is still unknown. The author believes that there is some very interesting mathematics hidden in these intriguing properties, and that the general machinery of quantum cohomology can be very useful in uncovering them.

One of the goals of this paper is to attract more people to work on this topic. Hence, to entertain the reader, the author will try to make some conjectures and proposals for general cases. Since this is a survey rather than a research paper, the author will concentrate on the ideas and techniques. The main reference for this article is [**LR**], where the reader can find more details and more complete references. The author apologizes for the lack of details.

The paper is organized as follows. In Section 2 we give a brief survey on quantum cohomology and the naturality problem, and fix notation. Then, in Section 3, we briefly outline Mori's minimal model program in birational geometry. We discuss the relation between the naturality problem and birational geometry in Section 4. The main theorem was proved by a degeneration technique developed

---

1991 *Mathematics Subject Classification*. Primary 58Dxx; Secondary 14E30.

Partially supported by an NSF grant and a Sloan fellowship

©1999 American Mathematical Society

independently by Li and Ruan [**LR**] and by Ionel and Parker [**IP**]. Only the approach of [**LR**] is presented in Section 5. We will discuss the possible generalization of Mori's program in the last section.

The author would like to thank H. Clemens for explaining to him the semistable degenerations. The author benefited from many interesting conversations with J. Kollár, An-Min Li and D. Morrison on the topics related to this article. He wishes to express his thanks to all of them.

2. Quantum cohomology and its naturality

Due to the efforts of many people, the mathematical theory of quantum cohomology is well understood. We refer to [**R6**] for a survey. Developing the theory was rather difficult and time consuming. Naturally, we are looking for some applications to justify the time and energy we spent in building the general machinery. If we expect quantum cohomology to be as useful as regular cohomology, the results so far are disappointing.

The fundamental reason why cohomology is so useful is its naturality. Namely, a continuous map induces a ring homomorphism on cohomology. A fundamental problem in quantum cohomology is the following question.

QUANTUM NATURALITY PROBLEM. *Define "morphisms" for symplectic manifolds so that quantum cohomology is natural.*

The author believes that understanding the naturality of quantum cohomology will be essential for the future success of quantum cohomology theory.

It has been known for a while that quantum cohomology is not natural, even with respect to holomorphic maps. A crucial calculation is the quantum cohomology of projective bundles by Qin and Ruan [**QR**]. This calculation clearly demonstrated that quantum cohomology is not natural for fibrations. The lack of naturality is very different from ordinary cohomology, and severely limits our efforts to develop some nice consequences of cohomology theory, such as characteristic classes. The Qin–Ruan result shows that possible "morphisms" must be very rigid. The existence of these rigid morphisms will set quantum cohomology apart from ordinary cohomology and give it its own identity. The author does not know the full story at present. In this article, we will describe a class of "morphisms" of symplectic manifolds related to birational geometry. This class of "morphisms" is a certain class of surgeries called "transitions".

Let us briefly review quantum cohomology and naturality. We start with the definition of GW (Gromov–Witten) invariants. Suppose that (M, ω) is a symplectic manifold of dimension $2n$ and J is a tamed almost complex structure. Let $\mathcal{M}_A(J, g, k)$ be the moduli space of genus g J-holomorphic maps with fundamental class A and k marked points in the stable range $2g + k \geq 3$. If $2g + k \leq 3$, we divide the space of J-holomorphic maps by the automorphism group. When $g = 0$, we assume that $A \neq 0$. Consider the space of corresponding stable maps $\overline{\mathcal{M}}_A(J, g, k)$, which compactifies $\mathcal{M}_A(J, g, k)$. There are several methods of defining GW-invariants ([**FO**], [**LT3**], [**R5**], [**S**]). One method ([**R5**], [**S**]) is to construct a "virtual neighborhood" (U_k, S_k, E_k) with the following properties:

(1) U_k is a smooth, oriented, open orbifold whose points are maps;
(2) E_k is a smooth, oriented orbifold bundle over U_k;
(3) S_k is a proper section of E_k such that $S^{-1}(0) = \overline{\mathcal{M}}_A(J, g, k)$.

There is a map

$$\Xi_k : U_k \to M^k$$

given by evaluating a map at the marked points. Let Θ be a Thom form of E_k supported in a small neighborhood of the zero section. We define

$$\Psi^M_{(A,k)} = \int_{U_k} S_k^*\Theta \wedge \Xi_k^* \prod_i \alpha_i \tag{2.1}$$

for $\alpha_i \in H^*(M, \mathbb{R})$. One can eliminate divisor classes $\alpha \in H^2(M, \mathbb{R})$ by the relation

$$\Psi^M_{(A,g,k+1)}(\alpha, \alpha_1, \ldots \alpha_k) = \alpha(A)\Psi^M_{(A,g,k)}(\alpha_1, \ldots, \alpha_k), \tag{2.2}$$

for $A \neq 0$.

The above invariants are only the *primitive* GW-invariants. In general, we can also pull back cohomology classes of $\overline{\mathcal{M}}_{g,k}$ and insert them in formula (2.1). But these more general invariants are conjectured ([**RT2**]) to be computable from primitive invariants. When $A = 0$, $g = 0$, $k \neq 3$, we define $\Psi^M_{(A,0,k)} = 0$ and $\Psi^M_{(0,0,3)}(\alpha_1, \alpha_2, \alpha_3) = \alpha_1 \cup \alpha_2 \cup \alpha_3[M]$.

Choose a basis $A_1, \ldots, A_k$ of $H_2(M, \mathbb{Z})$. For $A = \sum_i a_i A_i$, we define the formal product $q^A = (q_{A_1})^{a_1} \cdots (q_{A_k})^{a_k}$. For each cohomology class $w \in H^*(M)$ we define the quantum 3-point function

$$\Psi^M_w(\alpha_1, \alpha_2, \alpha_3) = \sum_k \frac{1}{k!} \sum_A \Psi^M_{(A,0,k+3)}(\alpha_1, \alpha_2, \alpha_3, w, \ldots, w) q^A, \tag{2.3}$$

where w appears k times. Here, we view Ψ^M as a power series in formal variables $p_i = q^{A_i}$. Clearly, a homomorphism on H_2 induces a change of variables p_i. To define the quantum product, we need to fix a symplectic class $[\omega]$ to define q^A as an element of the Novikov ring Λ_ω (see [**RT1**]). Formally, we can define quantum multiplication by the formula

$$\alpha \times^w_Q \beta \cup \gamma[M] = \Psi^M_w(\alpha, \beta, \gamma).$$

To define a homomorphism of the quantum product, we need to match Novikov rings and hence symplectic classes. Here, we view quantum cohomology as a theory of Gromov–Witten invariants instead of just the quantum product. Since the GW-invariants are invariant under deformations, the symplectic class is not a fundamental ingredient of the quantum cohomology. In fact, the symplectic class often obstructs our understanding of quantum cohomology. Clearly, the quantum 3-point function contains the same information as the quantum product. It is more convenient to work directly with the quantum 3-point function.

DEFINITION 2.1. Suppose that

$$\varphi : H_2(X, \mathbb{Z}) \to H_2(Y, \mathbb{Z}), \ H^{\text{even}}(Y, \mathbb{R}) \to H^{\text{even}}(X, \mathbb{R})$$

are group homomorphisms such that the maps on H_2, H^2 are dual to each other. We say that φ is *natural* with respect to (big) quantum cohomology if $\varphi^*\Psi^X_0 = \Psi^Y_0$ ($\varphi^*\Psi^X_{\varphi^* w} = \Psi^Y_w$) after the change of formal variable $q^A \to q^{\varphi(A)}$. If φ is also an isomorphism, we say φ induces an isomorphism on (big) quantum cohomology, or that X and Y have the same (big) quantum cohomology.

We will not distinguish two power series F, G such that $F = H + F'$, $G = H + G'$, where G' is an analytic continuation of F'. For example, we can expand

$\frac{1}{1-t} = \sum_{i=0} t^i$ at $t = 0$, or $\frac{1}{1-t} = \frac{1}{-t(1-t^{-1})} = -\sum_{i=0} t^{-i-1}$ at $t = \infty$. Hence, we will treat $\sum_{i=0} t^i$ and $\sum_{i=0} t^{-i-1}$ as the same power series.

When X, Y are 3-folds, such a φ is completely determined by maps on H_2. For example, the dual map of $\varphi : H_2(X,\mathbb{Z}) \to H_2(Y,\mathbb{Z})$ gives a map $H^2(Y,\mathbb{R}) \to H^2(X,\mathbb{R})$. A map $H^4(Y,\mathbb{R}) \to H^4(X,\mathbb{R})$ is Poincaré dual to a map $H_2(Y,\mathbb{R}) \to H_2(X,\mathbb{R})$. In the case of a flop, the natural map $H_2(X,\mathbb{Z}) \to H_2(Y,\mathbb{Z})$ is an isomorphism. Therefore, we can take the map $H_2(Y,\mathbb{R}) \to H_2(X,\mathbb{R})$ as its inverse.

GW-invariants are invariants under symplectic deformation. Hence, two symplectic deformation equivalent manifolds have isomorphic big quantum cohomology. Holomorphic symplectic varieties are such examples. But these are trivial examples. The only known nontrivial examples are given by the work of Li and Ruan [**LR**] (see Section 4).

From the physical point of view, it is also natural (perhaps better) to allow certain **mirror transformations**. The author does not know precisely what these mirror transformations are, but they appear naturally in mirror symmetry and in the quantum hyperplane section conjecture [**Kim**]. Judging from known examples, they must include a nonlinear changes of coordinates in $H^*(X,\mathbb{R}), H^*(Y,\mathbb{R})$ and a scalings of the 3-point function. The readers can find more motivations for mirror transformations in the next section.

DEFINITION 2.2. Under the assumptions of Definition 2.1, we say φ is p-natural with respect to (big) quantum cohomology if we allow a mirror transformation in Definition 2.1. In the same way, we say φ is a p-isomorphism of (big) quantum cohomology if φ is an isomorphism and p-natural with respect to (big) quantum cohomology.

Obviously, naturality (isomorphism) implies p-naturality (p-isomorphism). The author does not know any nontrivial example of p-naturality or p-isomorphism at present.

Another interesting notion is the notion of Frobenius manifold due to Dubrovin [**D**]. Here, we require the bilinear form $\langle \alpha, \beta \rangle = \int_X \alpha \wedge \beta$ to be preserved under φ. This is the case for blow-downs. But the intersection form may not be preserved under general transitions when we have nontrivial vanishing cycles (see the next section). All three constructions (quantum products, Frobenius manifolds, quantum 3-point functions) contain the same quantum information. Their differences lie in the classical information such as the symplectic class and bilinear form, which can be studied separately. The author believes that Frobenius manifolds may be useful in understanding mirror transformations.

3. Birational geometry

In this section, we review Mori's minimal model program. There are several excellent surveys on this topic [**K2**], [**K3**]. We shall sketch the ideas mainly for non-algebraic geometers. In the process, we introduce the surgeries we will need later. Birational geometry is the central topic in algebraic geometry. The goal of birational geometry is to classify algebraic varieties in the same birational class. Two projective algebraic varieties are birational to each other if and only if there is a rational map between them with rational inverse. We call such a map a *birational map*. Any birational map is an isomorphism between Zariski open sets. Note that a birational map is not necessarily defined everywhere. If it is defined everywhere, we call it a contraction. By definition, a contraction changes a lower dimensional subset only. Hence, we can view a contraction as a surgery. Intuitively, a contraction simplifies a variety. For algebraic curves, the birational equivalence is the same as the isomorphism. In dimension two, any two birational algebraic surfaces are related by blow-up and blow-down. The blow-down is a contraction. If the surface is neither rational nor ruled, one can always perform blow-down until one reaches a "minimal model" where blow-downs can no longer be applied. This is the reason why Mori's program is often called the minimal model program. We want to factor a birational map into a sequence of contractions and find the minimal ones in the same birational class. In dimension two, the minimal model is unique. In higher dimensions, it is much more difficult to carry out the minimal model program. The first difficulty is that the contraction is not unique. To overcome this difficulty, Mori found a beautiful correspondence between contractions and some combinatorial information about the algebraic manifold itself. This correspondence is called the Mori extremal ray theory.

To define extremal rays, consider the Mori effective cone

$$\begin{aligned}\overline{NE}(X) &= \{\sum_i a_iA_i, a_i \geq 0,\ A_i \text{ is represented by a holomorphic curve}\}\\ &\subset H_2(X, \mathbb{R})\end{aligned} \tag{3.1}$$

By definition, $\overline{NE}(X)$ is a closed cone. Let $K(X)$ be the ample cone. One of the nice properties of projective geometry is

$$\overline{NE}(X) = \overline{K(X)^*}. \tag{3.2}$$

In symplectic geometry, we do not know if this property holds. This lack of knowledge is one of the primary difficulties of symplectic geometry. An edge of $\overline{NE}(X)$ is called an extremal ray of X. Suppose that L is an extremal ray with $K_X \cdot L < 0$. Mori showed that an extremal ray is represented by rational curves. Each extremal ray L gives a nef class H_L such that $H_L \cdot C = 0$ if $[C] \in L$ and $H_L \cdot C > 0$ if $[C] \notin L$, where C is a holomorphic curve. In general, *a class H is said to be nef if and only if $H \cdot C \geq 0$ for any holomorphic curve C.* Then H_L defines a contraction $\phi_L : X \to Y$ by contracting every curve whose homology class is in L. Moreover, the Picard number satisfies the relation $\text{Pic}(X) = \text{Pic}(Y) + 1$. The contraction ϕ_L is called primitive. We abuse the notation and say that ϕ_L contracts L. One can contract several extremal rays simultaneously. Such a contraction corresponds to contracting an extremal face or other higher dimensional boundary. Conversely, if we have a contraction $\phi : X \to Y$ such that $\text{Pic}(X) = \text{Pic}(Y) + 1$, then $\phi = \phi_L$ for some extremal ray.

Once the contraction is found, an additional difficulty arises because Y could be singular even if X is smooth. Obviously, the topology of Y is simpler. However, if the singularities of Y become more and more complicated, we just trade one difficult problem for another equally difficult problem. The next idea is that we can fix the class of singularities. There are two commonly used classes of singularities: terminal singularities and canonical singularities. We refer readers to [**K2**] for the precise definitions. The terminal singularities comprise the minimal class of singularities in the minimal model theory. It has been shown that in complex dimension three, the minimal model program indeed can be carried out. Except for a well-understood exceptional class of varieties (uniruled varieties), the minimal model exists. A minimal model is, by definition, an algebraic variety X such that

(i) X has terminal singularities only, and

(ii) the canonical bundle K_X is nef.

One should mention that contractions are not sufficient to find the minimal model. A more difficult operation, "flip", is needed. We will not give any details about flip. However, its companion, "flop", is very important to us because it resolves another difficulty of the minimal model program. Unlike the dimension two case, the minimal model is not unique in dimension three or higher. However, different minimal models are related by a sequence of flops [**Ka**], [**K1**].

We now give a local description of a "simple flop". Let Y_s be a threefold with ordinary double points. Namely, the neighborhood of a singular point is complex analytically equivalent to a hypersurface of $\mathbb{C}^4$ defined by the equation

$$x^2 + y^2 + z^2 + w^2 = 0. \tag{3.3}$$

The origin is the only singular point. When we blow up the origin to obtain Y_b, we obtain an exceptional divisor $E \cong \mathbb{P}^1 \times \mathbb{P}^1$. Then, we contract one ruling to obtain Y containing a rational curve with normal bundle $O(-1) + O(-1)$. We can also contract another ruling to obtain $\tilde{Y}$ containing a rational curve with normal bundle $O(-1) + O(-1)$. Clearly, Y and $\tilde{Y}$ coincide locally. However, they could differ globally. The process of going from Y to $\tilde{Y}$ is called a simple flop. For smooth threefolds, a general flop can be deformed locally into the disjoint union of several simple flops.

The final difficulty is that Mori's minimal model program has only been carried out for complex dimension three. It is still an open problem for higher dimensions.

The fundamental surgeries in birational geometry are contractions, flips, and flops. Every smooth Calabi–Yau threefold is a minimal model by definition. For minimal models, we only have flops. But flops are not sufficient to classify smooth Calabi–Yau threefolds. We need to introduce another surgery, called *extremal transition* or *transition*. As we mentioned previously, a contraction could lead to a singular manifold. Sometimes, a contraction can result in a manifold with singularities beyond the class of terminal singularities. In fact, the flip was introduced precisely to deal with this problem, where it will improve the singularity. Smoothing is another well-known method to improve the singularity. A smoothing is defined as follows. Consider

$$\pi : U \to D(0, \epsilon) \subset \mathbb{C}, \tag{3.4}$$

where U is an analytic variety and π is holomorphic and of maximal rank everywhere except at zero. Then $X_z = \pi^{-1}(z)$ is smooth except for the central fiber X_0. In this situation, we say that X_z is a smoothing of X_0. If π is also of maximal rank at zero, X_0 is smooth and X_z is deformation equivalent to X_0. Hence, they have the same quantum cohomology. If U is also Kähler, we say (U, π) is a Kähler smoothing. By definition, *a transition is a composition of a contraction and a Kähler smoothing.* A contraction is said to be *small* if the exceptional locus is of complex codimension two or more. A transition is said to be *small* if the corresponding contraction is small. A flip-flop is a small operation in the sense that it only changes a subset of codimension two. Incidentally, flip-flop has been completely classified in dimension three. In higher dimensions, the classification is still an open question. It was conjectured that any two Calabi–Yau threefolds can be connected to each other by a sequence of flops or transitions and their inverses.

We first study the change of topology regarding flops and transitions, which is easy in dimension three. Suppose that $F : X \rightsquigarrow \tilde{X}$ is a flop, where X and $\tilde{X}$ are threefolds. In this case, the exceptional locus is of complex dimension one. Each $A \in H_2(X, \mathbb{Q})$ is represented by a pseudo-submanifold Σ. Using PL-transversality, we can assume that Σ is disjoint from the exceptional locus. Then Σ can also be viewed as a pseudo-submanifold of $\tilde{X}$. We can also reverse this process. Therefore, F induces an isomorphism on $H_2(X, \mathbb{Z})$, and hence $H^2 = \mathrm{Hom}(H_2, \mathbb{Q})$. Using Poincaré duality, it induces an isomorphism on H^4 as well. The maps on H^0, H^6 are obvious. The first important theorem is

THEOREM 3.1 (Li and Ruan). *Under the previous assumptions, F induces an isomorphism on quantum cohomology. In particular, any two three-dimensional smooth minimal models have isomorphic quantum cohomology, where the isomorphism is induced by flops.*

In dimension ≥ 4, flop is not completely understood. However, there are theorems of Batyrev [**Ba**] and Wang [**Wa**] that any two smooth minimal models have the same Betti number. We conjecture that

QUANTUM MINIMAL MODEL CONJECTURE. *Any two smooth minimal models in any dimension have isomorphic quantum cohomology.*

The Quantum Minimal Model Conjecture is true for holomorphic symplectic varieties as well. In this case, two birational holomorphic symplectic varieties are deformation equivalent. Hence, there is an abstract isomorphism of big quantum cohomology. However, it is still an interesting question to give a geometric construction of the isomorphism.

For transitions, let $c : X \to X_s$ be the contraction and $\pi : U \to D(0, \epsilon)$ be the Kähler smoothing such that X_s is the central fiber. There is a deformation retract $r : U \to X_s$. Therefore, there is a map $\delta : H^*(X_s, \mathbb{Q}) \to H^*(X_z, \mathbb{Q})$, where $X_z = \pi^{-1}(z)$ is the nearby smooth fiber. In the case of Kähler smoothing, the image of δ can be described in terms of monodromy. Readers can find more detailed references in [**C**]. The result is as follows. We first blow up the singular points of U along the central fiber to obtain a smooth complex manifold V whose central fiber is a union of smooth complex submanifolds intersecting transversally. Then we perform the base change (pull-back by a map $z \to z^k$ on the base) to obtain W. W has the additional property that every component of the central fiber has multiplicity one in the sense that the function

$$\tilde{\pi} : W \to V \to U \to D(0, \epsilon) \tag{3.5}$$

vanishes with order one on each component. This process is called *semi-stable reduction.* A consequence of semi-stable reduction is that the monodromy r of $\tilde{\pi}$ is unipotent. In this case, we can define

$$\log(r) : H^*(X_z, \mathbb{Q}) \to H^*(X_z, \mathbb{Q}). \tag{3.6}$$

The matrix $\log(r)$ is nilpotent, and $\ker \log(r)$ is precisely the space of r-invariant cocycles. The main result of Deligne, Schmid and Clemens [**C**] is that $\mathrm{im}(\delta)^* = \ker \log(r)$. There is a natural decomposition

$$H^*(X_z, \mathbb{Q}) = \ker \log(r) + \mathrm{im} \log(r). \tag{3.7}$$

The subspace $\operatorname{im}\log(r)$ is Poincaré dual to the space of vanishing cycles. Using the decomposition (3.7), we can define a map

$$H^*(X_z, \mathbb{Q}) \to H^*(X_s, \mathbb{Q}). \tag{3.8}$$

Let $\phi : H^*(X_s, \mathbb{Q}) \to H^*(X, \mathbb{Q})$ be its composition with c^*. Our second theorem is

THEOREM 3.2 (Li and Ruan). *Suppose that $T : X \to X_z$ is a small transition in complex dimension three as described previously. Then $\phi : H^*(X_z, \mathbb{R}) \to H^*(X, \mathbb{R})$ is a natural map for big quantum cohomology.*

The author conjectured the following:

QUANTUM NATURALITY CONJECTURE. *Theorem 3.2 is true in any dimension.*

One corollary of the Quantum Naturality Conjecture is as follows. Suppose that α is Poincaré dual to a vanishing cycle, i.e., $\alpha \in \operatorname{im}\log(r)$ is such that the central fiber X_s admits a small resolution. Then $\Psi^{X_z}_{(A,0,k)}(\alpha, \dots) = 0$.

The above results are not satisfactory, because they do not say anything about the easiest transition, namely blow-down. The calculation in [**RT1**] (Example 8.6) shows that blow-down is not natural for quantum cohomology. But this is not the end of the story. For Calabi–Yau threefolds, there are two other types of transitions which are not small. These are blow-down type surgeries. By the Mirror Surgery Conjecture [**LR**], there should be corresponding operations on the Hodge structures. The operation on the Hodge structure is obviously natural in algebraic coordinates. To compare with quantum cohomology, we should change from algebraic coordinates to the flat coordinates near the large complex structure limit (mirror transformation). This suggests

QUANTUM p-NATURALITY CONJECTURE. *Transition induces a p-natural map on big quantum cohomology.*

We hope to give such an example of p-naturality in [**LQR**].

For the same reason, we also have

p-QUANTUM MINIMAL MODEL CONJECTURE. *Any two smooth minimal models have p-isomorphic quantum cohomology.*

The Quantum Minimal Model Conjecture implies the p-Quantum Minimal Model Conjecture, but the converse could be false.

In Mori's program, it is essential to consider varieties with terminal singularities. It would be an important problem to work out the singular case in complex dimension three to see if the same phenomenon holds in the category of varieties with terminal singularities. Fortunately, a complete description of flops is also available in this case [**K3**].

4. Relative Gromov–Witten invariants and the degeneration formula

In this section, we describe the techniques used to prove Theorems 3.1 and 3.2. There are two slightly different approaches by Li and Ruan [**LR**] and by Ionel and Parker [**IP**]. Here, we present the Li–Ruan approach only. Readers should consult [**IP**] for their approach. Moreover, the author uses different terminology from the original paper to make it easier to the readers who are not familiar with contact geometry. The heart of the technique is a degeneration formula for GW-invariants under semi-stable degeneration such that the central fiber has only two components. The definition of semi-stable degeneration was already given in the previous section. In the case where the central fiber has only two components, the normal bundle of their intersection Z has opposite first Chern classes. During the last several years, semi-stable degeneration reappeared in symplectic geometry under the names "symplectic norm sum" and "symplectic cutting"; it plays an important role in some of the recent developments in symplectic geometry. The author should point out that the recent construction of symplectic geometry is independent from algebraic geometry. In fact, the symplectic construction is stronger because it shows that one can always produce a semi-stable degeneration with prescribed central fiber in the symplectic category. Of course, the analogous result is far from being true in the algebraic category. There are two symplectic constructions, norm sum and cutting, which are inverse to each other. Let me present the construction of symplectic cutting due to Lerman [**L**].

Suppose that $H : M \to \mathbb{R}$ is a periodic Hamiltonian function such that the Hamiltonian vector field X_H generates a circle action. By adding a constant, we can assume that 0 is a regular value. Then $H^{-1}(0)$ is a smooth submanifold preserved by the circle action. The quotient $Z = H^{-1}(0)/S^1$ is the famous symplectic reduction. Namely, it has an induced symplectic structure.

For simplicity, we assume that M has a global Hamiltonian circle action. Once we write down the construction, we observe that a local Hamiltonian circle action is enough to define symplectic cutting.

Consider the product manifold $(M \times \mathbb{C}, \omega \oplus -i\,dz \wedge d\bar{z})$. The moment map $H - |z|^2$ generates a Hamiltonian circle action $e^{i\theta}(x, z) = (e^{i\theta}x, e^{-i\theta}z)$. Zero is a regular value, and we have symplectic reduction

$$(4.1) \qquad \overline{M}^+ = \{H = |z|^2\}/S^1.$$

We have the decomposition

$$(4.2) \qquad \overline{M}^+ = \{H = |z|^2\}/S^1 = \{H = |z|^2 > 0\}/S^1 \cup H^{-1}(0)/S^1.$$

Furthermore,

$$(4.3) \qquad \phi^+ : \{H > 0\} \to \{H = |z|^2 > 0\}/S^1 \phi^+(x) = (x, \sqrt{H(x)})$$

is a symplectomorphism. Let

$$(4.4) \qquad M_b^+ = H^{-1}(\geq 0).$$

Then M_b^+ is a manifold with boundary, and there is a map

$$(4.5) \qquad M_b^+ \to \overline{M}^+.$$

Clearly, $\overline{M}^+$ is obtained by collapsing the S^1 action on $H^{-1}(0)$. It is obvious that we only need a local S^1 Hamiltonian action. To obtain $\overline{M}^-$, we consider the circle action $e^{i\theta}(x, z) = (e^{i\theta}x, e^{i\theta}z)$ with the moment map $H + |z|^2$. The manifolds $\overline{M}^+$ and $\overline{M}^-$ are called symplectic cuttings of M. We define M_b^- similarly. By the construction, $Z = H^{-1}(0)/S^1$, with induced symplectic structure embedded symplectically into $\overline{M}^\pm$. Moreover, its normal bundles have opposite first Chern classes.

Symplectic norm sum is an inverse operation of symplectic cutting. Gompf developed it before symplectic cutting appeared. Ionel observed that symplectic norm sum/symplectic cutting is the same as semi-stable degeneration such that $\overline{M}^+ \cup_Z \overline{M}^-$ is the central fiber.

The main result of this section is a degeneration formula for GW-invariants under semi-stable degeneration with the central fiber consisting of two components for a symplectic norm sum or a symplectic cutting. To simplify the notation, we use the term symplectic cutting only. The first step is to introduce relative GW-invariants of a pair (M, Z), where Z is a smooth codimension two symplectic submanifold. We can always choose an almost complex structure J such that Z is an almost complex submanifold. Hence, the algebraic geometer should view Z as a smooth divisor. The process of defining relative GW-invariants is similar to that of defining regular GW-invariants. First, we define relative stable maps. It is helpful to recall the definition of a stable map.

DEFINITION 4.1 ([**PW, Ye, KM**]). Let $(\Sigma, \{x_i\})$ be a stable Riemann surface. A stable holomorphic map (associated with $(\Sigma, \{x_i\})$) is an equivalence class of continuous maps f from Σ' to M such that f is smooth at smooth points of Σ'. Here the domain Σ' is obtained by joining chains of $\mathbb{P}^1$'s at some double points of Σ to separate the two components, and then attaching some trees of $\mathbb{P}^1$'s. We call the components of Σ *principal components*, and all others *bubble components*. Furthermore,

(1) If we attach a tree of $\mathbb{P}^1$ at a marked point x_i, then x_i will be replaced by a point different from intersection points on some component of the tree. Otherwise, the marked points do not change.
(2) The singularities of Σ' are normal crossing, and there are at most two components intersecting at one point.
(3) If the restriction of f to a bubble component is constant, then it has at least three special points (intersection points or marked points). We call this component *a ghost bubble* [**PW**].
(4) The restriction of f to each component is J-holomorphic.

Two such maps are equivalent if one is the composition of the other with an automorphism of the domain of f.

A stable map can be naturally decomposed into connected components lying outside of Z (*non-Z-factors*) or completely inside Z (*Z-factors*). Both are stable maps. They intersect each other at marked points. Let N be the projective completion of the normal bundle $E \to Z$, i.e., $N = P(E \oplus \mathbb{C})$. Then N has a zero section Z_0 and an infinity section Z_∞. We view Z in M as a zero section. A rough definition of a relative stable map is a stable map whose Z-component can be lifted to N in a prescribed way. More precisely,

DEFINITION 4.2. A relative stable map f satisfies the definition of stable map with the following additional data. We divide each Z-factor into a set of stable maps $\{f_\mu\}$ intersecting at marked points. Then, we associate an integer a_p to marked points of $\{f_\mu\}$ and non-Z factors with the following properties.

(5) $a_p \geq 0$ if p is a marked point of f. Otherwise, $a_p \neq 0$.
(6) Each f_μ can be lifted to a stable map $\tilde{f}_\mu$ into N such that (i) $\tilde{f}_\mu(p) \in Z_0$ with order a_p if and only if $a_p > 0$; (ii) $\tilde{f}_\mu(p) \in Z_\infty$ with order $|a_p|$ if and

only if $a_p < 0$; (iii) $\tilde{f}_\mu \in N - Z_0 \cup Z_\infty$ if and only if $a_p = 0$. Moreover, $\tilde{f}_\mu$ has no other intersections with Z_0 or Z_∞ except at marked points

(7) Each non-Z-factor intersects Z only at marked points. Moreover, the order of intersection at a marked point p is a_p.

(8) If f_μ, f_ν intersect at $f_\mu(p_i) = f_\nu(q_i)$ for $1 \le i \le l$, then (i) all a_{p_i} are > 0 or < 0 simultaneously for $1 \le i \le l$; (ii) $a_{p_i} = -a_{q_i}$.

(9) If a non-Z-factor intersects f_μ, the same compatibility condition as (7) is satisfied.

We call such a system of a_p a *compatible label.*

Note that the existence of $\tilde{f}_\mu$ implies that

$$\sum a_{p_i} = C_1(E)([f_\mu]). \tag{4.6}$$

Therefore, not every stable map may have a compatible label such that the lifting $\tilde{f}_\mu$ exists. This accounts for the difference between relative and absolute GW-invariants. However, it is easy to show that the lifting is unique if it exists.

By (5), there are two kinds of marked points of f determined by the value a_{x_i}. To simplify the notation, we distinguish two kinds of marked points. Suppose that the marked points are $x_1, \ldots, x_k, y_1, \ldots, y_m$ such that $a_{x_i} = 0$ and $a_{y_i} > 0$. We fix $\mathbf{k} = (a_{y_1}, \ldots, a_{y_m})$ and let $\overline{M}_A(Z, g, k, \mathbf{k}, J)$ be the moduli space of relative stable maps with fixed $g, k, \mathbf{k}$. Intuitively, $\overline{\mathcal{M}}_A(Z, g, k, \mathbf{k}, J)$ is the compactification of the space of holomorphic maps with prescribed tangency condition $\mathbf{k}$ at Z. We can use $\overline{\mathcal{M}}_A(Z, g, k, \mathbf{k}, J)$ to define the *relative GW-invariant* $\Psi^{(M,Z)}$. This invariant counts the number of relatively stable maps intersecting Z at finitely many points with prescribed tangency. More precisely, we have the maps

$$\Xi_{g,k} : \overline{\mathcal{M}}_A(Z, g, k, \mathbf{k}, J) \to M^k, \tag{4.7}$$

$$P : \overline{\mathcal{M}}_A(Z, g, k, \mathbf{k}, J) \to Z^m, \tag{4.8}$$

$$\chi_{g,k+m} : \overline{\mathcal{M}}_A(Z, g, k, \mathbf{k}, J) \to \overline{\mathcal{M}}_{g,k+m}. \tag{4.9}$$

Roughly, the relative GW-invariants are defined as

$$\begin{aligned} &\Psi^{(M,Z)}_{(A,g)}(K;\alpha_1, \ldots, \alpha_k; \beta_1, \ldots, \beta_m) \\ &= \int_{\overline{\mathcal{M}}_A(Z,g,k,\mathbf{k},J)} \chi^*_{g,k+m} K \wedge \Xi^*_{g,k} \prod_i \alpha_i \wedge P^* \prod_j \beta_j. \end{aligned} \tag{4.10}$$

To be precise, the virtual techniques developed by Fukaya and Ono [**FO**], Li and Tian [**LT3**], Ruan [**R5**], and Siebert [**S**] apply to this case. For example, we can construct a virtual neighborhood (7.1 in [**LR**]) of $\overline{\mathcal{M}}_A(Z, g, k, \mathbf{k}, J)$. Then we integrate (1.17) over the virtual neighborhood.

Clearly, there is a map

$$R : \overline{\mathcal{M}}_A(Z, g, k, \mathbf{k}, J) \to \overline{\mathcal{M}}_A(g, k+m, J). \tag{4.11}$$

However, it may not be surjective even if $\mathbf{k} = (1, \ldots, 1)$. Because of this, we remark that when $\mathbf{k} = \{1, \ldots, 1\}$, $\Psi^{(M,Z)}$ is usually different from the ordinary or absolute GW-invariant. So $\Psi^{(M,Z)}$ is not a "generalized" GW-invariant. The relation between relative and absolute GW-invariants is one of the main topics of [**LQR**].

THEOREM 4.3 (Theorem 7.6 in [**LR**]). (i) $\Psi^{(M,Z)}_{(A,g,k,\mathbf{k})}(K;\alpha_1,\dots,\alpha_k;\beta_1,\dots,\beta_m)$ *is well defined, multilinear and skew-symmetric.*

(ii) $\Psi^{(M,Z)}_{(A,g,k,\mathbf{k})}(K;\alpha_1,\dots,\alpha_k;\beta_1,\dots,\beta_m)$ *is independent of the choice of forms* K,α_i,β_j *representing the cohomology classes* $K,[\beta_j],[\alpha_i]$, *and independent of the choice of virtual neighborhoods.*

(iii) $\Psi^{(M,Z)}_{(A,g,k,\mathbf{k})}(K;\alpha_1,\dots,\alpha_k;\beta_1,\dots,\beta_m)$ *is independent of the choice of almost complex structures on* M, *where* Z *is an almost complex submanifold, and hence is an invariant of* (M,Z).

If $K=1$, we will drop K from the formula. This invariant is called a *primitive relative GW-invariant.* In this article, we will give a degeneration formula for primitive invariants and comment on how to modify it for non-primitive invariants.

Now we explain the degeneration formula of GW-invariants under symplectic cutting. First of all, symplectic cutting defines a map

$$\pi: M \to \overline{M}^+ \cup_Z \overline{M}^-, \tag{4.12}$$

where the right hand side is the central fiber of the degeneration. By construction, $\omega^+|_Z=\omega^-|_Z$. Hence, the pair (ω^+,ω^-) defines a cohomology class of $\overline{M}^+\cup_Z\overline{M}^-$, denoted by $\omega^+\cup_Z\omega^-$. It is easy to see that

$$\pi^*(\omega^+\cup_Z\omega^-)=\omega. \tag{4.13}$$

The map π induces the map

$$\pi_*: H_2(M,\mathbb{Z})\to H_2(\overline{M}^+\cup_Z\overline{M}^-,\mathbb{Z}). \tag{4.14}$$

Let $B\in\ker(\pi_*)$. By (4.13), $\omega(B)=0$. Define $[A]=A+\ker(\pi_*)$ and

$$\Psi^M_{([A],g,k)}=\sum_{B\in[A]}\Psi_{(B,g,k)}. \tag{4.15}$$

For $B,B'\in[A]$ we have $\omega(B)=\omega(B')$. By the Gromov compactness theorem, there are only finitely many such B represented by J-holomorphic stable maps. Therefore, the sum in (4.15) is finite. Moreover, the cohomology class α_i is not arbitrary. Namely, let $\alpha_i^{\pm}\in H^*(\overline{M}^{\pm},\mathbb{R})$ be such that $\alpha_i^+|_Z=\alpha_i^-|_Z$. It defines a class $\alpha_i^+\cup_Z\alpha_i^-\in H^*(\overline{M}^+\cup_Z\overline{M}^-,\mathbb{R})$. We choose $\alpha_i=\pi^*(\alpha^+\cup_Z\alpha^-)$. The degeneration formula is a large sum:

$$\Psi^M_{([A],g,k)}(\alpha_1,\dots,\alpha_k)=\sum_{\text{indices}}\text{terms}. \tag{4.16}$$

Next, we give a procedure to write down the terms on the right hand side.

Step (1). First we draw a graph representing the topological type of a degenerate Riemann surface in $\overline{M}^+\cup_Z\overline{M}^-$ with the following properties.

(i) Each component is completely inside either $\overline{M}^+$ or $\overline{M}^-$.

(ii) The components only intersect each other along Z.

(iii) No two components that are both in $\overline{M}^+$ or in $\overline{M}^-$ intersect each other.

(iv) The arithmetic genus is g and the number of marked points is k.

(v) The total homology class is $\pi_*([A])$.

(vi) Each intersection point carries a positive integer representing the order of tangency.

We use C to denote such a graph. C will be used as the index of summation, and (4.16) can be written as

$$\Psi^M_{([A],g,k)}(\alpha_1, \dots, \alpha_k) = \sum_C \Psi_C. \tag{4.17}$$

Step (2). Suppose that C has components $C_1, \dots, C_s$, and let (A_{C_i}, g_{C_i}) be the homology class and the genus of the component C_i. Then,

$$\Psi_C = r_C \prod_i \Psi^{(\overline{M}^?, Z)}_{(A_{C_i}, g_{C_i})}(\text{variables}), \tag{4.18}$$

where $\overline{M}^?$ is the one of $\overline{M}^{\pm}$ in which C_i lies and r_C is the product of certain numbers to be determined.

Step (3). The following two steps determine the variables of each term in (4.18) and the number r_C. If a marked point x_i appears in some component C_t, then $\alpha_i^{\pm}$ should be in the variable of $\Psi^{(\overline{M}^?, Z)}_{(A_{C_t}, g_{C_t})}(\text{variables})$. The sign $\pm$ depends on whether C_t lies in $\overline{M}^+$ or $\overline{M}^-$.

Step (4). $\Psi^{(\overline{M}^?, Z)}_{(A_{C_t}, g_{C_t})}(\text{variables})$ has also other variables associated to intersection points. Suppose that y is an intersection point of components C_i and C_j with tangency of order k_y. Let β_a be a basis of $H^*(Z, \mathbb{R})$ and $\eta^{ab} = \int_Z \beta_a \beta_b$. Let (η_{ab}) be the inverse matrix; it can be thought of as the intersection matrix of the Poincaré dual of β_a, β_b. Then y contributes a term β_a in $\Psi_{(A_{C_i}, g_{C_i})}$, β_b in $\Psi_{(A_{C_j}, g_{C_j})}$ and $k_y \eta_{ab}$ in r_C.

The four steps above will completely determine the formula (4.16).

THEOREM 4.1 ([**LR**], Theorem 7.9, 7.10). *There is a degeneration formula of GW-invariants under symplectic cutting described by Steps* (1)–(4).

An important comment is that different $\alpha_i^{\pm}$ may define the same α_i. Then we have different ways to express $\Psi^M(\dots, \alpha_i, \dots)$. This is very important in the application of the degeneration formula (4.16). For example, the Pioncaré dual of a point can be chosen to have its support completely inside $\overline{M}^+$ or $\overline{M}^-$.

To derive a degeneration formula with class K, let us consider its Poincaré dual K^*. Then we have to specify a degeneration K^*_∞ of K^* and look at what kind of graph C could appear in the degeneration of K^*. Namely, we look for those C for which we obtain an element of K^*_∞ after contracting all the unstable components of C. For example, suppose that $K^* = \{\Sigma\}$. We can require that Σ stay in the interior of $\overline{M}_{g,k}$. This requirement will force C to have one component the same as Σ in one of $\overline{M}^{\pm}$ and other components that are unstable rational components.

Using Theorem 4.1, we can prove Theorems 3.1 and 3.2. Consider a simple flop $X \rightsquigarrow \tilde{X}$. They have the same blow-up X_b. Take a trivial family of X and blow up the central fiber along the the rational curve L. This is a semi-stable degeneration whose central fiber is a union of X_b and Y_L, where Y_L is the projective completion of normal bundle $O(-1) + O(-1)$ of L intersecting with X_b along the projectivization of the normal bundle $P(O(-1)+O(-1)) \cong \mathbb{P}^1 \times \mathbb{P}^1$. The crucial information for the proof of Theorem 3.1 is that $C_1(Y_L) = 3Z_\infty$. An index calculation shows that the only nonzero term in (4.16) corresponds to C that lies completely in X_b or Y_L. We repeat the same argument for $\tilde{X}$ to express the invariant of $\tilde{X}$ in terms of that of X_b, Y_L. We then obtain a formula for the change of GW-invariants under a simple flop. To obtain the isomorphism of quantum cohomology, we need the amazing fact that the contribution of multiple cover maps to L cancels the error from the change of the classical triple product under a simple flop.

To prove Theorem 3.2, we consider the case where X_s has only an ordinary double point. We then do the semi-stable reduction to obtain a semi-stable degeneration. In this case, the central fiber has again only two components X_b, Y_s (a fact I learned from H. Clemens), where X_b is the blow-up along L and Y_s is a quadratic threefold intersecting with X_b along a hyperplane H. This is completely consistent with the symplectic cutting description in Section 2 of [**LR**]! Note that the first Chern class of a quadric threefold is $3H$. This is precisely what we need in the case of a flop. Using our degeneration formula, we conclude that the only nonzero term in (4.16) corresponds to C staying in either X_b or Y_s. But there is no holomophic curve in Y_s not intersecting H. Hence, it must be in X_b. Therefore, we get an expression (Theorem 8.1 in [**LR**]) of the invariant of X_z in terms of the relative invariant of X_b and hence X by the previous argument. The rest of the proof is just sorting out the formula for quantum threefold points.

For more general flops or small transitions, an argument of P. Wilson shows that one can reduce it to the above case by an almost complex deformation. Hence, our previous arguments work in all cases. If the reader is looking for an algebraic proof of our theorem, it seems that one can generalize our degeneration formula to semi-stable degeneration with an arbitrary number of components. Then one can use this more general degeneration formula to bypass the almost complex deformation. It is clear that such a general degeneration formula should also contain relative GW-invariants, but the combinatorics will be much more complicated.

5. Generalization of minimal model program

After going over technical mathematics in the previous section, it is time to have some fun and make some wild speculations. Mori's minimal model program can be viewed as a surgery theory of contractions and flip-flops. As we mentioned in section three, transitions play a crucial role in the classification of Calabi–Yau threefolds. The author hopes he has convinced the reader that they also play a crucial role in quantum cohomology. It is natural that we speculate that there should be a minimal model theory with transitions as fundamental surgeries. We call this proposed theory the "moduli minimal model program" because by definition it is concerned with the moduli space of complex structures. The moduli minimal model program is no longer in the category of birational geometry. Instead, it addresses the important problem of connecting different moduli spaces of complex manifolds. An interesting question is: what are the minimal models in the moduli minimal model program? Recall that in the Mori theory a minimal model has the nef K_X. We speculate that, for a minimal model in the moduli minimal model program, either we should have $\mathrm{Pic}(X) = 1$, or K_X should be ample. Similarly to the case of the minimal model program, there are exceptional cases. It is of great interest to study these exceptional cases as well. Clearly, transition improves singularities. The author does not have any feeling for what kind of singularities should be allowed in the moduli minimal model program.

An interesting aspect of the moduli minimal model program is its connection to symplectic geometry. Due to the topological nature of symplectic forms, symplectic geometry is an ideal place to study moduli problems. In my view, the guiding problem of symplectic geometry should be the classification of symplectic manifolds. The question is how to classify them. In topology, it is rare that we can label manifolds. An essential step is to establish some fundamental surgeries and classify a class of manifolds under such surgeries. These fundamental surgeries should simplify the manifolds. In dimension two, the fundamental surgery is the connected sum. In dimension three, the fundamental surgeries are connected sums over spheres and tori. In dimension four, people are still struggling to understand the fundamental surgeries. A natural question is "what are fundamental surgeries of symplectic manifolds?" I believe that transition is one of the fundamental surgeries of symplectic manifolds. In fact, if we believe that quantum cohomology is a fundamental invariant of a symplectic manifold, then any surgery that is natural with respect to quantum cohomology deserves our attention. This line of thought is very appealing because transitions have a beautiful interpretation in terms of classical symplectic geometry.

In a neighborhood of a singularity, the boundary has a natural contact structure. It is known classically that there are two ways to "fill" a contact manifold. A contact manifold could bound a resolution of a singularity or a neighborhood of the zero section of a cotangent bundle. In the case of a singular point, transition is a local duality which interchanges these two fillings. In the general case, we need to consider a fiberwise version of the above construction.

Therefore, it is natural to consider a "symplectic minimal model program" using transition as the fundamental surgery. As the author showed in [**R3**], one can generalize the Mori cone $NE(X)$ to symplectic manifolds. Transition simplifies a symplectic manifold X in the sense that it simplifies the Mori cone $NE(X)$. One important ingredient in Mori's program is the interpolation between the Mori cone and the ample cone, which are related by $NE(X) = \overline{K(X)^*}$. In symplectic geometry, we no longer have such a relation. So we encounter severe difficulties at the first step of our symplectic minimal model program. It is probably a long shot to establish such a program. But I have no doubt that much interesting mathematics will come out of our investigation. We end our discussion with following question: What is the minimal model in the symplectic minimal model program?

References

[Ba] V. Batyrev, *On the Betti number of birationally isomorphic projective varieties with trivial canonical bundles*, Preprint `math. AG/9710020`.

[C] H. Clemens, *Degeneration of Kahler manifolds*, Duke Math. J. **44** (1977), 215–290.

[D] B. Dubrovin, *Geometry and analytic theory of Frobenius manifolds*, Proc. Internat. Congress Math. (Berlin, 1998), Documenta Math., Vol. II, Berlin, 1998, pp. 315–327.

[FO] K. Fukaya and K. Ono, *Arnold conjecture and Gromov–Witten invariant*, Topology **38** (1999), 933–1048.

[Go2] R. Gompf, *A new construction of symplectic manifolds*, Ann. of Math. (2) **142** (1995), 527–595.

[Gr] M. Gromov, *Pseudo holomorphic curves in symplectic manifolds*, Invent. Math. **82** (1985), 307-347.

[IP] E. Ionel and T. Parker, *Gromov–Witten invariants of symplectic sums*, Preprint `math. SG/9806063`; Math. Res. Lett. **5** (1998), 563–576.

[Ka] Y. Kawamata, *Crepant blowing ups of three dimensional canonical singularities and applications to degeneration of surfaces*, Ann. of Math. (2) **119** (1984), 603–633.

[Kim] B. Kim, *Quantum hyperplane theorem for homogeneous spaces*, Preprint `alg-geom/9712008`.

[K1] J. Kollár, *Flops*, Nagoya Math. J. **113** (1989), 15–36.

[K2] ———, *The structure of algebraic threefolds—an introduction to Mori's program*, Bull. (N.S.) Amer. Math. Soc. **17** (1987), 211–273.

[K3] ———, *Flips, flops, minimal models, etc.*, Surveys in Differential Geometry, vol. 1, International Press, Cambridge, MA, 1991, pp. 113–199.

[KM] M. Kontsevich and Y. Manin, *GW-classes, quantum cohomology and enumerative geometry*, Comm. Math. Phys. **164** (1994), 525–562.

[L] E. Lerman, *Symplectic cuts*, Math. Res. Lett. **2** (1995), 247–258.

[LR] A. Li and Y. Ruan, *Symplectic surgery and GW-invariants of Calabi–Yau threefolds.* I, Preprint `math. AG/9803036`.

[LQR] A. Li, Z. Qin, and T. Ruan, *Symplectic surgeries and GW-invariants of Calabi–Yau threefolds.* II, in preparation.

[LT3] J. Li and G. Tian, *Virtual moduli cycles and Gromov–Witten invariants of general symplectic manifolds*, Topics in Symplectic 4-Manifolds (Irvine, CA, 1996), Internat. Press, Cambridge, MA, 1998, pp. 47–83.

[MW] G. McCarthy and J. Wolfson, *Symplectic normal connect sum*, Topology **33** (1994), 729–764.

[Mo1] D. Morrison, *Beyond the Kähler cone*, Proc. Hirzebruch 65 Conf. Algebraic Geometry (Ramat Gan, 1993), Israel Math. Conf. Proc., vol. 9, Bar-Ilan Univ., Ramat Gan, 1996, pp. 361–376.

[Mo2] ———, *Through the looking glass*, Mirror Symmetry, III (Montréal, 1995), AMS/IP Stud. Adv. Math., vol. 10, Amer. Math. Soc., Providence, RI, 1999, pp. 263–277.

[PW] T. Parker and J. Wolfson, *A compactness theorem for Gromov moduli space*, J. Geom. Anal. **3** (1993), 63–98.

[QR] Z. Qin and Y. Ruan, *Quantum cohomology of projective bundles over* $\mathbb{P}^n$, Trans. Amer. Math. Soc. **350** (1998), 3615–3638.

[R3] Y. Ruan, *Symplectic topology and extremal rays*, Geom. Funct. Anal. **3** (1993), 395–430.

[R5] ———, *Virtual neighborhoods and pseudo-holomorphic curves*, Preprint.

[R6] ———, *Quantum cohomology and its applications,*, Proc. Internat. Congress Math. (Berlin, 1998), Documenta Math., Vol. II, Berlin, 1998, pp. 411-420.

[RT1] Y. Ruan and G. Tian, *A mathematical theory of quantum cohomology*, J. Diff. Geom. **42** (1995), 259–367.

[RT2] ———, *Higher genus symplectic invariants and sigma model coupled with gravity*, Invent. Math. **130** (1997), 455–516.

[S] B. Siebert, *Gromov–Witten invariants for general symplectic manifolds*, Preprint.

[Wa] Chin-Lung Wang, *On the topology of birational minimal models*, J. Diff. Geom. **50** (1998), 129–146.

[Wi2] P. M. H. Wilson, *Symplectic deformations of Calabi–Yau threefolds*, J. Diff Geom. **45** (1997), 611–637.

[Wi3] ———, *Flops, type III contractions and GW-invariants of Calabi–Yau threefolds*, Preprint `alg-geom/9707008`.

[Ye] R. Ye, *Gromov's compactness theorem for pseudo-holomorphic curves*, Trans. Amer. Math. Soc. **340** (1994), 671–694.

Department of Mathematics, University of Wisconsin-Madison, Madison, WI 53706

Amer. Math. Soc. Transl.
(2) Vol. **196**, 1999

Quantum Products for Mapping Tori and the Atiyah–Floer Conjecture

Dietmar A. Salamon

1. Introduction

Let Σ be a compact oriented Riemann surface with area form ω, and let $f_0, f_1 \dots, f_n$ be orientation and area preserving diffeomorphisms of Σ such that

$$f_n \circ \cdots \circ f_0 = \mathrm{id}.$$

Suppose that $P \to \Sigma$ is a principal SO(3)-bundle with nonzero second Stiefel–Whitney class, and choose lifts

$$\begin{array}{ccc} P & \xrightarrow{\tilde{f}_j} & P \\ \downarrow & & \downarrow \\ \Sigma & \xrightarrow{f_j} & \Sigma \end{array}$$

to bundle automorphisms that also satisfy $\tilde{f}_n \circ \cdots \circ \tilde{f}_0 = \mathrm{id}$. These lifts induce symplectomorphisms $\phi_{\tilde{f}_j} : M_\Sigma \to M_\Sigma$ on the moduli space of flat connections on P. They also determine SO(3)-bundles $Q_{\tilde{f}_j} \to Y_{f_j}$ over mapping tori. There are natural product structures

$$\mathrm{HF}(Y_{f_0}, Q_{\tilde{f}_0}) \otimes \mathrm{HF}(Y_{f_1}, Q_{\tilde{f}_1}) \otimes \cdots \otimes \mathrm{HF}(Y_{f_n}, Q_{\tilde{f}_n}) \to \mathbb{Z} \tag{1}$$

in instanton Floer homology and

$$\mathrm{HF}^{\mathrm{symp}}(M_\Sigma, \phi_{\tilde{f}_0}) \otimes \mathrm{HF}^{\mathrm{symp}}(M_\Sigma, \phi_{\tilde{f}_1}) \otimes \cdots \otimes \mathrm{HF}^{\mathrm{symp}}(M_\Sigma, \phi_{\tilde{f}_n}) \to \mathbb{Z} \tag{2}$$

in symplectic Floer homology. Both were discovered by Donaldson. They are well defined up to an overall sign. The homomorphism (1) can be interpreted as a relative Donaldson invariant on a 4-manifold with boundary. In other words, this product is obtained by counting anti-self-dual connections over a 4-dimensional cobordism with $n+1$ cylindrical ends corresponding to Y_{f_j}. The second homomorphism is obtained by counting pseudo-holomorphic sections of a symplectic fibre bundle over the punctured sphere with fibre M_Σ and holonomies $\phi_{\tilde{f}_j}$ around the $n+1$ punctures. The next theorem is the main result of this paper. It can be viewed as an extension of the Atiyah–Floer conjecture to product structures (cf. [**1, 17**]).

1991 *Mathematics Subject Classification.* Primary 58E05; Secondary 57R57.

©1999 American Mathematical Society

THEOREM 1.1. *The natural isomorphism* $\mathrm{HF}(Y_f, Q_{\tilde f}) \to \mathrm{HF}^{\mathrm{symp}}(M_\Sigma, \phi_{\tilde f})$ *constructed in* [**6**] *intertwines the two product structures* (1) *and* (2) (*up to a sign*).

COROLLARY 1.2. *The quantum cohomology ring of* M_Σ *is isomorphic to the instanton Floer homology ring of* $S^1 \times \Sigma$.

PROOF. Just combine Theorem 1.1 with $\tilde f_0 = \tilde f_1 = \tilde f_2 = \mathrm{id}$ and Theorem 5.1 in [**16**]. □

REMARK 1.1. **(i)** The techniques of [**4, 5, 6**] suggest a general relation between holomorphic curves in M_Σ and anti-self-dual instantons on the 4-manifold $S \times \Sigma$ for any compact Riemann surface S. The technique of proof is an adiabatic limit argument where the metric on the fibre Σ converges to zero. In this limit the anti-self-dual instantons on the product $S \times \Sigma$ degenerate to holomorphic curves $S \to M_\Sigma$. The proof of Theorem 1.1 is the analogue of this argument for Riemann surfaces S with $n+1$ cylindrical ends. Instead of holomorphic maps $S \to M_\Sigma$ one obtains holomorphic sections of a fibre bundle $W \to S$ with fibres M_Σ and holonomies $\phi_{\tilde f_j}$ at the $n+1$ cylindrical ends. The comparison theorem for Floer homologies in [**6**] corresponds to the case of two cylindrical ends.

(ii) It is interesting to relate the results of this paper to the recent work of Donaldson about symplectic Lefschetz fibration (cf. [**2, 3**]). He proved that every symplectic manifold, after blowing up a suitable number of points, admits the structure of a topological Lefschetz fibration

$$\begin{array}{ccc} \Sigma & \hookrightarrow & X \\ & & \downarrow \\ & & S^2. \end{array}$$

The projection $\pi : X \to S^2$ has finitely many nondegenerate singular points (with distinct singular values) near which π is holomorphic. Moreover, each regular fibre of π is a symplectic submanifold of X. Thus the holonomy around each singular value is a positive Dehn twist f_j, and the composition of these Dehn twists is the identity. Cutting out neighbourhoods of the singular fibres, one obtains a fibration $X \to S$ over the punctured sphere. Now the neighbourhoods of the singular fibres determine natural Floer homology classes

$$a(\tilde f_j) \in \mathrm{HF}(Y_{f_j}, Q_{\tilde f_j}),$$

and their product under (1) is the degree-0 Donaldson invariant of X.

(iii) In his recent thesis [**11**] Handfield discusses the relation between anti-self-dual instantons on $S \times \Sigma$ and holomorphic curves $S \to M_\Sigma$ for a closed Riemann surface S. His work is based on the ideas developed in [**4, 5, 6**].

(iv) Corollary 1.2 was established independently by Munoz [**14, 15**], who computed both ring structures separately, without using the natural isomorphism.

2. The punctured sphere

Fix an integer $n \geq 2$, and let

$$S = \mathbb{C}P^1 \setminus \{z_0, \dots, z_n\}$$

be the Riemann sphere with punctures at $n+1$ distinct points $z_0, \dots, z_n$. We assume throughout that the points $z_0, \dots, z_n$ are ordered and in general position, i.e.

$$0 = |z_0| < |z_1| < \cdots < |z_{n-1}| < |z_n| = \infty.$$

It is convenient to use polar coordinates $z = e^{2\pi w}$. Let us fix lifts $w_j = s_j + it_j \in \mathbb{C}$ of the points z_j. Then

$$-\infty = s_0 < s_1 < \cdots < s_{n-1} < s_n = \infty,$$

and S can be identified with the quotient $S = U/\sim$ of the open set

$$U = \{s + it \in \mathbb{C} \mid \text{if } s = s_j \text{ then } t_j < t < t_j + 1\}$$

by the equivalence relation $(s, t) \sim (s, t + 1)$. It is convenient to choose a metric on S in which the punctures become cylindrical ends. Hence let

$$\lambda : \mathbb{C} \setminus \{w_j + ik \mid j, k \in \mathbb{Z}, \, 0 \leq j \leq n\} \to \mathbb{R}$$

be a smooth positive function such that $\lambda(s, t + 1) = \lambda(s, t)$ and, for some $T > 1$ and every $w = s + it \in U$,

$$\lambda(w) = \begin{cases} 1/2\pi|w - w_j| & \text{if } |w - w_j| \leq e^{-2\pi T} \text{ for some } j, \\ 1 & \text{if } |s| \geq T. \end{cases}$$

Consider the volume form

$$\mathrm{dvol}_S = \lambda^2 \mathrm{d}s \wedge \mathrm{d}t$$

on S, and note that the function $\zeta \mapsto w_j + e^{2\pi\zeta}$ is an isometry from the half cylinder $Z = (-\infty, -T) \times \mathbb{R}/\mathbb{Z}$ with the standard metric to the punctured ball $\{w \in \mathbb{C} \mid 0 < |w - w_j| < e^{-2\pi T}\}$ with the metric $\lambda^2(\mathrm{d}s^2 + \mathrm{d}t^2)$.

3. Fibre bundles

Throughout this paper, we identify $S^1 = \mathbb{R}/\mathbb{Z}$. Let Σ be a compact oriented Riemann surface with volume form ω. The mapping torus of a symplectomorphism $f \in \mathrm{Diff}(\Sigma, \omega)$ is a fibre bundle

$$Y_f \to S^1$$

with holonomy f. It is defined by $Y_f := \mathbb{R} \times \Sigma/\sim$, where

$$(t + 1, z) \sim (t, f(z)).$$

Let $f_0, f_1, \ldots, f_n \in \mathrm{Diff}(\Sigma, \omega)$ be such that

$$f_n \circ f_{n-1} \circ \cdots \circ f_0 = \mathrm{id}.$$

Then there is a natural fibre bundle

$$\begin{array}{ccc} \Sigma & \hookrightarrow & X \\ & & \downarrow \\ & & S \end{array}$$

over the punctured sphere with holonomy f_j around the jth puncture. Thus the restriction of the bundle to a circle around w_j is diffeomorphic to Y_{f_j}. To construct the bundle explicitly it is convenient to introduce the maps

$$g_j = f_j \circ f_{j-1} \circ \cdots \circ f_0$$

for $j = 0, 1, \ldots, n-1$. Then $X := U \times \Sigma/\sim$ under the equivalence relation

$$s_j < s < s_{j+1} \qquad \Longrightarrow \qquad (s, t + 1, z) \sim (s, t, g_j(z)).$$

For $j = 0, \ldots, n$ there is a natural embedding $\iota_j : (-\infty, -T) \times Y_{f_j} \hookrightarrow X$, defined by $\iota_0(s,t,z) = [s,t,z]$ and $\iota_n(s,t,z) = [-s,-t,z]$ for $j = 0, n$, and by

$$\iota_j(s,t,z) = \begin{cases} [w_j + e^{2\pi(s+it)}, {f_j}^k(z)] & \text{if } k - 1/4 < t < k + 3/4, \\ [w_j + i + e^{2\pi(s+it)}, {g_j}^{-1} \circ {f_j}^k(z)] & \text{if } k - 3/4 < t < k + 1/4, \end{cases}$$

for $j = 1, \ldots, n-1$. These maps are well defined and satisfy

$$\iota_j(s, t+1, z) = \iota_j(s, t, f_j(z))$$

for $j = 0, \ldots, n$.

Let $\mathcal{J}(\Sigma)$ denote the space of complex structures on Σ that are compatible with the orientation given by ω. A **vertical complex structure** on X is an almost complex structure on the vertical tangent bundle. Explicitly, such a complex structure can be represented by a smooth map $U \to \mathcal{J}(\Sigma) : (s,t) \mapsto J(s,t)$ that satisfies

$$s_j < s < s_{j+1} \qquad \Longrightarrow \qquad J(s,t+1) = {g_j}^* J(s,t).$$

The pullback $J_j = {\iota_j}^* J$ is a vertical complex structure on $(-\infty, -T) \times Y_{f_j}$. It is given by

$$J_j(s,t) = \begin{cases} ({f_j}^k)^* J(w_j + e^{2\pi(s+it)}) & \text{if } k - 1/4 < t < k + 3/4, \\ ({f_j}^k)^* {g_j}_* J(w_j + i + e^{2\pi(s+it)}) & \text{if } k - 3/4 < t < k + 1/4, \end{cases}$$

for $j = 1, \ldots, n-1$, and it satisfies $J_j(s,t+1) = {f_j}^* J_j(s,t)$. We assume throughout that $J_j(s,t)$ is independent of the s-variable for $-s$ sufficiently large. Let $*_{s,t}$ denote the Hodge $*$-operator on Σ corresponding to the metric $\omega(\cdot, J(s,t)\cdot)$, and $*_{j;s,t}$ the Hodge $*$-operator of the metric $\omega(\cdot, J_j(s,t)\cdot)$. Note that $*_{s,t}\alpha = -\alpha \circ J(s,t)$ for $\alpha \in \Omega^1(\Sigma, \mathfrak{g}_P)$.

4. Connections and gauge transformations

Let $P \to \Sigma$ be a principal bundle with structure group $\mathrm{G} = \mathrm{SO}(3)$ and nonzero second Stiefel–Whitney class. Denote by $\mathfrak{g} = \mathrm{Lie}(\mathrm{G}) = \mathfrak{so}(3)$ the Lie algebra of G, by $\mathcal{A}(P) \subset \Omega^1(P, \mathfrak{g})$ the space of connection 1-forms on P, and by $\mathcal{G}(P)$ the identity component in the group of gauge transformations (thought of as equivariant maps from P to G). A lift $\tilde{f} : P \to P$ of $f \in \mathrm{Diff}(\Sigma, \omega)$ determines a principal SO(3)-bundle

$$\begin{array}{ccc} P & \hookrightarrow & Q_{\tilde{f}} \\ & & \downarrow \\ & & Y_f \end{array}$$

given by the mapping torus $Q_{\tilde{f}} := \mathbb{R} \times Q/\sim$, where

$$(t+1, p) \sim (t, \tilde{f}(p)).$$

Choose $n+1$ such lifts $\tilde{f}_0, \ldots, \tilde{f}_n$ of the maps $f_0, \ldots, f_n$ such that

$$\tilde{f}_n \circ \cdots \circ \tilde{f}_0 = \mathrm{id}$$

and define $\tilde{g}_j = \tilde{f}_j \circ \cdots \circ \tilde{f}_0$. Then there is a principal SO(3)-bundle

$$\begin{array}{ccc} P & \hookrightarrow & Q \\ & & \downarrow \\ & & X \end{array}$$

defined as the quotient $Q := U \times P/\sim$ under the equivalence relation

$$s_j < s < s_{j+1} \qquad \Longrightarrow \qquad (s, t+1, p) \sim (s, t, \tilde{g}_j(p)).$$

The embedding ι_j lifts to a bundle map $\tilde{\iota}_j : (-\infty, -T) \times Q_{\tilde{f}_j} \to Q$, given by

$$\tilde{\iota}_j(s,t,p) = \begin{cases} [w_j + e^{2\pi(s+it)}, \tilde{f}_j{}^k(p)] & \text{if } k - 1/4 < t < k + 3/4, \\ [w_j + i + e^{2\pi(s+it)}, \tilde{g}_j{}^{-1} \circ \tilde{f}_j{}^k(p)] & \text{if } k - 3/4 < t < k + 1/4, \end{cases}$$

for $j = 1, \ldots, n-1$. A connection on Q is a 1-form

$$\Xi = A + \Phi\, \mathrm{d}s + \Psi\, \mathrm{d}t$$

on $U \times P$ such that the functions $A : U \to \mathcal{A}(P)$ and $\Phi, \Psi : U \to C^\infty(\Sigma, \mathfrak{g}_P)$ satisfy the periodicity conditions

$$(3) \qquad \begin{gathered} A(s,t+1) = \tilde{g}_j{}^* A(s,t), \\ \Phi(s,t+1) = \Phi(s,t) \circ \tilde{g}_j, \qquad \Psi(s,t+1) = \Psi(s,t) \circ \tilde{g}_j, \end{gathered}$$

for $s_j < s < s_{j+1}$. For $j = 1, \ldots, n-1$ the pullback connection

$$\Xi_j = \tilde{\iota}_j{}^* \Xi = A_j + \Phi_j\, \mathrm{d}s + \Psi_j\, \mathrm{d}t$$

on $(-\infty, -T) \times Q_{\tilde{f}_j}$ is given by

$$A_j(s,t) = \begin{cases} (\tilde{f}_j{}^k)^* A(w_j + e^{2\pi(s+it)}) & \text{if } k - 1/4 < t < k + 3/4, \\ (\tilde{g}_j{}^{-1} \circ \tilde{f}_j{}^k)^* A(w_j + i + e^{2\pi(s+it)}) & \text{if } k - 3/4 < t < k + 1/4, \end{cases}$$

$$\Phi_j(s,t) = 2\pi e^{2\pi s}\left(\cos(2\pi t)\Phi(w_j + e^{2\pi(s+it)}) + \sin(2\pi t)\Psi(w_j + e^{2\pi(s+it)})\right),$$

$$\Psi_j(s,t) = 2\pi e^{2\pi s}\left(\cos(2\pi t)\Psi(w_j + e^{2\pi(s+it)}) - \sin(2\pi t)\Phi(w_j + e^{2\pi(s+it)})\right)$$

for $-1/4 < t < 3/4$. For general t these functions are determined by the periodicity conditions $\Phi_j(s,t+1) = \Phi_j(s,t) \circ \tilde{f}_j$ and $\Psi_j(s,t+1) = \Psi_j(s,t) \circ \tilde{f}_j$. Let $\mathcal{A}(Q)$ denote the set of all connections Ξ that satisfy (3).

A gauge transformation of Q is a smooth function $u : U \to \mathcal{G}(P)$ that satisfies

$$s_j < s < s_{j+1} \qquad \Longrightarrow \qquad u(s,t+1) = u(s,t) \circ \tilde{g}_j.$$

The pullback gauge transformation $u_j = u \circ \tilde{\iota}_j : (-\infty, -T) \times Q_{\tilde{f}_j} \to \mathrm{G}$ is given by

$$u_j(s,t) = \begin{cases} u(w_j + e^{2\pi(s+it)}) \circ \tilde{f}_j{}^k & \text{if } k - 1/4 < t < k + 3/4, \\ u(w_j + i + e^{2\pi(s+it)}) \circ \tilde{g}_j{}^{-1} \circ \tilde{f}_j{}^k & \text{if } k - 3/4 < t < k + 1/4, \end{cases}$$

and satisfies $u_j(s,t+1) = u_j(s,t) \circ \tilde{f}_j$. Let $\mathcal{G}(Q)$ denote the group of gauge transformations such that $u_j(s,t) = \mathbb{1}$ for $-s$ sufficiently large. This group acts on $\mathcal{A}(Q)$ by

$$\begin{gathered} A \mapsto u^{-1}\mathrm{d}u + u^{-1}Au, \\ \Phi \mapsto u^{-1}\partial_s u + u^{-1}\Phi u, \qquad \Psi \mapsto u^{-1}\partial_t u + u^{-1}\Psi u. \end{gathered}$$

The action of u_j on Ξ_j is given by the same formulae.

5. Anti-self-dual instantons

The curvature of $\Xi = (A, \Phi, \Psi) \in \mathcal{A}(Q)$ is the 2-form

$$\begin{aligned} F_\Xi = {} & F_A + (\partial_s\Psi - \partial_t\Phi + [\Phi, \Psi])\, \mathrm{d}s \wedge \mathrm{d}t \\ & - (\partial_s A - \mathrm{d}_A\Phi) \wedge \mathrm{d}s - (\partial_t A - \mathrm{d}_A\Psi) \wedge \mathrm{d}t. \end{aligned}$$

Since

$$\begin{aligned} *F_\Xi = {} & \lambda^{-2} * (\partial_s\Psi - \partial_t\Phi + [\Phi, \Psi]) + \lambda^2(*F_A)\, \mathrm{d}s \wedge \mathrm{d}t \\ & - (*_{s,t}(\partial_t A - \mathrm{d}_A\Psi)) \wedge \mathrm{d}s + (*_{s,t}(\partial_s A - \mathrm{d}_A\Phi)) \wedge \mathrm{d}t, \end{aligned}$$

the connection Ξ is anti-self-dual if and only if

$$(4)\qquad \begin{aligned} (\partial_s A - \mathrm{d}_A\Phi) + *_{s,t}(\partial_t A - \mathrm{d}_A\Psi) &= 0, \\ \partial_s\Psi - \partial_t\Phi + [\Phi,\Psi] + \lambda^2 * F_A &= 0. \end{aligned}$$

Note that the Hodge $*$-operator on 1-forms depends on the complex structure $J(s,t)$ but is invariant under rescaling, while the Hodge $*$-operator on 2-forms depends only on the volume form. Note also that $\Xi_j = \tilde{\iota}_j{}^*\Xi$ satisfies the same equations with $\lambda = 1$ and $*_{s,t}$ replaced by $*_{j;s,t}$.

The Yang-Mills action of Ξ is given by

$$(5)\qquad \mathcal{YM}(\Xi) = \int_0^1\int_{-\infty}^{\infty}\left(\|\partial_s A - \mathrm{d}_A\Phi\|^2_{L^2(\Sigma,J)} + \lambda^2\|F_A\|^2_{L^2(\Sigma)}\right)\mathrm{d}s\mathrm{d}t.$$

If this action is finite then the connection has limits at the cylindrical ends. To be more precise, we shall assume that the connection is in temporal gauge on the cylindrical ends. This means that $\Phi_j(s,t) = 0$ for $-s$ sufficiently large. If this holds, and the flat connections on $Q_{\tilde{f}_j}$ are nondegenerate for all j, then there exist flat connections

$$a_j = A_j(t) + \Psi_j(t)\,\mathrm{d}t \in \mathcal{A}^{\mathrm{flat}}(Q_{\tilde{f}_j})$$

such that

$$(6)\qquad \lim_{s\to-\infty}(A_j(s,t) + \Psi_j(s,t)\,\mathrm{d}t) = a_j.$$

The convergence is exponential and in the C^∞-topology. Moreover,

$$\lim_{s\to-\infty}\partial_s A_j(s,t) = 0, \qquad \lim_{s\to-\infty}\partial_s\Psi_j(s,t) = 0.$$

Warning: When referring to (6), we shall always mean convergence in the C^1-norm together with uniform convergence of $\partial_s A_j$ and $\partial_s\Psi_j$ to zero and with $\partial_s\Phi_j = 0$ for $-s$ sufficiently large.

In general, the flat connections on the mapping tori $Q_{\tilde{f}_j}$ may occur in families, and, to obtain smooth moduli spaces, we must choose suitable perturbations. The quantum product structure (1) can then be obtained by counting the solutions of the perturbed instanton equations with prescribed limit connections.

6. Perturbations

Throughout this section let g denote the genus of Σ. For any loop $\gamma : \mathbb{R}/\mathbb{Z} \to P$ denote by $\rho_\gamma : \mathcal{A}(P) \to \mathrm{SU}(2)$ the holonomy along γ. Thus $\rho_\gamma(A) = u(1)$, where $u : [0,1] \to \mathrm{SU}(2)$ is the solution of the ordinary differential equation

$$\dot{u} + A(\dot{\gamma})u = 0, \qquad u(0) = 1\!\!1.$$

The differential of ρ_γ at A has the form

$$\rho_\gamma(A)^{-1}\mathrm{d}\rho_\gamma(A)\alpha = -\int_0^1 u(\theta)^{-1}\alpha(\dot{\gamma}(\theta))u(\theta)\,\mathrm{d}\theta$$

for $\alpha \in \Omega^1(\Sigma, \mathfrak{g}_P)$. Now fix a path of basepoints $[-1,1] \to P : \tau \mapsto p_\tau$ and $2g$ maps $[-1,1]\times\mathbb{R}/\mathbb{Z} \to P : (\tau,\theta) \mapsto \gamma_j(\tau,\theta)$ such that the projections $\pi\circ\gamma_j$ are orientation preserving embeddings, generate the fundamental group of Σ, and satisfy $\gamma_j(\tau,0) = p_\tau$. For $\tau \in [-1,1]$ let

$$\rho_\tau : \mathcal{A}(P) \to \mathrm{SU}(2)^{2g}$$

denote the holonomy along the loops $\theta \mapsto \gamma_j(\tau, \theta)$, $j = 1, \ldots, 2g$. Choose a smooth cutoff function $\beta : (-1, 1) \to \mathbb{R}$ with compact support and mean value 1. Then every smooth function $h : \mathrm{SU}(2)^{2g} \to \mathbb{R}$ that is invariant under conjugacy determines a Hamiltonian function

$$\mathcal{A}(P) \to \mathbb{R} : A \mapsto H(A) = \int_{-1}^{1} \beta(\tau) h(\rho_\tau(A))\, \mathrm{d}\tau.$$

The partial derivative of h with respect to the jth coordinate u_j can be represented by a function $\eta_j : \mathrm{SU}(2)^{2g} \to \mathfrak{su}(2)$ such that

$$\frac{\partial h}{\partial u_j}(u) u_j \xi = \langle \eta_j(u), \xi \rangle$$

for $u = (u_1, \ldots, u_{2g}) \in \mathrm{SU}(2)^{2g}$ and $\xi \in \mathfrak{g}$. Consider the vector field

$$\mathcal{A}(P) \to \Omega^1(\mathfrak{g}_P) : A \mapsto v(A) = \sum_{j=1}^{2g} v_j(A),$$

where $v_j(A) \in \Omega^1(\Sigma, \mathfrak{g}_P)$ is supported in $\pi^{-1}(\mathrm{im}\, \pi \circ \gamma_j)$ and satisfies

$$\gamma_j{}^* v_j(A) = \beta(\tau) u_j(\tau, \theta) \eta_j(\rho_\tau(A)) u_j(\tau, \theta)^{-1} \mathrm{d}\tau.$$

Here $\theta \mapsto u_j(\tau, \theta)$ is the holonomy of A along the loop $\theta \mapsto \gamma_j(\tau, \theta)$. Direct computation shows that

$$\mathrm{d}H(A)\alpha = \int_\Sigma \langle v(A) \wedge \alpha \rangle$$

for $A \in \mathcal{A}(P)$ and $\alpha \in \Omega^1(\Sigma, \mathfrak{g}_P)$. This means that v is the Hamiltonian vector field of H. Any such Hamiltonian function is invariant under $\mathcal{G}(P)$, and hence

$$H(u^* A) = H(A), \qquad v(u^* A) = u^{-1} v(A) u, \qquad \mathrm{d}_A v(A) = 0 \tag{7}$$

for $A \in \mathcal{A}(P)$ and $u \in \mathcal{G}(P)$. The vector fields v that arise from the holonomy are smooth with respect to the $W^{k,p}$-norm for all k and p.

Now let $U \times \mathcal{A}(P) \to \mathbb{R} : (s, t, A) \mapsto H_{s,t}(A)$ be a smooth family of such Hamiltonian functions, and let $U \times \mathcal{A}(P) \to \Omega^1(\Sigma, \mathfrak{g}_P) : (s, t, A) \mapsto v_{s,t}(A)$ be the corresponding family of Hamiltonian vector fields. Suppose that

$$H_{s,t}(A) = H_{s,t+1}(\tilde{g}_j{}^* A), \qquad \tilde{g}_j{}^* v_{s,t}(A) = v_{s,t+1}(\tilde{g}_j{}^* A)$$

for $s_j < s < s_{j+1}$. Let $K_{s,t}$ be another such family of Hamiltonian functions with corresponding vector fields $w_{s,t}$, and consider the perturbed anti-self-duality equations

$$\begin{aligned} (\partial_s A - \mathrm{d}_A \Phi - v_{s,t}(A)) + *_{s,t}(\partial_t A - \mathrm{d}_A \Psi - w_{s,t}(A)) &= 0, \\ \partial_s \Psi - \partial_t \Phi + [\Phi, \Psi] + \lambda^2 * F_A &= 0. \end{aligned} \tag{8}$$

The perturbed Yang-Mills action of a solution of (8) is given by

$$\mathcal{YM}(\Xi; H, K) = \int_0^1 \int_{-\infty}^{\infty} \Big(\|\partial_s A - \mathrm{d}_A \Phi - v_{s,t}(A)\|_{L^2(\Sigma_J)}^2 + \lambda^2 \|F_A\|_{L^2(\Sigma)}^2 \Big).$$

If this action is finite, then the limits (6) will exist in the case of a suitably chosen perturbation. To be more precise, consider the pullback perturbations under $\tilde{\iota}_j$. They are given by

$$\begin{aligned} v_{j;s,t} &= 2\pi e^{2\pi s}(\cos(2\pi t) v + \sin(2\pi t) w), \\ w_{j;s,t} &= 2\pi e^{2\pi s}(\cos(2\pi t) w - \sin(2\pi t) v), \end{aligned}$$

for $j = 1, \ldots, n-1$ and $-1/4 < t < 3/4$, where v and w are evaluated at the point $w_j + e^{2\pi(s+it)}$. These functions extend to $(-\infty, -T) \times \mathbb{R}$ via the periodicity conditions

$$\tilde{f}_j^{\,*} v_{j;s,t}(A) = v_{j;s,t+1}(\tilde{f}_j^{\,*} A), \qquad \tilde{f}_j^{\,*} w_{j;s,t}(A) = w_{j;s,t+1}(\tilde{f}_j^{\,*} A).$$

We shall assume throughout that $v_{j;s,t} = 0$ and $w_{j;s,t}$ is independent of s for $-s$ sufficiently large. Call $a_j = A_j(t) + \Psi_j(t)\,dt$ a **perturbed flat connection** for the jth cylindrical end if

$$A_j(t+1) = \tilde{f}_j^{\,*} A(t), \qquad \Psi_j(t+1) = \Psi_j(t) \circ \tilde{f}_j, \tag{9}$$

and

$$F_{A_j(t)} = 0, \qquad \dot{A}_j(t) - d_{A_j(t)}\Psi_j(t) - w_{j;-\infty,t}(A_j(t)) = 0. \tag{10}$$

Condition (9) asserts that a_j is a connection on $Q_{\tilde{f}_j}$. Denote the space of solutions of (9) and (10) by

$$\mathcal{A}^{\text{flat}}(Q_{\tilde{f}_j}, K_j) = \{a_j \in \mathcal{A}(Q_{\tilde{f}_j}) \,|\, (10)\}.$$

The gauge equivalence classes of such connections correspond naturally to the fixed points of a Hamiltonian deformation of the symplectomorphism $\phi_{\tilde{f}_j} : M_\Sigma \to M_\Sigma$ (defined below), and the perturbation K_j can be chosen such that these fixed points are all nondegenerate (cf. [**4, 5**]). Under this assumption the quotient space $\mathcal{A}^{\text{flat}}(Q_{\tilde{f}_j}, K_j)/\mathcal{G}(Q_{\tilde{f}_j})$ is a finite set and, for every solution Ξ of (8) with finite perturbed Yang-Mills energy, the limit (6) exists. Given $n+1$ perturbed flat connections

$$a_j \in \mathcal{A}^{\text{flat}}(Q_{\tilde{f}_j}, K_j), \qquad j = 0, \ldots, n,$$

we denote the moduli space of solutions of (8) with these limits by

$$\mathcal{M}_1(a_0, \ldots, a_n; H, K) = \frac{\{\Xi \in \mathcal{A}(Q) \,|\, (8), (6)\}}{\mathcal{G}(Q)}.$$

7. Pseudoholomorphic sections

Consider the moduli space

$$M_\Sigma = \mathcal{A}^{\text{flat}}(P)/\mathcal{G}(P).$$

This space is a smooth compact manifold of dimension $6g(\Sigma) - 6$, where $g(\Sigma)$ denotes the genus of Σ. It carries a natural symplectic form, and every complex structure $J \in \mathcal{J}(\Sigma)$ determines a complex structure on M_Σ via the Hodge $*$-operator. Each lift $\tilde{f} : P \to P$ of a symplectomorphism of Σ determines a symplectomorphism

$$\phi_{\tilde{f}} : M_\Sigma \to M_\Sigma$$

given by $\phi_{\tilde{f}}([A]) = [\tilde{f}^* A]$. As above, choose $n+1$ such lifts $\tilde{f}_0, \ldots, \tilde{f}_n$ such that

$$\tilde{f}_n \circ \cdots \circ \tilde{f}_0 = \mathrm{id}$$

and define $\tilde{g}_j = \tilde{f}_j \circ \cdots \circ \tilde{f}_0$. Then there is a fibre bundle

$$\begin{array}{ccc} M_\Sigma & \hookrightarrow & M_X \\ & & \downarrow \\ & & S \end{array}$$

over the punctured sphere with holonomy $\phi_{\tilde{f}_j}$ around the jth puncture. It is defined as the quotient $M_X = U \times M_\Sigma / \sim$ under the equivalence relation

$$s_j < s < s_{j+1} \qquad \Longrightarrow \qquad (s,t,[A]) \sim (s,t+1,[\tilde{g}_j{}^*A]).$$

The complex structure $J : U \to \mathcal{J}(\Sigma)$ determines a vertical complex structure on M_X. The (perturbed) holomorphic sections of M_X with respect to this complex structure can be expressed as connections $\Xi \in \mathcal{A}(Q)$ that satisfy the equations

$$(\partial_s A - \mathrm{d}_A\Phi - v_{s,t}(A)) + *_{s,t}(\partial_t A - \mathrm{d}_A\Psi - w_{s,t}(A)) = 0, \qquad F_A = 0. \tag{11}$$

As in [4], the additional terms Φ and Ψ are uniquely determined by the requirement that the 1-forms $\partial_s A - \mathrm{d}_A\Phi$ and $\partial_t A - \mathrm{d}_A\Psi$ in $\Omega^1(\Sigma, \mathfrak{g}_P)$ are harmonic with respect to A. The quantum product structure (2) in symplectic Floer homology is obtained by counting the solutions $\Xi \in \mathcal{A}(Q)$ of (11) that are in temporal gauge near the cylindrical ends and satisfy (6). Given $n+1$ connections $a_j \in \mathcal{A}^{\mathrm{flat}}(Q_{\tilde{f}_j}, K_j)$, denote by

$$\mathcal{M}_0(a_0, \ldots, a_n; H, K) = \frac{\{\Xi \in \mathcal{A}(Q) \,|\, (11), (6)\}}{\mathcal{G}(Q)}$$

the moduli space of gauge equivalence classes of pseudoholomorphic sections of M_X with prescribed limiting data at the cylindrical ends. The task at hand is to compare the solutions of (11) with those of (8).

A perturbation (H, K) is called **regular** if the perturbed flat connections corresponding to the $n+1$ cylindrical ends are all nondegenerate (see for example [4]) and if the linearized operator $\mathcal{D}_0(\Xi)$ (defined in Section 9 below) is surjective for all a_j and every solution Ξ of (11) and (6). The set of regular perturbations will be denoted by $\mathcal{HK}_{\mathrm{reg}}$. It was proved in [20] and [10] that $\mathcal{HK}_{\mathrm{reg}}$ is of the second Baire category in the space of all smooth perturbations.

8. Adiabatic limits

The strategy for the proof of Theorem 1.1 is, precisely as in [6], to show that the solutions of (4) degenerate to pseudoholomorphic sections in the limit where the metric on the fibre converges to zero. More precisely, if we multiply the metric on Σ by a factor ε^2, then the anti-self-duality equations for this metric take the form

$$\begin{aligned} (\partial_s A - \mathrm{d}_A\Phi - v_{s,t}(A)) + *_{s,t}(\partial_t A - \mathrm{d}_A\Psi - w_{s,t}(A)) &= 0, \\ \partial_s\Psi - \partial_t\Phi + [\Phi, \Psi] + \varepsilon^{-2}\lambda^2 * F_A &= 0. \end{aligned} \tag{12}$$

The moduli space of solutions of these equations with given limit connections a_j will be denoted by

$$\mathcal{M}_\varepsilon(a_0, \ldots, a_n; H, K) = \frac{\{\Xi \in \mathcal{A}(Q) \,|\, (12), (6)\}}{\mathcal{G}(Q)}.$$

As before we consider only solutions that are in temporal gauge near the cylindrical ends. The Yang-Mills action for the ε-dependent metric is given by

$$\mathcal{YM}_\varepsilon(\Xi) = \int_0^1 \int_{-\infty}^{\infty} \Big(\|\partial_s A - \mathrm{d}_A\Phi - v_{s,t}(A)\|^2_{L^2(\Sigma,J)} + \frac{\lambda^2}{\varepsilon^2} \|F_A\|^2_{L^2(\Sigma)} \Big). \tag{13}$$

It depends on the a_j, but not on ε.

THEOREM 8.1. *Suppose that the perturbed flat connections $a_j \in \mathcal{A}^{\mathrm{flat}}(Q_{\tilde{f}_j}, K_j)$ are nondenerate for $j = 0, \dots, n$. Then the moduli spaces $\mathcal{M}_0(a_0, \dots, a_n; H, K)$ and $\mathcal{M}_\varepsilon(a_0, \dots, a_n; H, K)$ have the same virtual dimensions.*

THEOREM 8.2. *Let $(H, K) \in \mathcal{HK}_{\mathrm{reg}}$ and suppose that the moduli space $\mathcal{M}_0(a_0, \dots, a_n; H, K)$ has virtual dimension zero. Then, for $\varepsilon > 0$ sufficiently small, there is a natural orientation preserving bijection*

$$\mathcal{T}_\varepsilon : \mathcal{M}_0(a_0, \dots, a_n; H, K) \to \mathcal{M}_\varepsilon(a_0, \dots, a_n; H, K).$$

Theorem 1.1 is an immediate corollary of these results. Theorem 8.1 will be proved in the next section. Theorem 8.2 then follows easily from the techniques developed in [**6**], and we shall summarize the main points in Section 10.

9. Proof of the index formula

The proof of Theorem 8.1 rests on estimates for the differential operators that arise from the linearized equations. To examine these operators in detail we introduce some further notation. Denote by $\mathcal{X}$ the space of triples $\xi = (\alpha, \phi, \psi)$, where $\alpha : U \to \Omega^1(\Sigma, \mathfrak{g}_P)$ and $\phi, \psi : U \to \Omega^0(U, \mathfrak{g}_P)$ are smooth functions that have compact support in $S = U/\sim$ and that, for $s_j < s < s_{j+1}$, satisfy the periodicity conditions

$$\begin{aligned} \alpha(s, t+1) &= \tilde{g}_j{}^*\alpha(s, t), \\ \phi(s, t+1) = \phi(s, t) \circ \tilde{g}_j, &\qquad \psi(s, t+1) = \psi(s, t) \circ \tilde{g}_j. \end{aligned} \tag{14}$$

Such a triple can be identified with a 1-form $\xi = \alpha + \phi\, \mathrm{d}s + \psi\, \mathrm{d}t \in \Omega^1(X, \mathfrak{g}_Q)$ with values in the Lie algebra bundle $\mathfrak{g}_Q$ associated to Q. The self-duality operator

$$\Omega^1(X, \mathfrak{g}_Q) \to \Omega^{2,+}(X, \mathfrak{g}_Q) \oplus \Omega^0(X, \mathfrak{g}_Q) : \xi \mapsto \mathcal{D}\xi = (\mathrm{d}_\Xi^+\xi, \mathrm{d}_\Xi^*\xi)$$

has the form

$$\mathrm{d}_\Xi^+\xi = -\tilde{\alpha} \wedge \mathrm{d}s + *_{s,t}\tilde{\alpha} \wedge \mathrm{d}t + \tilde{\psi}\, \mathrm{d}s \wedge \mathrm{d}t + \lambda^{-2}\tilde{\psi}\omega, \qquad \mathrm{d}_\Xi^*\xi = -\lambda^{-2}\tilde{\phi},$$

where $\tilde{\xi} = (\tilde{\alpha}, \tilde{\phi}, \tilde{\psi})$ is given by

$$\begin{aligned} \tilde{\alpha} &= \nabla_s\alpha - \mathrm{d}_A\phi - \mathrm{d}v_{s,t}(A)\alpha + *_{s,t}(\nabla_t\alpha - \mathrm{d}_A\psi - \mathrm{d}w_{s,t}(A)\alpha), \\ \tilde{\phi} &= \nabla_s\phi + \nabla_t\psi + \lambda^2 * \mathrm{d}_A *_{s,t} \alpha, \\ \tilde{\psi} &= \nabla_s\psi - \nabla_t\phi + \lambda^2 * \mathrm{d}_A\alpha. \end{aligned}$$

The L^p and $W^{1,p}$-norms of the 1-form $\xi = \alpha + \phi\, \mathrm{d}s + \psi\, \mathrm{d}t$ are given by

$$\begin{aligned} \|\xi\|_{L^p}^p &= \int_0^1 \int_{-\infty}^\infty \left(\lambda^2 \|\alpha\|_{L^p(\Sigma,J)}^p + \lambda^{2-p} \|\phi\|_{L^p(\Sigma)}^p + \lambda^{2-p} \|\psi\|_{L^p(\Sigma)}^p\right), \\ \|\xi\|_{W^{1,p}}^p &= \int_0^1 \int_{-\infty}^\infty \Big(\lambda^2 \|\alpha\|_{W^{1,p}(\Sigma,J)}^p \\ &\quad + \lambda^{2-p} \|\nabla_s\alpha\|_{L^p(\Sigma,J)}^p + \lambda^{2-p} \|\nabla_t\alpha\|_{L^p(\Sigma,J)}^p \\ &\quad + \lambda^{2-p} \|\phi\|_{W^{1,p}(\Sigma,J)}^p + \lambda^{2-p} \|\psi\|_{W^{1,p}(\Sigma,J)}^p \\ &\quad + \lambda^{2-2p} \|\nabla_s\phi\|_{L^p(\Sigma)}^p + \lambda^{2-2p} \|\nabla_s\psi\|_{L^p(\Sigma)}^p \\ &\quad + \lambda^{2-2p} \|\nabla_t\phi\|_{L^p(\Sigma)}^p + \lambda^{2-2p} \|\nabla_t\psi\|_{L^p(\Sigma)}^p\Big), \end{aligned}$$

where $\nabla_s = \partial_s + \Phi$, $\nabla_t = \partial_t + \Psi$, and

$$\|\phi\|^p_{W^{1,p}(\Sigma,J)} = \|\phi\|^p_{L^p(\Sigma)} + \|\mathrm{d}_A\phi\|^p_{L^p(\Sigma,J)},$$

$$\|\alpha\|^p_{W^{1,p}(\Sigma,J)} = \|\alpha\|^p_{L^p(\Sigma,J)} + \|\mathrm{d}_A\alpha\|^p_{L^p(\Sigma)} + \|\mathrm{d}_A(\alpha\circ J)\|^p_{L^p(\Sigma)}.$$

Thus the $W^{1,p}$-norm depends on a connection $\Xi = (A, \Phi, \Psi)$ which should be chosen to be in temporal gauge near the cylindrical ends. We assume further that the limits (6) exist. The L^p-norm of the triple $\tilde{\xi} \cong \mathcal{D}\xi$ is given by

$$\|\tilde{\xi}\|^p_{\widetilde{L}^p} = \int_0^1\int_{-\infty}^{\infty}\left(\lambda^{2-p}\|\tilde{\alpha}\|^p_{L^p(\Sigma,J)} + \lambda^{2-2p}\|\tilde{\phi}\|^p_{L^p(\Sigma)} + \lambda^{2-2p}\|\tilde{\psi}\|^p_{L^p(\Sigma)}\right).$$

For $1 < p < \infty$ we denote by L^p, $W^{1,p}$, and $\widetilde{L}^p$ the completions of $\mathcal{X}$ with respect to these norms. They are independent of the choice of the connection Ξ. Note that $\tilde{\xi} \in \widetilde{L}^p$ if and only if $\lambda^{-1}\tilde{\xi} \in L^p$.

Linearizing the equations (12), we obtain the ε-dependent self-duality operator

$$\mathcal{D}_\varepsilon = \mathcal{D}_\varepsilon(\Xi) : W^{1,p} \to \widetilde{L}^p,$$

given by

$$\begin{aligned} \mathcal{D}_\varepsilon\xi &= \tilde{\xi} = (\tilde{\alpha}, \tilde{\phi}, \tilde{\psi}), \\ \tilde{\alpha} &= \nabla_s\alpha - \mathrm{d}_A\phi - \mathrm{d}v_{s,t}(A)\alpha + *_{s,t}(\nabla_t\alpha - \mathrm{d}_A\psi - \mathrm{d}w_{s,t}(A)\alpha), \\ \tilde{\phi} &= \nabla_s\phi + \nabla_t\psi + (\lambda/\varepsilon)^2 * \mathrm{d}_A *_{s,t} \alpha, \\ \tilde{\psi} &= \nabla_s\psi - \nabla_t\phi + (\lambda/\varepsilon)^2 * \mathrm{d}_A\alpha. \end{aligned} \tag{15}$$

Next we discuss the Cauchy–Riemann operator along the section $S \to \mathcal{M} : (s,t) \mapsto [A(s,t)]$. The pullback vertical tangent bundle under this section is the bundle $\mathrm{H}_A \to S$ whose fibre over $s + it$ is the space

$$\mathrm{H}^1_{A(s,t)} = \ker \mathrm{d}_{A(s,t)} \cap \ker \mathrm{d}_{A(s,t)}*_{s,t}$$

of harmonic 1-forms with respect to $A(s,t)$ and $*_{s,t}$. Let

$$\pi_A = \pi_{A(s,t)} : L^2(\Sigma, T^*\Sigma \otimes \mathfrak{g}_P) \to \mathrm{H}^1_{A(s,t)}$$

denote the L^2 orthogonal projection onto the harmonic part. We introduce the subspaces

$$L^p(\mathrm{H}_A) \subset L^p, \qquad W^{1,p}(\mathrm{H}_A) \subset W^{1,p}, \qquad \widetilde{L}^p(\mathrm{H}_A) \subset \widetilde{L}^p,$$

of those triples (α, ϕ, ψ) that satisfy $\phi = \psi = 0$ and $\pi_A(\alpha) = \alpha$. The operator

$$\mathcal{D}_0 = \mathcal{D}_0(\Xi) : W^{1,p}(\mathrm{H}_A) \to \widetilde{L}^p(\mathrm{H}_A)$$

is given by

$$\mathcal{D}_0\alpha_0 = \pi_A\big(\nabla_s\alpha_0 - \mathrm{d}v_{s,t}(A)\alpha_0 + *_{s,t}\nabla_t\alpha_0 - *_{s,t}\mathrm{d}w_{s,t}(A)\alpha_0\big).$$

Note that $\tilde{\alpha}_0 = \mathcal{D}_0\alpha_0$ can be interpreted as a $(0,1)$-form $\tilde{\alpha}_0 \wedge \mathrm{d}s - *_{s,t}\tilde{\alpha}_0 \wedge \mathrm{d}t$ on S with values in the bundle $\mathrm{H}_A \to S$. The L^p-norm of this 1-form is precisely the norm of the space $\widetilde{L}^p$.

Both operators $\mathcal{D}_\varepsilon(\Xi)$ and $\mathcal{D}_0(\Xi)$ are Fredholm whenever the limit connections $a_0, \ldots, a_n$ of Ξ are all nondegenerate. To examine the relation between these operators it will be convenient to introduce the ε-dependent norms

$$\begin{aligned}
\|\xi\|_{0,p,\varepsilon}^p &= \int_0^1 \int_{-\infty}^{\infty} \left(\lambda^2 \|\alpha\|_{L^p(\Sigma,J)}^p + \varepsilon^p \lambda^{2-p} \|\phi\|_{L^p(\Sigma)}^p + \varepsilon^p \lambda^{2-p} \|\psi\|_{L^p(\Sigma)}^p\right), \\
\|\xi\|_{1,p,\varepsilon}^p &= \int_0^1 \int_{-\infty}^{\infty} \left(\lambda^2 \|\alpha\|_{W^{1,p}(\Sigma,J)}^p\right. \\
&\quad + \varepsilon^p \lambda^{2-p} \|\nabla_s \alpha\|_{L^p(\Sigma,J)}^p + \varepsilon^p \lambda^{2-p} \|\nabla_t \alpha\|_{L^p(\Sigma,J)}^p \\
&\quad + \varepsilon^p \lambda^{2-p} \|\phi\|_{W^{1,p}(\Sigma,J)}^p + \varepsilon^p \lambda^{2-p} \|\psi\|_{W^{1,p}(\Sigma,J)}^p \\
&\quad + \varepsilon^{2p} \lambda^{2-2p} \|\nabla_s \phi\|_{L^p(\Sigma)}^p + \varepsilon^{2p} \lambda^{2-2p} \|\nabla_s \psi\|_{L^p(\Sigma)}^p \\
&\quad \left. + \varepsilon^{2p} \lambda^{2-2p} \|\nabla_t \phi\|_{L^p(\Sigma)}^p + \varepsilon^{2p} \lambda^{2-2p} \|\nabla_t \psi\|_{L^p(\Sigma)}^p\right), \\
\|\tilde{\xi}\|_{\tilde{0},p,\varepsilon}^p &= \int_0^1 \int_{-\infty}^{\infty} \left(\lambda^{2-p} \|\tilde{\alpha}\|_{L^p(\Sigma,J)}^p + \varepsilon^p \lambda^{2-2p} \|\tilde{\phi}\|_{L^p(\Sigma)}^p + \varepsilon^p \lambda^{2-2p} \|\tilde{\psi}\|_{L^p(\Sigma)}^p\right).
\end{aligned}$$

PROPOSITION 9.1. *Fix a connection $\Xi = (A, \Phi, \Psi) \in \mathcal{A}(Q)$ that is in temporal gauge near the cylindrical ends and satisfies $F_{A(s,t)} = 0$ for all s and t. Suppose that the limits (6) exist. Then, for every $p \geq 2$, there exist constants $\varepsilon_0 > 0$ and $c \geq 1$ such that, for $0 < \varepsilon < \varepsilon_0$ and $\xi \in \mathcal{X}$,*

$$\|\xi\|_{1,p,\varepsilon} \leq c\left(\varepsilon \|\mathcal{D}_\varepsilon \xi\|_{\tilde{0},p,\varepsilon} + \|\pi_A(\xi)\|_{L^p}\right), \tag{16}$$

$$\|\xi - \pi_A(\xi)\|_{1,p,\varepsilon} \leq c\varepsilon \left(\|\mathcal{D}_\varepsilon \xi\|_{\tilde{0},p,\varepsilon} + \|\pi_A(\xi)\|_{L^p}\right), \tag{17}$$

$$\|\pi_A(\mathcal{D}_\varepsilon \xi) - \mathcal{D}_0 \pi_A(\xi)\|_{\tilde{L}^p} \leq c \|\xi - \pi_A(\xi)\|_{0,p,\varepsilon}. \tag{18}$$

PROOF. See Appendix D. (For $p = 2$ see [**6**, Lemmata 4.2 and 4.3].) □

PROOF OF THEOREM 8.1. Let us choose an L^2-orthonormal basis $\alpha_1, \ldots, \alpha_k \in W^{1,2}(\mathrm{H}_A)$ of the kernel of $\mathcal{D}_0$ and an $\widetilde{L}^2$-orthonormal basis $\beta_1, \ldots, \beta_\ell \in \widetilde{L}^2(\mathrm{H}_A)$ of the cokernel of $\mathcal{D}_0$. Define the operator

$$\widehat{\mathcal{D}}_\varepsilon : W^{1,2} \oplus \mathbb{R}^\ell \to \widetilde{L}^2 \oplus \mathbb{R}^k$$

by

$$\widehat{\mathcal{D}}_\varepsilon \hat{\xi} := \left(\mathcal{D}_\varepsilon \xi + \sum_{j=1}^{\ell} \lambda_j \beta_j, \langle \alpha_1, \pi_A(\xi)\rangle, \ldots, \langle \alpha_k, \pi_A(\xi)\rangle\right)$$

for $\hat{\xi} = (\xi, \lambda_1, \ldots, \lambda_\ell) \in W^{1,2} \oplus \mathbb{R}^\ell$. We prove that $\widehat{\mathcal{D}}_\varepsilon$ is injective. To see this, note first that there is a constant $c_0 \geq 1$ such that

$$\|\pi_A(\xi)\|_{L^2} + \sum_{j=1}^{\ell} |\lambda_j| \leq c_0 \left\|\mathcal{D}_0 \pi_A(\xi) + \sum_{j=1}^{\ell} \lambda_j \beta_j\right\|_{\widetilde{L}^2}$$

for all $\hat{\xi} = (\xi, \lambda_1, \dots, \lambda_\ell) \in W^{1,2} \oplus \mathbb{R}^\ell$. Hence it follows from (18) and (17) with $p = 2$ that

$$\begin{aligned}
&\|\pi_A(\xi)\|_{L^2} + \sum_{j=1} |\lambda_j| \\
&\quad \le c_0 \left\| \pi_A(\mathcal{D}_\varepsilon \xi) + \sum_{j=1}^{\ell} \lambda_j \beta_j \right\|_{\widetilde{L}^2} + c_0 \left\| \pi_A(\mathcal{D}_\varepsilon \xi) - \mathcal{D}_0 \pi_A(\xi) \right\|_{\widetilde{L}^2} \\
&\quad \le c_0 \left\| \mathcal{D}_\varepsilon \xi + \sum_{j=1}^{\ell} \lambda_j \beta_j \right\|_{\widetilde{0},2,\varepsilon} + c_0 c \left\| \xi - \pi_A(\xi) \right\|_{0,2,\varepsilon} \\
&\quad \le c_0 \left\| \mathcal{D}_\varepsilon \xi + \sum_{j=1}^{\ell} \lambda_j \beta_j \right\|_{\widetilde{0},2,\varepsilon} + c_0 c^2 \varepsilon \left(\|\mathcal{D}_\varepsilon \xi\|_{\widetilde{0},2,\varepsilon} - \|\pi_A(\xi)\|_{L^2} \right) \\
&\quad \le c_0(1 + c^2\varepsilon) \left\| \mathcal{D}_\varepsilon \xi + \sum_{j=1}^{\ell} \lambda_j \beta_j \right\|_{\widetilde{0},2,\varepsilon} + c_0 c^2 \varepsilon \left(\|\pi_A(\xi)\|_{L^2} + \sum_{j=1}^{\ell} |\lambda_j| \right).
\end{aligned}$$

With $c_0 c^2 \varepsilon \le 1/2$ we obtain

$$\|\pi_A(\xi)\|_{L^2} + \sum_{j=1}^{\ell} |\lambda_j| \le 3c_0 \|\widehat{\mathcal{D}}_\varepsilon \hat{\xi}\|_{\widetilde{0},2,\varepsilon}.$$

Combining this inequality with (16) gives

$$\begin{aligned}
\|\hat{\xi}\|_{1,2,\varepsilon} &= \sqrt{\|\xi\|_{1,2,\varepsilon}^2 + \sum_{j=1}^{\ell} \lambda_j{}^2} \\
&\le \|\xi\|_{1,2,\varepsilon} + \sum_{j=1}^{\ell} |\lambda_j| \\
&\le c \left(\varepsilon \|\mathcal{D}_\varepsilon \xi\|_{\widetilde{0},2,\varepsilon} + \|\pi_A(\xi)\|_{L^2} \right) + \sum_{j=1}^{\ell} |\lambda_j| \\
&\le c\varepsilon \left\| \mathcal{D}_\varepsilon \xi + \sum_{j=1}^{\ell} \lambda_j \beta_j \right\|_{\widetilde{0},2,\varepsilon} + c' \left(\|\pi_A(\xi)\|_{L^2} + \sum_{j=1}^{\ell} |\lambda_j| \right) \\
&\le c'' \|\widehat{\mathcal{D}}_\varepsilon \hat{\xi}\|_{\widetilde{0},2,\varepsilon}.
\end{aligned}$$

This shows that $\widehat{\mathcal{D}}_\varepsilon$ is injective for $\varepsilon > 0$ sufficiently small. To prove surjectivity we examine the adjoint operator

$$\mathcal{D}_\varepsilon{}' : W^{1,2} \to \widetilde{L}^2.$$

It is defined by the identity

$$\langle \lambda^{-1} \mathcal{D}_\varepsilon{}' \xi', \xi \rangle_{0,2,\varepsilon} = \langle \xi', \lambda^{-1} \mathcal{D}_\varepsilon \xi \rangle_{0,2,\varepsilon}$$

for $\xi, \xi' \in \mathcal{X}$. Direct computation shows that $\mathcal{D}_\varepsilon{}' \xi' = \tilde{\xi}'$ is given by

$$\begin{aligned}
\tilde{\alpha}' &= -\widetilde{\nabla}_s \alpha' - \mathrm{d}_A \phi' - \mathrm{d}v_{s,t}(A)\alpha' + *_{s,t}(\widetilde{\nabla}_t \alpha' - \mathrm{d}_A \psi - \mathrm{d}w_{s,t}(A)\alpha'), \\
\tilde{\phi}' &= -\widetilde{\nabla}_s \phi' + \widetilde{\nabla}_t \psi' + (\lambda/\varepsilon)^2 * \mathrm{d}_A *_{s,t} \alpha',
\end{aligned}$$

$$\widetilde{\psi}' = -\widetilde{\nabla}_s\psi' - \widetilde{\nabla}_t\phi' + (\lambda/\varepsilon)^2 * \mathrm{d}_A\alpha',$$

where

$$\widetilde{\nabla}_s\alpha' = \nabla_s\alpha' + \lambda^{-1}(\partial_s\lambda)\alpha' + \alpha' \circ (J\partial_s J), \qquad \widetilde{\nabla}_t\alpha' = \nabla_s\alpha' + \lambda^{-1}(\partial_t\lambda)\alpha',$$

and

$$\widetilde{\nabla}_s\phi' = \nabla_s\phi' - \lambda^{-1}(\partial_s\lambda)\phi', \qquad \widetilde{\nabla}_t\phi' = \nabla_t\phi' - \lambda^{-1}(\partial_t\lambda)\phi'.$$

Similarly, the operator

$$\mathcal{D}_0' : W^{1,2}(\mathrm{H}_A) \to \widetilde{L}^2(\mathrm{H}_A)$$

is given by

$$\mathcal{D}_0'\alpha_0' = \pi_A\Big(\widetilde{\nabla}_s\alpha_0' - \mathrm{d}v_{s,t}(A)\alpha_0' + *_{s,t}\widetilde{\nabla}_t\alpha_0' - *_{s,t}\mathrm{d}w_{s,t}(A)\alpha_0'\Big).$$

Note that there exists a constant $c_1 > 0$ such that

$$|\partial_s J| + |\lambda^{-1}\partial_s\lambda| + |\lambda^{-1}\partial_t\lambda| \leq c_1\lambda.$$

Hence it follows from Proposition 9.1 that the adjoint operator $\mathcal{D}_\varepsilon{}'$ satisfies the same estimates as $\mathcal{D}_\varepsilon$. Namely, for every $p \geq 2$, there exist constants $\varepsilon_0 > 0$ and $c \geq 1$ such that, for $0 < \varepsilon < \varepsilon_0$ and $\xi' \in \mathcal{X}$,

$$\|\xi'\|_{1,p,\varepsilon} \leq c\left(\varepsilon\left\|\mathcal{D}_\varepsilon{}'\xi'\right\|_{\tilde{0},p,\varepsilon} + \|\pi_A(\xi')\|_{L^p}\right),$$

$$\|\xi' - \pi_A(\xi')\|_{1,p,\varepsilon} \leq c\varepsilon\left(\left\|\mathcal{D}_\varepsilon{}'\xi'\right\|_{\tilde{0},p,\varepsilon} + \|\pi_A(\xi')\|_{L^p}\right),$$

$$\left\|\pi_A(\mathcal{D}_\varepsilon{}'\xi') - \mathcal{D}_0{}'\pi_A(\xi')\right\|_{\widetilde{L}^p} \leq c\,\|\xi' - \pi_A(\xi')\|_{0,p,\varepsilon}.$$

Now it follows by the same arguments as above that the operator $\widehat{\mathcal{D}}_\varepsilon{}'$ is injective. This shows that $\widehat{\mathcal{D}}_\varepsilon$ is surjective. Hence

$$\operatorname{index}\mathcal{D}_\varepsilon = k - \ell = \operatorname{index}\mathcal{D}_0.$$

This proves the theorem. □

The Fredholm index will be denoted by

$$\mu(a_0, \ldots, a_n; H, K) = \operatorname{index}\mathcal{D}_\varepsilon = \operatorname{index}\mathcal{D}_0.$$

This index depends on the limit connections a_j but not on the connection Ξ used to define it. It is the virtual dimension of the moduli spaces $\mathcal{M}_0(a_0, \ldots, a_n; H, K)$ and $\mathcal{M}_\varepsilon(a_0, \ldots, a_n; H, K)$. As pointed out above, the space $\mathcal{HK}_{\mathrm{reg}}$ of regular perturbations consists of those pairs (H, K) such that the operator $\mathcal{D}_0(\Xi)$ is surjective for all solutions Ξ of (11) and (6). This set is a countable intersection of open and dense sets in the space of all smooth perturbations (cf. [**20, 10**]). If $(H, K) \in \mathcal{HK}_{\mathrm{reg}}$, then $\mathcal{M}_0(a_0, \ldots, a_n; H, K)$ is a smooth orientable manifold of dimension

$$\dim\mathcal{M}_0(a_0, \ldots, a_n; H, K) = \mu(a_0, \ldots, a_n; H, K).$$

If, moreover, $\mu(a_0, \ldots, a_n; H, K) = 0$, then there exists a constant $\varepsilon_0 > 0$ such that $\mathcal{M}_\varepsilon(a_0, \ldots, a_n; H, K)$ is a zero-dimensional manifold for $0 < \varepsilon \leq \varepsilon_0$.

REMARK 9.1. The same argument as in the proof of Theorem 8.1 shows that if $\mathcal{D}_0$ is injective then so is $\mathcal{D}_\varepsilon$ for $\varepsilon > 0$ sufficiently small, and similarly for surjectivity. The argument also provides a uniform estimate for the left, respectively right, inverse of $\mathcal{D}_\varepsilon$ (cf. [**6**]).

10. Proof of the main result

To restate Theorem 8.2 more precisely we shall need some notation. Fix a reference connection

$$\hat{\Xi} = \hat{A} + \hat{\Phi}\,ds + \hat{\Psi}\,dt \in \mathcal{A}(Q)$$

such that the pullback connections on the cylindrical ends $(-\infty, -T) \times Q_{\tilde{f}_j}$ satisfy (6) and

$$\partial_s \hat{A}_j = 0, \qquad \hat{\Phi}_j = 0, \qquad \partial_s \hat{\Psi}_j = 0$$

for $-s$ sufficiently large. Then denote

$$\mathcal{A}^{1,p}(a_0, \ldots, a_n) = \{\hat{\Xi} + \xi \,|\, \xi \in W^{1,p}\}.$$

Likewise, denote by $\mathcal{G}^{2,p}$ the completion of the group $\mathcal{G}(Q)$ (of gauge transformations that are equal to $\mathbb{1}$ near the cylindrical ends) with respect to the $W^{1,p}$-norm on $u^*\hat{\Xi}$. Next define

$$\mathcal{A}_0^{1,p}(a_0, \ldots, a_n; H, K) = \left\{\Xi \in \mathcal{A}^{1,p}(a_0, \ldots, a_n) \,|\, (11)\right\},$$

and

$$\mathcal{A}_\varepsilon^{1,p}(a_0, \ldots, a_n; H, K) = \left\{\Xi \in \mathcal{A}^{1,p}(a_0, \ldots, a_n) \,|\, (12)\right\}.$$

Thus, for $\varepsilon \geq 0$,

$$\mathcal{M}_\varepsilon(a_0, \ldots, a_n; H, K) \cong \frac{\mathcal{A}_\varepsilon^{1,p}(a_0, \ldots, a_n; H, K)}{\mathcal{G}^{2,p}}.$$

THEOREM 10.1. *Assume $(H, K) \in \mathcal{HK}_{\mathrm{reg}}$ and $\mu(a_0, \ldots, a_n; H, K) = 0$. Then there exist constants $\varepsilon_0 > 0$ and $c > 0$ such that, for $0 < \varepsilon < \varepsilon_0$, there exists a map*

$$\mathcal{T}_\varepsilon : \mathcal{A}_0^{1,p}(a_0, \ldots, a_n; H, K) \to \mathcal{A}_\varepsilon^{1,p}(a_0, \ldots, a_n; H, K)$$

that satisfies the following conditions.

(i) *If $\Xi_0 \in \mathcal{A}_0^{1,p}(a_0, \ldots, a_n; H, K)$, then $\Xi_\varepsilon = \mathcal{T}_\varepsilon(\Xi_0)$ satisfies*[1]

$$d_{\Xi_0}^{*_\varepsilon}(\Xi_\varepsilon - \Xi_0) = 0, \tag{19}$$

$$\|\Xi_\varepsilon - \Xi_0\|_{1,p,\varepsilon} \leq c\varepsilon^2. \tag{20}$$

Here the $1, p, \varepsilon$-norm is the one determined by Ξ_0.

(ii) *$\mathcal{T}_\varepsilon$ is equivariant under the action of $\mathcal{G}^{2,p}$, i.e.*

$$\mathcal{T}_\varepsilon(u^*\Xi_0) = u^*\mathcal{T}_\varepsilon(\Xi_0)$$

for every $\Xi_0 \in \mathcal{A}_0^{1,p}(a_0, \ldots, a_n; H, K)$ and every $u \in \mathcal{G}^{2,p}$. The induced map of the moduli spaces will also be denoted by $\mathcal{T}_\varepsilon$.

(iii) *$\mathcal{T}_\varepsilon$ is injective.*

(iv) *$\mathcal{T}_\varepsilon$ is surjective.*

---

[1]Here $d_\Xi^{*_\varepsilon} : \Omega^1(X, \mathfrak{g}_Q) \to \Omega^0(X, \mathfrak{g}_Q)$ is the adjoint of the covariant differential $d_\Xi : \Omega^0(X, \mathfrak{g}_Q) \to \Omega^1(X, \mathfrak{g}_Q)$ with respect to the ε-dependent L^2-norm on 1-forms. In explicit terms,

$$d_\Xi^{*_\varepsilon}\xi = - * d_A *_{s,t} \alpha - (\varepsilon/\lambda)^2 \nabla_s \phi - (\varepsilon/\lambda)^2 \nabla_t \psi.$$

The first assertion of Theorem 10.1 is a refined version of the implicit function theorem and follows from a Newton type iteration argument. The details are word for word the same as in the proof of Theorem 5.1 in [**6**], and will be omitted. The next two assertions follow from standard arguments in gauge theory, e.g. the gauge equivariance follows from the uniqueness part of the implicit function theorem (Proposition 10.2 below), and injectivity follows from the standard observation that the $W^{1,p}$-norms of two gauge equivalent connections control the $W^{2,p}$-norm of the gauge transformation by which they are related (cf. Proposition 5.7 in [**6**] for the present context). The hardest part of the proof of Theorem 10.1 is surjectivity. It relies on the following four propositions.

PROPOSITION 10.2. *Assume $(H,K) \in \mathcal{HK}_{\mathrm{reg}}$ and $\mu(a_0,\ldots,a_n;H,K) = 0$. Then there exist constants $\delta > 0$ and $\varepsilon_0 > 0$ such that the following holds. If $\Xi_0 \in \mathcal{A}_0^{1,p}(a_0,\ldots,a_n;H,K)$ and $\Xi \in \mathcal{A}_\varepsilon^{1,p}(a_0,\ldots,a_n;H,K)$ for some $\varepsilon \in (0,\varepsilon_0)$ and (19) is satisfied and*

$$\|\Xi - \Xi_0\|_{1,p,\varepsilon} \le \delta\varepsilon^{1/2+2/p},$$

then $\Xi = \mathcal{T}_\varepsilon(\Xi_0)$.

PROPOSITION 10.3. *Assume $(H,K) \in \mathcal{HK}_{\mathrm{reg}}$ and $\mu(a_0,\ldots,a_n;H,K) = 0$. Then there exist constants $\varepsilon_0 > 0$ and $\delta > 0$ such that the following holds. If $\Xi_0 \in \mathcal{A}_0^{1,p}(a_0,\ldots,a_n;H,K)$ and $\Xi \in \mathcal{A}_\varepsilon^{1,p}(a_0,\ldots,a_n;H,K)$ for some $\varepsilon \in (0,\varepsilon_0)$ and*

$$\|\Xi - \Xi_0\|_{1,p,\varepsilon} \le \delta\varepsilon^{1/2+2/p}, \tag{21}$$

then there exists a $u \in \mathcal{G}^{2,p}$ such that $u^*\Xi = \mathcal{T}_\varepsilon(\Xi_0)$.

As pointed out above, Proposition 10.2 is the uniqueness part of the implicit function theorem (compare with [**6**, Theorem 5.2]). Proposition 10.3 is an easy consequence of Proposition 10.2. One can use the gauge freedom to achieve condition (19), and one has to show that this can be done without destroying the estimate (21). (See [**6**, Section 6] for details.)

PROPOSITION 10.4. *Assume $(H,K) \in \mathcal{HK}_{\mathrm{reg}}$ and $\mu(a_0,\ldots,a_n;H,K) = 0$. Then, for every constant $c_0 > 0$, there exists a constant $\varepsilon_0 > 0$ such that the following holds. If $\Xi \in \mathcal{A}_\varepsilon^{1,p}(a_0,\ldots,a_n;H,K)$ for some $\varepsilon \in (0,\varepsilon_0)$ and*

$$\sup_{s,t}\left(\varepsilon^{-2}\|F_A\|_{L^\infty(\Sigma)} + \lambda^{-1}\|\partial_t A - d_A\Psi\|_{L^\infty(\Sigma,J)}\right) \le c_0, \tag{22}$$

then there exists a connection $\Xi_0 \in \mathcal{A}_0^{1,p}(a_0,\ldots,a_n;H,K)$ such that $\Xi = \mathcal{T}_\varepsilon(\Xi_0)$.

The proof of this result relies on Proposition 10.3. The key idea is to project the first component of $\Xi = (A,\Phi,\Psi)$ onto the moduli space M_Σ of flat connections and then prove estimates for the resulting section A' of the bundle $\mathcal{M}$. First. one can show that A' is approximately holomorphic in the sense that the complex anti-linear part of its differential is bounded by a constant times ε^2. Second, one can control the $W^{1,p}$ norm of this section. Hence it can be approximated by a pseudo-holomorphic section A'' (Theorem 2.5 in [**6**]). A further modification of A'' (by a gauge transformation, pointwise for all s and t) then gives rise to a connection $\Xi_0 \in \mathcal{A}_0^{1,p}(a_0,\ldots,a_n;H,K)$ such that

$$d_{A(s,t)} *_{s,t} (A(s,t) - A_0(s,t)) = 0$$

for all s and t. This connection satisfies an estimate

$$\|\Xi - \Xi_0\|_{1,2,\varepsilon} \le c\varepsilon^{1+2/p},$$

where c is independent of Ξ. With $c\varepsilon^{1/2} \le \delta$ it follows from Proposition 10.3 that Ξ is gauge equivalent to $\mathcal{T}_\varepsilon(\Xi_0)$. The details are carried out in [**6**, Sections 7 and 8] and will not be reproduced here.

PROPOSITION 10.5. *Assume $(H, K) \in \mathcal{HK}_{\mathrm{reg}}$ and $\mu(a_0, \ldots, a_n; H, K) = 0$. Then there exist constants $c_0 > 0$ and $\varepsilon_0 > 0$ such that, for every $\varepsilon \in (0, \varepsilon_0)$ and every $\Xi \in \mathcal{A}^{1,p}_\varepsilon(a_0, \ldots, a_n; H, K)$,*

$$\sup_{s,t} \left(\varepsilon^{-2} \|F_A\|_{L^\infty(\Sigma)} + \lambda^{-1} \|\partial_t A - d_A\Psi\|_{L^\infty(\Sigma,J)}\right) \le c_0.$$

To prove this one argues by contradiction. If the result were false, then bubbling would have to occur, and this would give rise to nonempty moduli spaces of negative virtual dimension, contradicting the assumption $(H, K) \in \mathcal{HK}_{\mathrm{reg}}$. For more details see [**6**, Section 9].

PROOF OF THEOREM 10.1 (IV). Choose c_0 and ε_0 as in Proposition 10.5. Making ε_0 smaller, if necessary, we may assume that Proposition 10.4 holds with these constants. Now let $\Xi \in \mathcal{A}^{1,p}_\varepsilon(a_0, \ldots, a_n; H, K)$ and suppose that $0 < \varepsilon < \varepsilon_0$. Then, by Proposition 10.5, Ξ satisfies (22). Hence, by Proposition 10.4, Ξ lies in the image of $\mathcal{T}_\varepsilon$. Hence $\mathcal{T}_\varepsilon$ is surjective for $0 < \varepsilon < \varepsilon_0$. □

PROOF OF THEOREM 8.2. By Theorem 10.1, there is a natural bijection

$$\mathcal{T}_\varepsilon : \mathcal{M}_0(a_0, \ldots, a_n; H, K) \to \mathcal{M}_\varepsilon(a_0, \ldots, a_n; H, K).$$

It remains to show that this bijection preserves orientations. To see this, consider the space

$$\mathcal{A}^{1,p}_\Sigma(a_0, \ldots, a_n) = \left\{\Xi \in \mathcal{A}^{1,p}(a_0, \ldots, a_n; H, K) \,|\, F_{A(s,t)} = 0 \,\forall\, s, t\right\}.$$

Since $\mathcal{A}^{\mathrm{flat}}(P)$ is connected and simply connected, this space is nonempty. For every $\Xi \in \mathcal{A}^{1,p}_\Sigma(a_0, \ldots, a_n)$, there are two Fredholm operators, $\mathcal{D}_0(\Xi)$ and $\mathcal{D}_\varepsilon(\Xi)$ for $\varepsilon > 0$. Their determinants give rise to two line bundles

$$\mathcal{L}_0 \to \mathcal{A}^{1,p}_\Sigma(a_0, \ldots, a_n), \qquad \mathcal{L}_\varepsilon \to \mathcal{A}^{1,p}_\Sigma(a_0, \ldots, a_n).$$

The argument in the proof of Theorem 8.1 establishes, for every Ξ, a natural identification of $\det(\mathcal{D}_0(\Xi))$ with $\det(\mathcal{D}_\varepsilon(\Xi))$ for sufficiently small $\varepsilon > 0$. Hence the line bundles $\mathcal{L}_0$ and $\mathcal{L}_\varepsilon$ are isomorphic along any given loop in $\mathcal{A}^{1,p}_\Sigma(a_0, \ldots, a_n)$ for ε sufficiently small. Since $\mathcal{L}_\varepsilon$ extends to the affine space $\mathcal{A}^{1,p}(a_0, \ldots, a_n)$, it follows that both bundles are orientable and there is a natural bijection from the (two element) set $\mathrm{Or}_0(a_0, \ldots, a_n)$ of orientations of $\mathcal{L}_0$ to the set $\mathrm{Or}_\varepsilon(a_0, \ldots, a_n)$ of orientations of $\mathcal{L}_\varepsilon$. Denote this bijection by

$$\tau_\varepsilon(a_0, \ldots, a_n) : \mathrm{Or}_0(a_0, \ldots, a_n) \to \mathrm{Or}_\varepsilon(a_0, \ldots, a_n). \tag{23}$$

Now suppose that we are given a system of orientations

$$\sigma_0(a_0, \ldots, a_n) \in \mathrm{Or}_0(a_0, \ldots, a_n), \tag{24}$$

one for each $(n+1)$-tuple of perturbed flat connections $a_j \in \mathcal{A}^{\text{flat}}(Q_{\tilde{f}_j}, K_j)$. Suppose further that we are given $n+1$ systems of coherent orientations for the $n+1$ (symplectic) Floer homologies:

$$\sigma_{\tilde{f}_j}(a_j, b_j) \in \mathrm{Or}_{\tilde{f}_j}(a_j, b_j)$$

(cf. Floer–Hofer [**9**]). The orientations (24) are called *compatible with the coherent orientations* if all the Floer gluing maps are orientation preserving. Given the coherent orientations of the Floer chain complexes, there exists a compatible collection of orientations (24), and such a collection is uniquely determined up to an overall sign. The same holds with the subscript 0 replaced by $\varepsilon > 0$.

Now one can show, as in [**6**, Proposition 10.3], that the maps (23) commute with the Floer gluing maps. Hence the maps τ_ε map any compatible collection of orientations (for the symplectic Floer theory) to a compatible collection

$$\sigma_\varepsilon(a_0, \ldots, a_n) \in \mathrm{Or}_\varepsilon(a_0, \ldots, a_n) \tag{25}$$

for the corresponding instanton theory.

Let us fix two such compatible collections of orientations, which are related by τ_ε. Then the sign $\nu_0(\Xi_0)$ of a solution $\Xi_0 \in \mathcal{A}_0^{1,p}(a_0, \ldots, a_n; H, K)$ is determined by comparing the natural orientation of $\det(\mathcal{D}_0(\Xi_0)) \cong \mathbb{R}$ with the orientation given by (24). Similarly for $\Xi_\varepsilon \in \mathcal{A}_\varepsilon^{1,p}(a_0, \ldots, a_n; H, K)$. But since $\sigma_\varepsilon(a_0, \ldots, a_n)$ is the image of $\sigma_0(a_0, \ldots, a_n)$ under $\tau_\varepsilon(a_0, \ldots, a_n)$, it follows that the two orientations of $\det(\mathcal{D}_0(\Xi_0))$ agree if and only if the two orientations of $\det(\mathcal{D}_\varepsilon(\Xi_0))$ agree. Since $\Xi_\varepsilon = \mathcal{T}_\varepsilon(\Xi_0)$ is close to Ξ_0, it follows that

$$\nu_\varepsilon(\mathcal{T}_\varepsilon(\Xi_0)) = \nu_0(\Xi_0)$$

for $\varepsilon > 0$ sufficiently small. Hence the map

$$\mathcal{T}_\varepsilon : \mathcal{M}_0(a_0, \ldots, a_n; H, K) \to \mathcal{M}_\varepsilon(a_0, \ldots, a_n; H, K)$$

is orientation preserving. This completes the proof of Theorem 8.2 □

PROOF OF THEOREM 1.1. The Floer homology groups $\mathrm{HF}^{\text{symp}}(\phi_{\tilde{f}})$ and $\mathrm{HF}(Q_{\tilde{f}})$ are both derived from the chain complex

$$\mathrm{CF}(\tilde{f}, K) = \bigoplus_{[a] \in \mathcal{A}^{\text{flat}}(Q_{\tilde{f}}, K)/\mathcal{G}(Q_{\tilde{f}})} \mathbb{Z}a$$

for a regular perturbation K. Given $n+1$ automorphisms $\tilde{f}_0, \ldots, \tilde{f}_n$ such that $\tilde{f}_n \circ \cdots \circ \tilde{f}_0 = \mathrm{id}$ and a regular perturbation $(H, K) \in \mathcal{HK}_{\text{reg}}$, there are two homomorphisms

$$\psi_0, \psi_\varepsilon : \mathrm{CF}(\tilde{f}_0, K_0) \otimes \mathrm{CF}(\tilde{f}_1, K_1) \otimes \cdots \otimes \mathrm{CF}(\tilde{f}_n, K_n) \to \mathbb{Z}$$

defined by

$$\psi_0(a_0, \ldots, a_n) = \sum_{[\Xi_0] \in \mathcal{M}_0(a_0, \ldots, a_n; H, K)} \nu_0(\Xi_0),$$

$$\psi_\varepsilon(a_0, \ldots, a_n) = \sum_{[\Xi_\varepsilon] \in \mathcal{M}_\varepsilon(a_0, \ldots, a_n; H, K)} \nu_\varepsilon(\Xi_\varepsilon),$$

whenever $\mu(a_0, \ldots, a_n; H, K) = 0$, and $\psi_0(a_0, \ldots, a_n) = \psi_\varepsilon(a_0, \ldots, a_n) = 0$ whenever $\mu(a_0, \ldots, a_n; H, K) \neq 0$. Here the definition of the signs $\nu_0(\Xi_0)$ and $\nu_\varepsilon(\Xi_\varepsilon)$ requires the choice of two compatible collections of orientations (24) and (25).

Changing an overall sign, if necessary, we may assume that these systems of orientations are related by the map (23). Under this assumption Theorem 8.2 asserts that $\nu_\varepsilon(\mathcal{T}_\varepsilon(\Xi_0)) = \nu_0(\Xi_0)$ for $\varepsilon > 0$ sufficiently small. Hence $\psi_0 = \psi_\varepsilon$ for $\varepsilon > 0$ sufficiently small. This proves that the product structures (1) and (2) agree up to a sign. □

Appendix A. Cauchy–Riemann operators in Hilbert space

Let V and H be separable Hilbert spaces and $V \hookrightarrow H$ be a compact linear inclusion with a dense image. Throughout we identify H with its dual space H^* via the Riesz representation theorem. Then the adjoint of the inclusion $V \hookrightarrow H$ is an inclusion $H \hookrightarrow V^*$ which is again compact and has a dense image. Thus there are two inclusions

$$V \subset H \subset V^*,$$

and we shall think of V as a subset of H and of H as a subset of V^*. Let $\mathcal{L}(H)$ denote the space of bounded linear operators on H and $\mathcal{L}(V,H)$ the space of bounded linear operators from $V \to H$. Denote by $\mathcal{P}(H)$ the set of self-adjoint Hilbert space isomorphisms $Q : H \to H$ such that $QV = V$ and

$$0 < \inf_{0 \neq x \in H} \frac{\langle x, Qx\rangle_H}{\|x\|_H^2} \leq \sup_{0 \neq x \in H} \frac{\langle x, Qx\rangle_H}{\|x\|_H^2} < \infty.$$

Every operator $Q \in \mathcal{P}(H)$ determines an inner product

$$\langle x, y\rangle_Q := \langle x, Qy\rangle_H$$

such that the corresponding norm is equivalent to the standard norm on H. Given $Q \in \mathcal{P}(H)$, we denote by $\mathcal{J}(H;Q)$ the set of Hilbert space automorphisms $J : H \to H$ that satisfy $JV = V$ and

$$J^{-1} = Q^{-1}J^*Q = -J.$$

The first identity means that J is an isometry for the Q-inner product. An operator $D \in \mathcal{L}(V,H)$ is called **Q-symmetric** if

$$\langle Dx, Qy\rangle_H = \langle x, QDy\rangle_H$$

for $x, y \in V$. It is called **Q-self-adjoint** if it is Q-symmetric and, for every $x \in H$,

$$x \in V \quad \Longleftrightarrow \quad \sup_{0 \neq y \in V} \frac{|\langle x, QDy\rangle_H|}{\|y\|_H} < \infty. \tag{26}$$

Let $\mathcal{S}(V,H;Q)$ denote the set of Q-self-adjoint operators $D \in \mathcal{L}(V,H)$ which satisfy the estimate

$$\|x\|_V \leq c\,(\|Dx\|_H + \|x\|_H) \tag{27}$$

for some constant $c > 0$ and all $x \in V$. For every $D \in \mathcal{S}(V,H;Q)$ we denote by

$$\pi_D : H \to H$$

the Q-orthogonal projection onto the kernel of D.

LEMMA A.1. *Suppose $D \in \mathcal{L}(V,H)$ is Q-symmetric. Then the following are equivalent.*

(i) $D \in \mathcal{S}(V,H;Q)$.

(ii) *There exists a constant $c_0 > 0$ such that, for every $x \in V$,*

$$\|x\|_V \leq c_0\,(\|Dx\|_H + \|\pi_D(x)\|_H)\,. \tag{28}$$

Moreover, if $x \in H$ satisfies $\langle x, QDy\rangle = 0$ for every $y \in V$, then $x \in V$.

(iii) *D is a Fredholm operator of index zero.*

PROOF. We prove that (i) implies (ii). Suppose, for a contradiction, that (28) does not hold. Then there exists a sequence $x_n \in H$ such that

$$\|x_n\|_V = 1, \qquad \|Dx_n\|_H + \|\pi_D(x_n)\|_H \le 1/n.$$

Since the inclusion $V \hookrightarrow H$ is compact, there exists a subsequence x_{n_k} that converges in H to a vector x. By (27), x_{n_k} is a Cauchy sequence in V. Since V is complete, it follows that $x \in V$ and $\lim_{k\to\infty} \|x_{n_k} - x\|_V = 0$. Hence

$$\|x\|_V = 1, \qquad Dx = 0, \qquad \pi_D(x) = 0.$$

Such a vector cannot exist, and this contradiction proves (28). The second assertion in (ii) is an obvious consequence of (26).

We prove that (ii) implies (iii). By assumption, the composition of $\pi_D : H \to H$ with the inclusion $V \hookrightarrow H$ is a compact operator. Hence it follows from standard arguments in functional analysis that every operator that satisfies (28) has a finite dimensional kernel and a closed range (e.g. [**19**, Appendix A]). Now the second assertion in (ii) shows that the Q-orthogonal complement of the image of D agrees with the kernel of D. Hence the cokernel of D is finite dimensional and has the same dimension as the kernel of D. This proves (iii).

We prove that (iii) implies (i). If D is Fredholm, then the operator $V \to H \oplus \ker D : x \mapsto (Dx, \pi_D(x))$ is injective and has a closed range. Hence (28) follows from the open mapping theorem. Hence D satisfies (27). To prove (26), suppose first that D is bijective. Suppose that $x \in H$ satisfies

$$\sup_{0 \ne y \in V} \frac{|\langle x, QDy\rangle_H|}{\|y\|_H} < \infty$$

and choose $w \in H$ such that $\langle x, QDy\rangle_H = \langle w, Qy\rangle_H$ for $y \in V$. Define $\xi = D^{-1}w \in V$. Since D is Q-symmetric, we have

$$\langle x - \xi, QDy\rangle_H = \langle x, QDy\rangle_H - \langle D\xi, Qy\rangle_H = \langle x, QDy\rangle_H - \langle w, Qy\rangle_H = 0$$

for every $y \in V$. Since D is surjective, it follows that $x = \xi \in V$. This proves (i) under the assumption that D is bijective. In general, the kernel of D is contained in the Q-orthogonal complement of the image of D. Since D has index zero, both must be equal. Hence the identity on H is an isomorphism from the kernel of D to a complement of the image. This implies that $D + \varepsilon\mathbb{1} : V \to H$ is bijective for $\varepsilon > 0$ sufficiently small. By what we have just proved, the operator $D + \varepsilon\mathbb{1}$ satisfies (26), and hence so does D. This proves the lemma. □

The previous lemma shows that the set $\mathcal{S}(V, H; Q)$ is open (with respect to the operator norm) in the Banach space of Q-symmetric operators in $\mathcal{L}(V, H)$. Moreover, the inequality (28) is stable under small perturbations. Namely, if $D \in \mathcal{S}(V, H; Q)$ satisfies (28), then every operator $D' \in \mathcal{S}(V, H; Q)$ that is sufficiently close to D in the operator norm satisfies (28) with c_0 replaced by $2c_0$. (The proof is an exercise.) This shows that for every compact subset $\mathcal{K} \subset \mathcal{S}(V, H; Q)$ there exists a constant $c_0 > 0$ such that (28) holds for every $D \in \mathcal{K}$.

If $Q \in \mathcal{P}(H)$ then the restriction of Q to V will still be denoted by Q. Its adjoint is an operator on V^* and is an extension of the original operator under the inclusion $H \hookrightarrow V^*$. This extension will also be denoted by Q. If $D \in \mathcal{S}(V, H; Q)$, then the dual operator D^* can be thought of as an operator from $H \to V^*$. Since

D is Q-symmetric, the operator $Q^{-1}D^*Q : H \to V^*$ is an extension of $D : V \to H$. Let us denote this extension again by

$$D = Q^{-1}D^*Q : H \to V^*.$$

Thus D can be thought of both as an operator from $V \to H$ and as an operator from $H \to V^*$. The self-adjoint property (26) can then be expressed by saying that $x \in V$ if and only if $x \in H$ and $Dx \in H$.

PROPOSITION A.2. *Let $V \subset H$ be Hilbert spaces with a compact dense inclusion of norm less than or equal to 1. Suppose that the functions*

$$Q, J : \mathbb{R}^2 \to \mathcal{L}(H), \qquad D : \mathbb{R}^2 \to \mathcal{L}(V, H)$$

are continuously differentiable with respect to the strong operator topology and satisfy $Q(s,t) \in \mathcal{P}(H)$, $J(s,t) \in \mathcal{J}(H; Q(s,t))$, $D(s,t) \in \mathcal{S}(V, H; Q(s,t))$, and

$$D(s,t)J(s,t) + J(s,t)D(s,t) = 0$$

for all $(s,t) \in \mathbb{R}^2$. Suppose further that there exist constants $c_0, c_1, c_2 \geq 1$ and $\delta > 0$ such that, for every $(s,t) \in \mathbb{R}^2$, (28) holds for $D = D(s,t)$ and

$$\delta \|x\|_H^2 \leq \langle x, Qx \rangle_H \leq \delta^{-1} \|x\|_H^2, \tag{29}$$

$$\|Dx\|_H + \|(\partial_s D)x\|_H + \|(\partial_t D)x\|_H \leq c_1 \|x\|_V, \tag{30}$$

$$\|(\partial_s J)x\|_H + \|(\partial_t J)x\|_H + \|(\partial_s Q)x\|_H + \|(\partial_t Q)x\|_H \leq c_2 \|x\|_H. \tag{31}$$

Then, for every $p \geq 2$, there exist constants $c \geq 1$ and $\varepsilon_0 > 0$ (depending only on δ, c_0, c_1, c_2, and p) such that the following holds. If $\xi \in L^p(\mathbb{R}^2, V) \cap W^{1,p}(\mathbb{R}^2, H)$ and $\tilde{\xi} \in L^p(\mathbb{R}^2, H)$ satisfy

$$\partial_s \xi(s,t) + J(\varepsilon s, \varepsilon t)\partial_t \xi(s,t) + D(\varepsilon s, \varepsilon t)\xi(s,t) = \tilde{\xi}(s,t) \tag{32}$$

for some $\varepsilon \in (0, \varepsilon_0)$, then

$$\int_{\mathbb{R}^2} \|\xi\|_H^p \leq c \int_{\mathbb{R}^2} \left(\|\tilde{\xi}\|_H^p + \|\pi_D(\xi)\|_H^p \right). \tag{33}$$

PROOF. Suppose, without loss of generality, that $\xi : \mathbb{R}^2 \to V$ is twice continuously differentiable and has compact support. Then the function $\tilde{\xi} : \mathbb{R}^2 \to H$, given by (32), is continuously differentiable and has compact support. It satisfies the equation

$$\partial_s \partial_s \xi + \partial_t \partial_t \xi - D^2 \xi = f, \tag{34}$$

where $f : \mathbb{R}^2 \to V^*$ is given by

$$f = \partial_s \tilde{\xi} - J\partial_t \tilde{\xi} - D\tilde{\xi} + \varepsilon(\partial_t J)\partial_s \xi - \varepsilon(\partial_s J)\partial_t \xi + \varepsilon(\partial_t(JD) - \partial_s D)\xi. \tag{35}$$

Here J and D are evaluated at the point $(\varepsilon s, \varepsilon t) \in \mathbb{R}^2$. Take the inner product with $\|\xi\|_Q^{p-2} Q\xi$ and compute

$$\begin{aligned}
\int_{\mathbb{R}^2} \|\xi\|_Q^{p-2} \langle Q\xi, f\rangle_{V,V^*} &= \int_{\mathbb{R}^2} \|\xi\|_Q^{p-2} \langle Q\xi, D^2\xi - \partial_s\partial_s\xi - \partial_t\partial_t\xi\rangle_{V,V^*} \\
&= \int_{\mathbb{R}^2} \|\xi\|_Q^{p-2} \left(\|D\xi\|_Q^2 + \|\partial_s\xi\|_Q^2 + \|\partial_t\xi\|_Q^2\right) \\
&\quad + \varepsilon \int_{\mathbb{R}^2} \|\xi\|_Q^{p-2} \left(\langle(\partial_s Q)\xi, \partial_s\xi\rangle_H + \langle(\partial_t Q)\xi, \partial_t\xi\rangle_H\right) \\
&\quad + \int_{\mathbb{R}^2} \left(\partial_s \|\xi\|_Q^{p-2}\right) \langle \xi, \partial_s\xi\rangle_Q \\
&\quad + \int_{\mathbb{R}^2} \left(\partial_t \|\xi\|_Q^{p-2}\right) \langle \xi, \partial_t\xi\rangle_Q \\
&= \int_{\mathbb{R}^2} \|\xi\|_Q^{p-2} \left(\|D\xi\|_Q^2 + \|\partial_s\xi\|_Q^2 + \|\partial_t\xi\|_Q^2\right) \\
&\quad + \varepsilon \int_{\mathbb{R}^2} \|\xi\|_Q^{p-2} \left(\langle(\partial_s Q)\xi, \partial_s\xi\rangle_H + \langle(\partial_t Q)\xi, \partial_t\xi\rangle_H\right) \\
&\quad + (p-2) \int_{\mathbb{R}^2} \|\xi\|_Q^{p-4} \left(\langle \xi, \partial_s\xi\rangle_Q^2 + \langle \xi, \partial_t\xi\rangle_Q^2\right) \\
&\quad + \frac{\varepsilon(p-2)}{2} \int_{\mathbb{R}^2} \|\xi\|_Q^{p-4} \langle(\partial_s Q)\xi, \xi\rangle_H \langle \xi, \partial_s\xi\rangle_Q \\
&\quad + \frac{\varepsilon(p-2)}{2} \int_{\mathbb{R}^2} \|\xi\|_Q^{p-4} \langle(\partial_t Q)\xi, \xi\rangle_H \langle \xi, \partial_t\xi\rangle_Q .
\end{aligned}$$

Let $c_3 := (p-1)c_2/\delta^2$. Then, by (29) and (31),

$$\begin{aligned}
&\int_{\mathbb{R}^2} \|\xi\|_Q^{p-2} \left(\|D\xi\|_Q^2 + \|\partial_s\xi\|_Q^2 + \|\partial_t\xi\|_Q^2\right) \\
&\qquad \le \int_{\mathbb{R}^2} \|\xi\|_Q^{p-2} \langle Q\xi, \tilde{\xi}\rangle_{V,V^*} + \varepsilon c_3 \int_{\mathbb{R}^2} \|\xi\|_Q^{p-1} \left(\|\partial_s\xi\|_Q + \|\partial_t\xi\|_Q\right).
\end{aligned}$$

Since $\|x\|_H \le \|x\|_V$ for every $x \in V$, it follows from (28) and (29) that

$$\|\xi\|_Q \le \delta^{-1} \|\xi\|_V \le c_0\delta^{-2} \left(\|D\xi\|_Q + \|\pi_D(\xi)\|_Q\right). \tag{36}$$

Let $c_4 := c_0c_3/\delta^2$. Then the last two inequalities show that

$$\begin{aligned}
&\int_{\mathbb{R}^2} \|\xi\|_Q^{p-2} \left(\|D\xi\|_Q^2 + \|\partial_s\xi\|_Q^2 + \|\partial_t\xi\|_Q^2\right) \\
&\qquad \le \int_{\mathbb{R}^2} \|\xi\|_Q^{p-2} \langle Q\xi, f\rangle_{V,V^*} \\
&\qquad\quad + \varepsilon c_4 \int_{\mathbb{R}^2} \|\xi\|_Q^{p-2} \left(\|\partial_s\xi\|_Q + \|\partial_t\xi\|_Q\right) \left(\|D\xi\|_Q + \|\pi_D(\xi)\|_Q\right) \\
&\qquad \le \int_{\mathbb{R}^2} \|\xi\|_Q^{p-2} \langle Q\xi, f\rangle_{V,V^*} \\
&\qquad\quad + \varepsilon c_4 \int_{\mathbb{R}^2} \|\xi\|_Q^{p-2} \left(\|D\xi\|_Q^2 + \|\partial_s\xi\|_Q^2 + \|\partial_t\xi\|_Q^2 + \|\pi_D(\xi)\|_Q^2\right).
\end{aligned}$$

With $\varepsilon c_4 \le 1/2$ this gives

$$\begin{aligned}&\int_{\mathbb{R}^2} \|\xi\|_Q^{p-2} \left(\|D\xi\|_Q^2 + \|\partial_s\xi\|_Q^2 + \|\partial_t\xi\|_Q^2\right) \\ &\qquad \le 2\int_{\mathbb{R}^2} \|\xi\|_Q^{p-2} \left(\langle Q\xi, f\rangle_{V,V^*} + \varepsilon c_4 \|\pi_D(\xi)\|_Q^2\right).\end{aligned} \tag{37}$$

Now recall from (35) that $f = g + h$, where $g : \mathbb{R}^2 \to V^*$ is given by

$$g = \partial_s\tilde{\xi} - J\partial_t\tilde{\xi} - D\tilde{\xi}$$

and $h : \mathbb{R}^2 \to H$ is given by

$$h = \varepsilon((\partial_t J)\partial_s\xi - (\partial_s J)\partial_t\xi + (\partial_t J)D\xi + J(\partial_t D)\xi - (\partial_s D)\xi).$$

Let $c_5 := (c_2 + c_0 c_1)/\delta^2$. Then, by (29), (30), (31), and (36),

$$\begin{aligned}\|h\|_Q &\le \varepsilon c_2\delta^{-2}\left(\|D\xi\|_Q + \|\partial_s\xi\|_Q + \|\partial_t\xi\|_Q\right) + \varepsilon c_1\delta^{-1}\|\xi\|_V \\ &\le \varepsilon c_5\left(\|D\xi\|_Q + \|\partial_s\xi\|_Q + \|\partial_t\xi\|_Q + \|\pi_D(\xi)\|_Q\right).\end{aligned}$$

Hence, by (36),

$$\begin{aligned}&\int_{\mathbb{R}^2} \|\xi\|_Q^{p-2}\langle Q\xi, h\rangle_H \\ &\qquad \le \int_{\mathbb{R}^2} \|\xi\|_Q^{p-1}\|h\|_Q \\ &\qquad \le \varepsilon c_5 \int_{\mathbb{R}^2} \|\xi\|_Q^{p-1}\left(\|D\xi\|_Q + \|\partial_s\xi\|_Q + \|\partial_t\xi\|_Q + \|\pi_D(\xi)\|_Q\right) \\ &\qquad \le \varepsilon c_0 c_5\delta^{-2}\int_{\mathbb{R}^2} \|\xi\|_Q^{p-2}\left(\|D\xi\|_Q + \|\pi_D(\xi)\|_Q\right) \\ &\qquad\qquad \times \left(\|D\xi\|_Q + \|\partial_s\xi\|_Q + \|\partial_t\xi\|_Q + \|\pi_D(\xi)\|_Q\right) \\ &\qquad \le 3\varepsilon c_0 c_5\delta^{-2}\int_{\mathbb{R}^2} \|\xi\|_Q^{p-2}\left(\|D\xi\|_Q^2 + \|\partial_s\xi\|_Q^2 + \|\partial_t\xi\|_Q^2 + \|\pi_D(\xi)\|_Q^2\right).\end{aligned}$$

Suppose that $3\varepsilon c_0 c_5\delta^{-2} \le 1/4$, and let $c_6 := c_4 + 3c_0c_5/\delta^2$. Then, inserting the last inequality into (37), we obtain

$$\begin{aligned}&\int_{\mathbb{R}^2} \|\xi\|_Q^{p-2} \left(\|D\xi\|_Q^2 + \|\partial_s\xi\|_Q^2 + \|\partial_t\xi\|_Q^2\right) \\ &\qquad \le 4\int_{\mathbb{R}^2} \|\xi\|_Q^{p-2} \left(\langle Q\xi, g\rangle_{V,V^*} + \varepsilon c_6 \|\pi_D(\xi)\|_Q^2\right).\end{aligned} \tag{38}$$

Now

$$\begin{aligned}\int_{\mathbb{R}^2} \|\xi\|_Q^{p-2}\langle Q\xi, g\rangle_{V,V^*} &= \int_{\mathbb{R}^2} \|\xi\|_Q^{p-2}\langle Q\xi, -D\tilde{\xi} + \partial_s\tilde{\xi} - J\partial_t\tilde{\xi}\rangle_{V,V^*} \\ &= \int_{\mathbb{R}^2} \|\xi\|_Q^{p-2}\left(-\langle D\xi, \tilde{\xi}\rangle_Q - \langle\partial_s\xi, \tilde{\xi}\rangle_Q + \langle\partial_t\xi, J\tilde{\xi}\rangle_Q\right) \\ &\quad - \varepsilon\int_{\mathbb{R}^2} \|\xi\|_Q^{p-2}\langle(\partial_s Q)\xi, \tilde{\xi}\rangle_H \\ &\quad + \varepsilon\int_{\mathbb{R}^2} \|\xi\|_Q^{p-2}\left(\langle(\partial_t Q)\xi, J\tilde{\xi}\rangle_H - \langle\xi, (\partial_t J)\tilde{\xi}\rangle_Q\right)\end{aligned}$$

$$
\begin{aligned}
&- \int_{\mathbb{R}^2} \left(\partial_s \|\xi\|_Q^{p-2}\right) \langle \xi, \tilde{\xi}\rangle_Q \\
&+ \int_{\mathbb{R}^2} \left(\partial_t \|\xi\|_Q^{p-2}\right) \langle \xi, J\tilde{\xi}\rangle_Q \\
&= \int_{\mathbb{R}^2} \|\xi\|_Q^{p-2} \left(-\langle D\xi, \tilde{\xi}\rangle_Q - \langle \partial_s\xi, \tilde{\xi}\rangle_Q + \langle \partial_t\xi, J\tilde{\xi}\rangle_Q\right) \\
&- \varepsilon \int_{\mathbb{R}^2} \|\xi\|_Q^{p-2} \langle (\partial_s Q)\xi, \tilde{\xi}\rangle_H \\
&+ \varepsilon \int_{\mathbb{R}^2} \|\xi\|_Q^{p-2} \left(\langle (\partial_t Q)\xi, J\tilde{\xi}\rangle_H + \langle \xi, (\partial_t J)\tilde{\xi}\rangle_Q\right) \\
&- (p-2) \int_{\mathbb{R}^2} \|\xi\|_Q^{p-4} \langle \xi, \partial_s\xi\rangle_Q \langle \xi, \tilde{\xi}\rangle_Q \\
&- \frac{\varepsilon(p-2)}{2} \int_{\mathbb{R}^2} \|\xi\|_Q^{p-4} \langle (\partial_s Q)\xi, \xi\rangle_H \langle \xi, \tilde{\xi}\rangle_Q \\
&+ (p-2) \int_{\mathbb{R}^2} \|\xi\|_Q^{p-4} \langle \xi, \partial_t\xi\rangle_Q \langle \xi, J\tilde{\xi}\rangle_Q \\
&+ \frac{\varepsilon(p-2)}{2} \int_{\mathbb{R}^2} \|\xi\|_Q^{p-4} \langle (\partial_t Q)\xi, \xi\rangle_H \langle \xi, J\tilde{\xi}\rangle_Q.
\end{aligned}
$$

Let $c_7 := (p+1)c_2/\delta^2$. Then, by (29) and (31),

$$
\begin{aligned}
\int_{\mathbb{R}^2} \|\xi\|_Q^{p-2} \langle Q\xi, g\rangle_{V,V^*} &\le (p-1) \int_{\mathbb{R}^2} \|\xi\|_Q^{p-2} \|\tilde{\xi}\|_Q \left(\|D\xi\|_Q + \|\partial_s\xi\|_Q + \|\partial_t\xi\|_Q\right) \\
&+ \varepsilon c_7 \int_{\mathbb{R}^2} \|\xi\|_Q^{p-1} \|\tilde{\xi}\|_Q.
\end{aligned}
$$

Inserting this inequality into (38) and using (36), we obtain

$$
\begin{aligned}
&\int_{\mathbb{R}^2} \|\xi\|_Q^{p-2} \left(\|D\xi\|_Q^2 + \|\partial_s\xi\|_Q^2 + \|\partial_t\xi\|_Q^2\right) \\
&\le 4(p-1) \int_{\mathbb{R}^2} \|\xi\|_Q^{p-2} \|\tilde{\xi}\|_Q \left(\|D\xi\|_Q + \|\partial_s\xi\|_Q + \|\partial_t\xi\|_Q\right) \\
&\quad + 4\varepsilon c_7 \int_{\mathbb{R}^2} \|\xi\|_Q^{p-1} \|\tilde{\xi}\|_Q + 4\varepsilon c_6 \int_{\mathbb{R}^2} \|\xi\|_Q^{p-2} \|\pi_D(\xi)\|_Q^2 \\
&\le 4(p-1) \int_{\mathbb{R}^2} \|\xi\|_Q^{p-2} \|\tilde{\xi}\|_Q \left(\|D\xi\|_Q + \|\partial_s\xi\|_Q + \|\partial_t\xi\|_Q\right) \\
&\quad + 4\varepsilon c_0 c_7 \delta^{-2} \int_{\mathbb{R}^2} \|\xi\|_Q^{p-2} \|\tilde{\xi}\|_Q \left(\|D\xi\|_Q + \|\pi_D(\xi)\|_Q\right) \\
&\quad + 4\varepsilon c_6 \int_{\mathbb{R}^2} \|\xi\|_Q^{p-2} \|\pi_D(\xi)\|_Q^2.
\end{aligned}
$$

Suppose that $\varepsilon c_0 c_7/\delta^2 \le 1$ and let $c_8 := 4c_0c_7/\delta^2$. Then

$$
\begin{aligned}
&\int_{\mathbb{R}^2} \|\xi\|_Q^{p-2} \left(\|D\xi\|_Q^2 + \|\partial_s\xi\|_Q^2 + \|\partial_t\xi\|_Q^2\right) \\
&\le 4p \int_{\mathbb{R}^2} \|\xi\|_Q^{p-2} \|\tilde{\xi}\|_Q \left(\|D\xi\|_Q + \|\partial_s\xi\|_Q + \|\partial_t\xi\|_Q\right) \\
&\quad + \frac{c_8}{2} \int_{\mathbb{R}^2} \|\xi\|_Q^{p-2} \|\tilde{\xi}\|_Q^2 + \frac{8\varepsilon c_6 + \varepsilon^2 c_8}{2} \int_{\mathbb{R}^2} \|\xi\|_Q^{p-2} \|\pi_D(\xi)\|_Q^2
\end{aligned}
$$

$$\leq \frac{1}{2}\int_{\mathbb{R}^2} \|\xi\|_Q^{p-2}\left(\|D\xi\|_Q^2 + \|\partial_s\xi\|_Q^2 + \|\partial_t\xi\|_Q^2\right)$$
$$+ \frac{c_8 + 48p^2}{2}\int_{\mathbb{R}^2} \|\xi\|_Q^{p-2}\|\tilde{\xi}\|_Q^2 + \frac{8\varepsilon c_6 + \varepsilon^2 c_8}{2}\int_{\mathbb{R}^2} \|\xi\|_Q^{p-2}\|\pi_D(\xi)\|_Q^2$$

Let $c_9 := \max\{c_8 + 48p^2, 8c_6 + 4\}$. Since $\varepsilon c_8 \leq 4$, we obtain

$$\int_{\mathbb{R}^2} \|\xi\|_Q^{p-2}\|D\xi\|_Q^2 \leq c_9 \int_{\mathbb{R}^2} \|\xi\|_Q^{p-2}\left(\|\tilde{\xi}\|_Q^2 + \varepsilon\|\pi_D(\xi)\|_Q^2\right). \tag{39}$$

Let $c_{10} := 2(c_0/\delta^2)^2(c_9 + 1)$. Then, by (36) and (39),

$$\begin{aligned}\int_{\mathbb{R}^2} \|\xi\|_Q^p &\leq 2(c_0\delta^{-2})^2 \int_{\mathbb{R}^2} \|\xi\|_Q^{p-2}\left(\|D\xi\|_Q^2 + \|\pi_D(\xi)\|_Q^2\right)\\ &\leq c_{10}\int_{\mathbb{R}^2} \|\xi\|_Q^{p-2}\left(\|\tilde{\xi}\|_Q^2 + \|\pi_D(\xi)\|_Q^2\right).\end{aligned}$$

By Hölder's inequality, $\int_{\mathbb{R}^2} \|\xi\|_Q^{p-2}\|\tilde{\xi}\|_Q^2 \leq (\int_{\mathbb{R}^2}\|\xi\|_Q^p)^{1-2/p}(\int_{\mathbb{R}^2}\|\tilde{\xi}\|_Q^p)^{2/p}$. Hence

$$\left(\int_{\mathbb{R}^2}\|\xi\|_Q^p\right)^{2/p} \leq c_{10}\left(\left(\int_{\mathbb{R}^2}\|\tilde{\xi}\|_Q^p\right)^{2/p} + \left(\int_{\mathbb{R}^2}\|\pi_D(\xi)\|_Q^p\right)^{2/p}\right).$$

This proves the proposition. □

Appendix B. Estimates on a Riemann surface

In this section we collect some standard estimates for connections over Riemann surfaces. We include the proofs for the sake of completeness. Let Σ be a compact oriented Riemann surface with volume form ω. Denote by $\mathcal{J}(\Sigma)$ the space of complex structures on Σ compatible with the orientation. Let $P \to \Sigma$ be a principal $\mathrm{SO}(3)$-bundle with nonzero second Stiefel–Whitney class. Let $\mathcal{A}(P) \subset \Omega^1(\Sigma, \mathfrak{g}_P)$ denote the space of connection 1-forms on P, $\mathcal{A}^{\mathrm{flat}}(P) \subset \mathcal{A}(P)$ the submanifold of flat connections, and $\mathcal{G}(P) \subset \mathrm{Map}(P, \mathrm{G})$ the identity component of the gauge group.

LEMMA B.1. *Fix a complex structure $J_0 \in \mathcal{J}(\Sigma)$ and a connection $A_0 \in \mathcal{A}(P)$. Then, for every $C > 0$ and every $p > 1$, there exists a constant $c = c(C, p, J_0, A_0) \geq 1$ such that, if $J \in \mathcal{J}(\Sigma)$ and $A \in \mathcal{A}(P)$ satisfy*

$$\|J\|_{C^1(\Sigma,J_0)} + \|A - A_0\|_{L^\infty(\Sigma,J_0)} \leq C, \tag{40}$$

then, for every $\alpha \in \Omega^1(\Sigma, \mathfrak{g}_P)$,

$$c^{-1}\|\alpha\|_{L^p(\Sigma,J_0)} \leq \|\alpha\|_{L^p(\Sigma,J)} \leq c\|\alpha\|_{L^p(\Sigma,J_0)}, \tag{41}$$

$$\|\nabla_{A_0}\alpha\|_{L^p(\Sigma,J_0)} \leq c\left(\|d_A\alpha\|_{L^p(\Sigma)} + \|d_A(\alpha\circ J)\|_{L^p(\Sigma)} + \|\alpha\|_{L^p(\Sigma,J)}\right). \tag{42}$$

Here ∇_{A_0} denotes the covariant derivative with respect to the connection on $T^*\Sigma\otimes\mathfrak{g}_P$ determined by A_0 and the metric $\omega(\cdot, J_0\cdot)$.

PROOF. For a fixed complex structure $J = J_1$ the estimate (41) is obvious, and it holds with a uniform constant c in some C^0-neighbourhood of J_1. By the Arzéla–Ascoli theorem, we can cover the set of all $J \in \mathcal{J}(\Sigma)$ with $\|J\|_{C^1(\Sigma,J_0)} \leq C$ by finitely many such neighbourhoods, and this proves (41).

We prove (42). For $A = A_0$ and a fixed complex structure $J = J_1$ this follows from the Calderón–Zygmund inequality. Now, for every $J_1 \in \mathcal{J}(\Sigma)$ there exists a constant $c_1 > 0$ such that, for every $\alpha \in \Omega^1(\Sigma, \mathfrak{g}_P)$,

$$\begin{aligned}\|\mathrm{d}_{A_0}(\alpha \circ J) - \mathrm{d}_{A_0}(\alpha \circ J_1)\|_{L^p(\Sigma)} &\le c_1 \|J - J_1\|_{C^1(\Sigma,J_0)} \|\alpha\|_{L^p(\Sigma,J_0)} \\ &\quad + c_1 \|J - J_1\|_{L^\infty(\Sigma,J_0)} \|\nabla_{A_0}\alpha\|_{L^p(\Sigma,J_0)}.\end{aligned}$$

Hence there exist constants $\delta = \delta(J_1) > 0$ and $c_2 = c_2(J_1) > 0$ such that, if $\|J - J_1\|_{L^\infty(\Sigma,J_0)} < \delta$, then

$$\begin{aligned}\|\nabla_{A_0}\alpha\|_{L^p(\Sigma,J_0)} &\le c_2 \left(\|\mathrm{d}_{A_0}\alpha\|_{L^p(\Sigma)} + \|\mathrm{d}_{A_0}(\alpha \circ J)\|_{L^p(\Sigma)}\right) \\ &\quad + c_2 \left(1 + \|J - J_1\|_{C^1(\Sigma,J_0)}\right) \|\alpha\|_{L^p(\Sigma,J_0)}.\end{aligned}$$

By the Arzéla–Ascoli theorem, we can cover the set of all $J \in \mathcal{J}(\Sigma)$ with $\|J\|_{C^1(\Sigma,J_0)} \le C$ by finitely many such δ-neighbourhoods. This proves (42) for $A = A_0$. Since $\mathrm{d}_A\alpha - \mathrm{d}_{A_0}\alpha = [(A - A_0) \wedge \alpha]$, the estimate (42) holds, with a larger constant c, for every pair (A, J) that satisfies (40). This proves the lemma. □

LEMMA B.2. *Fix a complex structure* $J_0 \in \mathcal{J}(\Sigma)$ *and a connection* $A_0 \in \mathcal{A}(P)$. *Then, for every* $\delta > 0$, $C > 0$, *and* $p \ge 2$, *there is a constant* $c = c(\delta, C, p, J_0, A_0) \ge 1$ *such that, if* $J \in \mathcal{J}(\Sigma)$ *and* $A \in \mathcal{A}(P)$ *satisfy (40), then, for every* $\phi \in \Omega^0(\Sigma, \mathfrak{g}_P)$ *and every* $\alpha \in \Omega^1(\Sigma, \mathfrak{g}_P)$,

$$\|\phi\|_{L^p(\Sigma)}^p \le \delta \|\mathrm{d}_A\phi\|_{L^p(\Sigma,J)}^p + c\|\phi\|_{L^2(\Sigma)}^p, \tag{43}$$

$$\|\alpha\|_{L^p(\Sigma,J)}^p \le \delta \left(\|\mathrm{d}_A\alpha\|_{L^p(\Sigma)}^p + \|\mathrm{d}_A(\alpha \circ J)\|_{L^p(\Sigma)}^p\right) + c\|\alpha\|_{L^2(\Sigma,J)}^p. \tag{44}$$

PROOF. For $p = 2$ there is nothing to prove. For $p > 2$ this is a standard estimate in partial differential equations. For the sake of completeness we give a proof. We show first that, if $u : \mathbb{R}^2 \to \mathbb{R}$ is continuously differentiable, then

$$|u(x)| \le \frac{\varepsilon^{1-2/p}}{(2\pi)^{1/p}} \left(\frac{p-1}{p-2}\right)^{1-1/p} \|\nabla u\|_{L^p(B_\varepsilon)} + \frac{2}{\pi\varepsilon^2}\|u\|_{L^1(B_\varepsilon)}, \tag{45}$$

where

$$B_\varepsilon = \left\{y \in \mathbb{R}^2 \,|\, |y - x| < \varepsilon\right\}.$$

To see this observe that, for $|\xi| = 1$ and $0 < t \le \varepsilon$,

$$\begin{aligned}|u(x)| &\le |u(x + t\xi)| + \int_0^\varepsilon |\nabla u(x + s\xi)|\,\mathrm{d}s \\ &\le |u(x + t\xi)| + \left(\int_0^\varepsilon s^{-1/(p-1)}\,\mathrm{d}s\right)^{1-1/p} \left(\int_0^\varepsilon |\nabla u(x + s\xi)|^p\, s\,\mathrm{d}s\right)^{1/p} \\ &= |u(x + t\xi)| + \varepsilon^{1-2/p}\left(\frac{p-1}{p-2}\right)^{1-1/p} \left(\int_0^\varepsilon |\nabla u(x + s\xi)|^p\, s\,\mathrm{d}s\right)^{1/p}.\end{aligned}$$

Here the second estimate follows from Hölder's inequality. Now integrate over $\xi \in S^1$ and use Hölder's inequality again to obtain

$$|u(x)| \le \frac{1}{2\pi}\int_0^{2\pi} |u(x + te^{i\theta})|\,\mathrm{d}\theta + \frac{\varepsilon^{1-2/p}}{(2\pi)^{1/p}}\left(\frac{p-1}{p-2}\right)^{1-1/p} \|\nabla u\|_{L^p(B_\varepsilon)}.$$

Now integrate over the interval $\varepsilon/2 \le t \le \varepsilon$ to obtain (45). It follows from (45) that there exists a constant $c_0 > 0$ such that

$$\|\phi\|_{L^p(\Sigma)} \le \varepsilon^{1-2/p} \|\mathrm{d}_{A_0}\phi\|_{L^p(\Sigma,J_0)} + c_0\varepsilon^{-1} \|\phi\|_{L^2(\Sigma)}, \tag{46}$$

$$\|\alpha\|_{L^p(\Sigma,J_0)} \le \varepsilon^{1-2/p} \|\nabla_{A_0}\alpha\|_{L^p(\Sigma,J_0)} + c_0\varepsilon^{-1} \|\alpha\|_{L^2(\Sigma)} \tag{47}$$

for $\phi \in \Omega^0(\Sigma, \mathfrak{g}_P)$, $\alpha \in \Omega^1(\Sigma, \mathfrak{g}_P)$, and $0 < \varepsilon \le 1$. Now (44) follows from (47) and (42) if $\varepsilon > 0$ is chosen sufficiently small. To prove (43) note that, by Lemma B.1,

$$\|\mathrm{d}_{A_0}\phi\|_{L^p(\Sigma,J_0)} \le c\,\|\mathrm{d}_{A_0}\phi\|_{L^p(\Sigma,J)} \le c\,\|\mathrm{d}_A\phi\|_{L^p(\Sigma,J)} + c'\,\|\phi\|_{L^p(\Sigma)}$$

Combining this with (46) gives (43), provided that $\varepsilon > 0$ is chosen sufficiently small. This proves the lemma. □

LEMMA B.3. *Fix a complex structure $J_0 \in \mathcal{J}(\Sigma)$. Then, for every $C > 0$ and every $p \ge 2$, there exists a constant $c = c(C, p, J_0) \ge 1$ such that, if $J \in \mathcal{J}(\Sigma)$ satisfies*

$$\|J\|_{C^1(\Sigma,J_0)} \le C, \tag{48}$$

and $A \in \mathcal{A}^{\mathrm{flat}}(P)$, then, for every $\phi \in \Omega^0(\Sigma, \mathfrak{g}_P)$ and every $\alpha \in \Omega^1(\Sigma, \mathfrak{g}_P)$,

$$\|\phi\|_{L^p(\Sigma)} \le c\,\|\mathrm{d}_A\phi\|_{L^p(\Sigma,J)}, \tag{49}$$

$$\|\alpha\|_{L^p(\Sigma,J)} \le c\left(\|\mathrm{d}_A\alpha\|_{L^p(\Sigma)} + \|\mathrm{d}_A(\alpha\circ J)\|_{L^p(\Sigma)} + \|\pi_A(\alpha)\|_{L^p(\Sigma,J)}\right). \tag{50}$$

PROOF. We prove (49). If $A_1 \in \mathcal{A}^{\mathrm{flat}}(P)$, then $\mathrm{d}_{A_1} : \Omega^0(\Sigma, \mathfrak{g}_P) \to \Omega^1(\Sigma, \mathfrak{g}_P)$ is injective. Hence there exists a constant $c = c(A_1) > 0$ such that (49) holds for $A = A_1$ and $J = J_0$. Hence

$$\|\phi\|_{L^p(\Sigma)} \le c\left(\|\mathrm{d}_A\phi\|_{L^p(\Sigma,J_0)} + \|A - A_1\|_{L^\infty(\Sigma,J_0)}\,\|\phi\|_{L^p(\Sigma)}\right).$$

This shows that (49) holds, with $J = J_0$ and a uniform constant c, in some C^0-neighbourhood of A_1. Cover the set

$$\mathcal{A}^{\mathrm{flat}}_C(P) = \left\{\mathcal{A}^{\mathrm{flat}}(P) \mid \|A - A_0\|_{C^2(\Sigma,J_0)} \le C\right\}$$

by finitely many such neighbourhoods to obtain (49) with $J = J_0$ for some constant $c > 0$, every $A \in \mathcal{A}^{\mathrm{flat}}_C(P)$, and every $\phi \in \Omega^0(\Sigma, \mathfrak{g}_p)$. Now use (41) to obtain (49) for any $A \in \mathcal{A}^{\mathrm{flat}}_C(P)$ and any $J \in \mathcal{J}(\Sigma)$ that satisfies (48). If C is sufficiently large then every flat connection is gauge equivalent to one in $\mathcal{A}^{\mathrm{flat}}_C(P)$, and this proves (49).

Next we prove (50) for $p = 2$. Write

$$\alpha = \pi_A(\alpha) + \mathrm{d}_A\zeta - (\mathrm{d}_A\eta)\circ J$$

for $\eta, \zeta \in \Omega^0(\Sigma, \mathfrak{g}_P)$. Since the three terms on the right are pairwise L^2-orthogonal, with respect to the metric determined by J, we have

$$\|\alpha\|^2_{L^2(\Sigma,J)} = \|\mathrm{d}_A\zeta\|^2_{L^2(\Sigma,J)} + \|\mathrm{d}_A\eta\|^2_{L^2(\Sigma,J)} + \|\pi_A(\alpha)\|^2_{L^2(\Sigma,J)}.$$

Since A is flat, we have $\mathrm{d}_A{}^{*_J}\mathrm{d}_A\eta = -*\mathrm{d}_A\alpha$. Taking the L^2-inner product with η and using the Cauchy–Schwarz inequality, we obtain

$$\|\mathrm{d}_A\eta\|^2_{L^2(\Sigma,J)} \le \|\eta\|_{L^2(\Sigma)}\,\|\mathrm{d}_A\alpha\|_{L^2(\Sigma)} \le c\,\|\mathrm{d}_A\eta\|_{L^2(\Sigma,J)}\,\|\mathrm{d}_A\alpha\|_{L^2(\Sigma)}.$$

The last inequality follows from (49) with $p = 2$. Hence

$$\|\mathrm{d}_A\eta\|_{L^2(\Sigma,J)} \le c\,\|\mathrm{d}_A\alpha\|_{L^2(\Sigma)}, \qquad \|\mathrm{d}_A\zeta\|_{L^2(\Sigma,J)} \le c\,\|\mathrm{d}_A(\alpha\circ J)\|_{L^2(\Sigma)}.$$

This proves (50) for $p=2$. If $A \in \mathcal{A}_C^{\mathrm{flat}}(P)$, then (50) for general p follows from the case $p=2$ and (44). Since there exists a constant $C>0$ such that every flat connection is gauge equivalent to one in $\mathcal{A}_C^{\mathrm{flat}}(P)$, this proves the lemma. □

LEMMA B.4. *Fix a complex structure $J_0 \in \mathcal{J}(\Sigma)$. Then, for every $C>0$ and every $p \geq 2$, there exists a constant $c = c(C,p,J_0) \geq 1$ such that, if $J \in \mathcal{J}(\Sigma)$ satisfies (48) and $A \in \mathcal{A}^{\mathrm{flat}}(P)$ then, for every $\phi \in \Omega^0(\Sigma, \mathfrak{g}_P)$ and every $\alpha \in \Omega^1(\Sigma, \mathfrak{g}_P)$,*

$$\|\mathrm{d}_A\phi\|_{L^p(\Sigma,J)} \leq c\left(\|\alpha\|_{L^p(\Sigma,J)} + \|\mathrm{d}_A((\mathrm{d}_A\phi)\circ J + \alpha)\|_{L^p(\Sigma)}\right). \tag{51}$$

PROOF. By (41) in Lemma B.1, it suffices to prove the inequality

$$\|\mathrm{d}_A\phi\|_{L^p(\Sigma,J_0)} \leq c\left(\|\alpha\|_{L^p(\Sigma,J_0)} + \|\mathrm{d}_A((\mathrm{d}_A\phi)\circ J + \alpha)\|_{L^p(\Sigma)}\right) \tag{52}$$

instead of (51). Let $\psi := *\mathrm{d}_A((\mathrm{d}_A\phi)\circ J + \alpha) \in \Omega^0(\Sigma, \mathfrak{g}_P)$. Then

$$\mathrm{d}_A{}^{*_J}\mathrm{d}_A\phi = \psi - *\mathrm{d}_A\alpha.$$

For $p=2$ the estimate follows by taking the inner product with ϕ and using (49). The general case can be reduced to $p=2$ via the Calderón–Zygmund inequality and (43). Hence, for every pair (J,A), there exists a constant $c = c(J,A) \geq 1$ such that (52) holds for every $\phi \in \Omega^0(\Sigma, \mathfrak{g}_P)$ and every $\alpha \in \Omega^1(\Sigma, \mathfrak{g}_P)$. We prove that the constant can be chosen independent of A. To see this, note that

$$\begin{aligned}
&\mathrm{d}_A((\mathrm{d}_A\phi)\circ J) - \mathrm{d}_{A_1}((\mathrm{d}_{A_1}\phi)\circ J) \\
&\quad = [(A-A_1)\wedge((\mathrm{d}_A\phi)\circ J)] + [(A-A_1)\circ J \wedge [(A-A_1),\phi]] \\
&\qquad + [\mathrm{d}_{A_1}((A-A_1)\circ J),\phi] - [(A-A_1)\circ J \wedge \mathrm{d}_A\phi].
\end{aligned}$$

Let $c_1 = c(J, A_1)$ and use (52) with $A = A_1$ to obtain

$$\begin{aligned}
\|\mathrm{d}_A\phi\|_{L^p(\Sigma,J_0)} &\leq \|A-A_1\|_{L^\infty(\Sigma,J_0)}\|\phi\|_{L^p(\Sigma)} \\
&\quad + c_1\left(\|\alpha\|_{L^p(\Sigma,J_0)} + \|\mathrm{d}_{A_1}((\mathrm{d}_{A_1}\phi)\circ J + \alpha)\|_{L^p(\Sigma)}\right) \\
&\leq c_1\left(1 + \|A-A_1\|_{L^\infty(\Sigma,J_0)}\right)\|\alpha\|_{L^p(\Sigma,J_0)} + c_1\|\mathrm{d}_A((\mathrm{d}_A\phi)\circ J + \alpha)\|_{L^p(\Sigma)} \\
&\quad + c_2\left(\|A-A_1\|^2_{L^\infty(\Sigma,J_0)} + \|A-A_1\|_{C^1(\Sigma,J_0)}\right)\|\mathrm{d}_A\phi\|_{L^p(\Sigma,J_0)}.
\end{aligned}$$

Here we have used the inequality $\|\phi\|_{L^p(\Sigma)} \leq c\|\mathrm{d}_A\phi\|_{L^p(\Sigma,J_0)}$. It follows that every flat connection A_1 has a C^1-neighbourhood in which (52) holds with $c = 2c_1(J,A_1)$. By the Arzéla–Ascoli theorem, cover the set $\mathcal{A}_C^{\mathrm{flat}}(P)$ by finitely many such neighbourhoods. Since (52) is gauge invariant, and every flat connection is gauge equivalent to one in $\mathcal{A}_C^{\mathrm{flat}}(P)$, there exists, for every $J \in \mathcal{J}(\Sigma)$, a constant $c_3 = c_3(J) > 0$ such that (52) holds with $c = c_3(J)$ for every flat connection A. Now apply (52) with $J = J_1$ to the pair $(\phi, \alpha + (\mathrm{d}_A\phi)\circ(J-J_1))$ to obtain

$$\begin{aligned}
\|\mathrm{d}_A\phi\|_{L^p(\Sigma,J_0)} \;&\leq\; c_3(J_1)\left(\|\alpha\|_{L^p(\Sigma,J_0)} + \|\mathrm{d}_A((\mathrm{d}_A\phi)\circ J + \alpha)\|_{L^p(\Sigma)}\right) \\
&\quad + c_3(J_1)\|J-J_1\|_{L^\infty(\Sigma,J_0)}\|\mathrm{d}_A\phi\|_{L^p(\Sigma,J_0)}.
\end{aligned}$$

Hence (52) holds with $c = 2c_3(J_1)$ whenever $c_3(J_1)\|J-J_1\|_{L^\infty(\Sigma,J_0)} \leq 1/2$. By the Arzéla–Ascoli theorem, cover the set of all $J \in \mathcal{J}(\Sigma)$ that satisfy (48) by finitely many such neighbourhoods. This proves the lemma. □

Appendix C. An L^p estimate

Continue the notation of the previous section. Let

$$\lambda : \mathbb{R}^2 \to (0,\infty), \qquad J : \mathbb{R}^2 \to \mathcal{J}(\Sigma), \qquad A : \mathbb{R}^2 \to \mathcal{A}(P)$$

be smooth functions. Suppose that, for some $A_0 \in \mathcal{A}(P)$ and $J_0 \in \mathcal{J}(\Sigma)$,

$$\sup_{s,t} \left(|\lambda(s,t)| + |1/\lambda(s,t)| + |\partial_s\lambda(s,t)| + |\partial_t\lambda(s,t)|\right) < \infty, \tag{53}$$

$$\sup_{s,t} \left(\|J(s,t)\|_{C^2(\Sigma)} + \|\partial_s J(s,t)\|_{C^1(\Sigma)} + \|\partial_t J(s,t)\|_{C^1(\Sigma)}\right) < \infty, \tag{54}$$

$$\sup_{s,t} \left(\|A(s,t) - A_0\|_{C^1(\Sigma)} + \|\partial_s A(s,t)\|_{L^\infty(\Sigma)} + \|\partial_s A(s,t)\|_{L^\infty(\Sigma)}\right) < \infty. \tag{55}$$

Here all norms are to be understood with respect to the metric induced by J_0. For a triple $\xi = (\alpha, \phi, \psi)$ of smooth functions $\phi, \psi : \mathbb{R}^2 \to \Omega^0(\Sigma, \mathfrak{g}_P)$ and $\alpha : \mathbb{R}^2 \to \Omega^1(\Sigma, \mathfrak{g}_P)$ with compact support we consider the equations

$$\begin{aligned} \tilde{\alpha} &= \partial_s\alpha - d_{A(\varepsilon s,\varepsilon t)}\phi + *_{\varepsilon s,\varepsilon t}(\partial_t\alpha - d_{A(\varepsilon s,\varepsilon t)}\psi), \\ \tilde{\phi} &= \partial_s\phi + \partial_t\psi + \lambda(\varepsilon s,\varepsilon t)^2 * d_{A(\varepsilon s,\varepsilon t)} *_{\varepsilon s\ \varepsilon t} \alpha, \\ \tilde{\psi} &= \partial_s\psi - \partial_t\phi + \lambda(\varepsilon s,\varepsilon t)^2 * d_{A(\varepsilon s,\varepsilon t)}\alpha. \end{aligned} \tag{56}$$

Here $*_{s,t} : \Omega^1(\Sigma, \mathfrak{g}_P) \to \Omega^1(\Sigma, \mathfrak{g}_P)$ and $* : \Omega^2(\Sigma, \mathfrak{g}_P) \to \Omega^0(\Sigma, \mathfrak{g}_P)$ denote the Hodge $*$-operators of the metric determined by ω and $J(s,t)$.

PROPOSITION C.1. *Let $\lambda : \mathbb{R}^2 \to (0,\infty)$, $J : \mathbb{R}^2 \to \mathcal{J}(\Sigma)$, and $A : \mathbb{R}^2 \to \mathcal{A}(P)$ be continuously differentiable functions that satisfy (53), (54), and (55). Then, for every $p \geq 2$, there exist constants $\varepsilon_0 > 0$ and $c_1 > 0$ such that the following holds. If ξ and $\tilde{\xi}$ have compact support and satisfy (56) for some $\varepsilon \in (0, \varepsilon_0)$, then*

$$\begin{aligned} \int_{\mathbb{R}^2} \Big(&\|d_A\alpha\|^p_{L^p(\Sigma)} + \|d_A(\alpha \circ J)\|^p_{L^p(\Sigma)} + \|\partial_s\alpha\|^p_{L^p(\Sigma,J)} + \|\partial_t\alpha\|^p_{L^p(\Sigma,J)} \\ &+ \|d_A\phi\|^p_{L^p(\Sigma,J)} + \|\partial_s\phi\|^p_{L^p(\Sigma)} + \|\partial_t\phi\|^p_{L^p(\Sigma)} \\ &+ \|d_A\psi\|^p_{L^p(\Sigma,J)} + \|\partial_s\psi\|^p_{L^p(\Sigma)} + \|\partial_t\psi\|^p_{L^p(\Sigma)}\Big) \\ \leq c_1 \int_{\mathbb{R}^2} &\left(\|\tilde{\alpha}\|^p_{L^p(\Sigma)} + \|\tilde{\phi}\|^p_{L^p(\Sigma)} + \|\tilde{\psi}\|^p_{L^p(\Sigma)} + \|\pi_A(\alpha)\|^p_{L^p(\Sigma,J)}\right). \end{aligned} \tag{57}$$

Here we abbreviate $A = A(\varepsilon s, \varepsilon t)$ and $J = J(\varepsilon s, \varepsilon t)$. Moreover, we use the identity $*_{\varepsilon s,\varepsilon t} = -\alpha \circ J$, and $\pi_{A(\varepsilon s,\varepsilon t)} : \Omega^1(\Sigma, \mathfrak{g}_P) \to \mathrm{H}^1_{A(\varepsilon s,\varepsilon t)}$ denotes the L^2-orthogonal projection with respect to the metric $*_{\varepsilon s,\varepsilon t}$.

LEMMA C.2. *Suppose λ, J, and A satisfy* (53), (54), and (55). *Then, for every $p \geq 2$, there exist constants $\varepsilon_0 > 0$ and $c_2 > 0$ such that the following holds. If ξ and $\tilde{\xi}$ have compact support and satisfy* (56) *for some $\varepsilon \in (0, \varepsilon_0)$, then*

$$\begin{aligned} &\int_{\mathbb{R}^2} \left(\|\alpha\|^p_{L^2(\Sigma,J)} + \|\phi\|^p_{L^2(\Sigma)} + \|\psi\|^p_{L^2(\Sigma)}\right) \\ &\qquad \leq c_2 \int_{\mathbb{R}^2} \left(\|\tilde{\alpha}\|^p_{L^2(\Sigma,J)} + \|\tilde{\psi}\|^p_{L^2(\Sigma)} + \|\tilde{\psi}\|^p_{L^2(\Sigma)} + \|\pi_A(\alpha)\|^p_{L^2(\Sigma,J)}\right). \end{aligned}$$

PROOF. Define

$$\begin{aligned} H &= L^2(\Sigma, T^*\Sigma \otimes \mathfrak{g}_P) \oplus L^2(\Sigma, \otimes\mathfrak{g}_P) \oplus L^2(\Sigma, \otimes\mathfrak{g}_P), \\ V &= W^{1,2}(\Sigma, T^*\Sigma \otimes \mathfrak{g}_P) \oplus W^{1,2}(\Sigma, \otimes\mathfrak{g}_P) \oplus W^{1,2}(\Sigma, \otimes\mathfrak{g}_P). \end{aligned}$$

Suppose that the inner product on H is determined by the reference metric $\langle\cdot,\cdot\rangle = \omega(\cdot, J_0\cdot)$, where $J_0 = J(0,0)$. Then the operators

$$\mathbf{Q}(s,t) = \begin{pmatrix} -*_0 *_{s,t} & 0 & 0 \\ 0 & \lambda(s,t)^{-2} & 0 \\ 0 & 0 & \lambda(s,t)^{-2} \end{pmatrix},$$

$$\mathbf{J}(s,t) = \begin{pmatrix} *_{s,t} & 0 & 0 \\ 0 & 0 & 1 \\ 0 & -1 & 0 \end{pmatrix},$$

$$\mathbf{D}(s,t) = \begin{pmatrix} 0 & -\mathrm{d}_{A(s,t)} & -*_{s,t}\,d_{A(s,t)} \\ \lambda(s,t)^2 * \mathrm{d}_{A(s,t)} *_{s,t} & 0 & 0 \\ \lambda(s,t)^2 * \mathrm{d}_{A(s,t)} & 0 & 0 \end{pmatrix}$$

satisfy the requirements of Proposition A.2. Explicitly, the inner product on H determined by $\mathbf{Q}(s,t)$ agrees with the one given by $\mathbf{J}(s,t)$ on 1-forms and with the standard metric multiplied by $\lambda(s,t)^{-2}$ on functions. Direct computation shows that

$$\mathbf{J}\mathbf{D} + \mathbf{D}\mathbf{J} = 0, \qquad \mathbf{Q}\mathbf{J} + \mathbf{J}^*\mathbf{Q} = 0, \qquad \mathbf{Q}\mathbf{D} - \mathbf{D}^*\mathbf{Q} = 0,$$

where the adjoint is to be understood with respect to the metric determined by ω and J_0. The kernel of $\mathbf{D}(s,t)$ consists of all triples

$$\xi = (\alpha, \phi, \psi) \in \Omega^1(\Sigma, \mathfrak{g}_P) \oplus \Omega^0(\Sigma, \mathfrak{g}_P) \oplus \Omega^0(\Sigma, \mathfrak{g}_P)$$

such that

$$\phi = \psi = 0, \qquad \alpha = \pi_{A(s,t)}(\alpha),$$

i.e., α is harmonic with respect to $A(s,t)$ and $*_{s,t}$. The conditions (53), (54), and (55) guarantee that that $\mathbf{Q}$, $\mathbf{J}$, and $\mathbf{D}$ satisfy (28), (29), (30), and (31). Note that ξ and $\tilde{\xi}$ satisfy (56) if and only if

$$\partial_s\xi + \mathbf{J}\partial_t\xi + \mathbf{D}\xi = \tilde{\xi}.$$

Here $\mathbf{J}$ and $\mathbf{D}$ are evaluated at the point $(\varepsilon s, \varepsilon t)$. Hence the assertion follows from Proposition A.2. □

LEMMA C.3. *Suppose λ, J, and A satisfy* (53), (54), *and* (55). *Then, for every $p \geq 2$, there exist constants $\varepsilon_0 > 0$ and $c_3 > 0$ such that the following holds. If ξ and $\tilde{\xi}$ have compact support and satisfy* (56) *for some $\varepsilon \in (0, \varepsilon_0)$, then (with $A = A(\varepsilon s, \varepsilon t)$ and $J = J(\varepsilon s, \varepsilon t)$)*

$$\begin{aligned}
&\int_{\mathbb{R}^2} \left(\|\mathrm{d}_A\alpha\|^p_{L^p(\Sigma)} + \|\mathrm{d}_A(\alpha\circ J)\|^p_{L^p(\Sigma)} + \|\partial_s\alpha\|^p_{L^p(\Sigma,J)} + \|\partial_t\alpha\|^p_{L^p(\Sigma,J)} \right) \\
&\quad + \int_{\mathbb{R}^2} \left(\|\mathrm{d}_A\phi\|^p_{L^p(\Sigma,J)} + \|\partial_s\phi\|^p_{L^p(\Sigma)} + \|\partial_t\phi\|^p_{L^p(\Sigma)} \right) \\
&\quad + \int_{\mathbb{R}^2} \left(\|\mathrm{d}_A\psi\|^p_{L^p(\Sigma,J)} + \|\partial_s\psi\|^p_{L^p(\Sigma)} + \|\partial_t\psi\|^p_{L^p(\Sigma)} \right) \\
&\leq c_3 \int_{\mathbb{R}^2} \left(\|\tilde{\alpha}\|^p_{L^p(\Sigma,J)} + \|\tilde{\phi}\|^p_{L^p(\Sigma)} + \|\tilde{\psi}\|^p_{L^p(\Sigma)} \right) \\
&\quad + c_3 \int_{\mathbb{R}^2} \left(\|\alpha\|^p_{L^p(\Sigma,J)} + \|\phi\|^p_{L^p(\Sigma)} + \|\psi\|^p_{L^p(\Sigma)} \right).
\end{aligned} \tag{58}$$

PROOF. The proof consists of three steps.

Step 1. *For every $C > 0$, $r > 0$, $p \ge 2$, $A_0 \in \mathcal{A}(P)$, and $J_0 \in \mathcal{J}(\Sigma)$, there exists a constant $c_4 = c_4(C, r, p, A_0, J_0) > 0$ such the following holds. Suppose that $\lambda > 0$, $J \in \mathcal{J}(\Sigma)$, and $A \in \mathcal{A}(P)$ satisfy*

$$\lambda + 1/\lambda + \|J\|_{C^2(\Sigma,J_0)} + \|A - A_0\|_{C^1(\Sigma,J_0)} \le C. \tag{59}$$

Suppose further that ξ and $\tilde{\xi}$ vanish for $s^2+t^2 \ge r^2$ and satisfy (56) with $A(s,t) \equiv A$, $J(s,t) \equiv J$, $\lambda(s,t) \equiv \lambda$. Then (58) holds with c_3 replaced by c_4.

For every fixed triple (λ_1, J_1, A_1) the estimate follows from the Calderón–Zygmund inequality. Moreover, if $\tilde{\xi}$ is given by (56), then

$$\begin{aligned}
&\tilde{\alpha} - ((\partial_s\alpha - d_{A_1}\phi) - (\partial_t\alpha - d_{A_1}\psi) \circ J_1) \\
&\qquad = [(A_1 - A), \phi] + [(A - A_1) \circ J_1, \psi] + (\partial_t\alpha - d_A\psi) \circ (J_1 - J),
\end{aligned}$$

$$\begin{aligned}
&\tilde{\phi} - \big(\partial_s\phi + \partial_t\psi - {\lambda_1}^2 * d_{A_1}(\alpha \circ J_1)\big) \\
&\qquad = ({\lambda_1}^2 - \lambda^2) * d_{A_1}(\alpha \circ J_1) \\
&\qquad\quad + \lambda^2 * d_{A_1}(\alpha \circ (J_1 - J)) + \lambda^2 * [(A_1 - A) \wedge (\alpha \circ J)],
\end{aligned}$$

$$\begin{aligned}
&\tilde{\psi} - \big(\partial_s\psi - \partial_t\phi - {\lambda_1}^2 * d_{A_1}\alpha\big) \\
&\qquad = ({\lambda_1}^2 - \lambda^2) * d_{A_1}\alpha + \lambda^2 * [(A_1 - A) \wedge \alpha].
\end{aligned}$$

Hence, by Lemma B.1, every triple (λ_1, J_1, A_1) has a neighbourhood (with respect to the C^1-topology on J and the C^0-topology on A) in which (58) holds with c_3 replaced by a uniform constant $c_4(\lambda_1, J_1, A_1)$. By the Arzéla–Ascoli theorem, the set of all triples (λ, J, A) that satisfy (59) can be covered by finitely many such neighbourhoods. This proves Step 1.

Step 2. *For every $r > 0$ there exist constants $c_5 > 0$ and $\varepsilon_0 > 0$ such that every ξ with support in $B_r(s_1, t_1) \times \Sigma$ satisfies (58) with c_3 replaced by c_5 and $0 < \varepsilon < \varepsilon_0$.*

Let $A_1 := A(s_1, t_1)$, $J_1 := J(s_1, t_1)$, and $\lambda_1 := \lambda(s_1, t_1)$. Consider the identities in the proof of Step 1 with $\tilde{\xi}$ given by (56). Since $A = A(\varepsilon s, \varepsilon t)$, $J = J(\varepsilon s, \varepsilon t)$, and $\lambda = \lambda(\varepsilon s, \varepsilon t)$, it follows from (53), (54), and (55) that there exists a constant $c_6 > 0$ such that, for every $\varepsilon > 0$ and every ξ with support in $B_r(s_1, t_1) \times \Sigma$,

$$\begin{aligned}
&\int_{\mathbb{R}^2} \Big(\|\tilde{\alpha} - ((\partial_s\alpha - d_{A_1}\phi) - (\partial_t\alpha - d_{A_1}\psi) \circ J_1)\|_{L^p(\Sigma,J_0)} \\
&\qquad + \|\tilde{\phi} - \big(\partial_s\phi + \partial_t\psi - {\lambda_1}^2 * d_{A_1}(\alpha \circ J_1)\big)\|_{L^p(\Sigma)} \\
&\qquad + \|\tilde{\psi} - \big(\partial_s\psi - \partial_t\phi - {\lambda_1}^2 * d_{A_1}\alpha\big)\|_{L^p(\Sigma)}\Big) \\
&\le c_6 r\varepsilon \int_{\mathbb{R}^2} \big(\|\alpha\|_{L^p(\Sigma,J_0)} + \|\partial_t\alpha\|_{L^p(\Sigma,J_0)} + \|\nabla_{A_0}\alpha\|_{L^p(\Sigma,J_0)} \\
&\qquad + \|\phi\|_{L^p(\Sigma)} + \|\psi\|_{L^p(\Sigma)} + \|d_A\psi\|_{L^p(\Sigma,J_0)}\big).
\end{aligned}$$

Hence the assertion of Step 2 follows from Step 1 and Lemma B.1.

Step 3. *We prove the lemma.*

Choose a partition of unity $\rho_j : \mathbb{R}^2 \to [0, 1]$, indexed by $j = (j_1, j_2) \in \mathbb{Z}^2$, such that each ρ_j is supported in the square $[j_1 - 1, j_1 + 2] \times [j_2 - 1, j_2 + 2]$ and $\rho_j(s, t) = \rho_0(s - j_1, t - j_2)$. Then Step 2 shows that the estimate (58) holds for the nine functions $\xi_i = \sum_j \rho_{i+3j}\xi$, where $i = (i_1, i_2) \in \mathbb{Z}^2$ with $0 \le i_1, i_2 \le 2$. Hence $\xi = \sum_{0\le i_1,i_2\le 2} \xi_i$ satisfies (58). This proves the lemma. □

PROOF OF PROPOSITION C.1. Let c_3 be the constant of Lemma C.3, and choose $\delta > 0$ such that $\delta c_3 \le 1/2$. Next choose $C > 0$ such that

$$\|J(s,t)\|_{C^2(\Sigma,J_0)} + \|A(s,t) - A_0\|_{C^1(\Sigma,J_0)} \le C$$

for all $(s,t) \in \mathbb{R}^2$, and let c be the constant of Lemma B.2 corresponding to δ and C. Then, by (43), (44), and (58),

$$\begin{aligned}
&\int_{\mathbb{R}^2} \left(\|\mathrm{d}_A\alpha\|^p_{L^p(\Sigma)} + \|\mathrm{d}_A(\alpha\circ J)\|^p_{L^p(\Sigma)} + \|\partial_s\alpha\|^p_{L^p(\Sigma,J)} + \|\partial_t\alpha\|^p_{L^p(\Sigma,J)}\right) \\
&\quad + \int_{\mathbb{R}^2} \left(\|\mathrm{d}_A\phi\|^p_{L^p(\Sigma,J)} + \|\partial_s\phi\|^p_{L^p(\Sigma)} + \|\partial_t\phi\|^p_{L^p(\Sigma)}\right) \\
&\quad + \int_{\mathbb{R}^2} \left(\|\mathrm{d}_A\psi\|^p_{L^p(\Sigma,J)} + \|\partial_s\psi\|^p_{L^p(\Sigma)} + \|\partial_t\psi\|^p_{L^p(\Sigma)}\right) \\
&\quad \le 2c_3 \int_{\mathbb{R}^2} \left(\|\tilde\alpha\|^p_{L^p(\Sigma,J)} + \|\tilde\phi\|^p_{L^p(\Sigma)} + \|\tilde\psi\|^p_{L^p(\Sigma)}\right) \\
&\quad + 2cc_3 \int_{\mathbb{R}^2} \left(\|\alpha\|^p_{L^2(\Sigma,J)} + \|\phi\|^p_{L^2(\Sigma)} + \|\psi\|^p_{L^2(\Sigma)}\right).
\end{aligned}$$

Hence (57) follows from Lemma C.2. □

Appendix D. Proof of Proposition 9.1

Let us return to the notation of Sections 3, 4, and 5. Fix a perturbation H, K as in Section 6, and $n+1$ perturbed flat connections $a_j \in \mathcal{A}^{\mathrm{flat}}(Q_{\tilde f_j}, K_j)$. Let $\Xi = (A, \Phi, \Psi) \in \mathcal{A}(Q)$ be a connection that is in temporal gauge near the cylindrical ends and satisfies (6). Suppose further that $F_{A(s,t)} = 0$ for all s and t, and let $\mathcal{D}_\varepsilon = \mathcal{D}_\varepsilon(\Xi) : W^{1,p} \to \widetilde{L}^p$ be the operator introduced in Section 9. Recall from (15) that $\tilde\xi = \mathcal{D}_\varepsilon\xi$ if and only if

$$\begin{aligned}
\tilde\alpha &= \nabla_s\alpha - \mathrm{d}_A\phi - \mathrm{d}v_{s,t}(A)\alpha + *_{s,t}(\nabla_t\alpha - \mathrm{d}_A\psi - \mathrm{d}w_{s,t}(A)\alpha), \\
\tilde\phi &= \nabla_s\phi + \nabla_t\psi + (\lambda/\varepsilon)^2 * \mathrm{d}_A *_{s,t} \alpha, \\
\tilde\psi &= \nabla_s\psi - \nabla_t\phi + (\lambda/\varepsilon)^2 * \mathrm{d}_A\alpha.
\end{aligned}$$

LEMMA D.1. *For every Ξ as above and every $p \ge 2$ there exist constants $c > 0$ and $\varepsilon_0 > 0$ such that the following holds. If $\xi = (\alpha, \phi, \psi) \in W^{1,p}$ and $\varepsilon \in (0, \varepsilon_0)$,* *then*

$$
\begin{aligned}
&\int_0^1\int_{-\infty}^{\infty}\lambda^2\left(\|\mathrm{d}_A\alpha\|^p_{L^p(\Sigma)}+\|\mathrm{d}_A(\alpha\circ J)\|^p_{L^p(\Sigma)}\right)\\
&\qquad+\int_0^1\int_{-\infty}^{\infty}\varepsilon^p\lambda^{2-p}\left(\|\nabla_s\alpha\|^p_{L^p(\Sigma,J)}+\|\nabla_t\alpha\|^p_{L^p(\Sigma,J)}\right)\\
&\qquad+\int_0^1\int_{-\infty}^{\infty}\varepsilon^p\lambda^{2-p}\left(\|\mathrm{d}_A\phi\|^p_{L^p(\Sigma,J)}+\|\mathrm{d}_A\psi\|^p_{L^p(\Sigma,J)}\right)\\
&\qquad+\int_0^1\int_{-\infty}^{\infty}\varepsilon^{2p}\lambda^{2-2p}\left(\|\nabla_s\phi\|^p_{L^p(\Sigma)}+\|\nabla_t\phi\|^p_{L^p(\Sigma)}\right)\\
&\qquad+\int_0^1\int_{-\infty}^{\infty}\varepsilon^{2p}\lambda^{2-2p}\left(\|\nabla_s\psi\|^p_{L^p(\Sigma)}+\|\nabla_t\psi\|^p_{L^p(\Sigma)}\right)\\
&\qquad\le c\int_0^1\int_{-\infty}^{\infty}\Big(\varepsilon^p\lambda^2\,\|\pi_A(\alpha)\|^p_{L^p(\Sigma,J)}+\varepsilon^p\lambda^{2-p}\,\|\tilde\alpha\|^p_{L^p(\Sigma,J)}\\
&\qquad\qquad+\varepsilon^{2p}\lambda^{2-2p}\|\tilde\phi\|^p_{L^p(\Sigma)}+\varepsilon^{2p}\lambda^{2-2p}\|\tilde\psi\|^p_{L^p(\Sigma)}\Big).
\end{aligned}\tag{60}
$$

where $\tilde\xi=(\tilde\alpha,\tilde\phi,\tilde\psi)\in\widetilde{L}^p$ *is given by* $\tilde\xi=\mathcal{D}_\varepsilon\xi$.

Proof. The proof consists of six steps. The first two steps are proved by direct calculation.

Step 1. *Assume* $v=w=0$. *Define* $\Xi'=(A',\Phi',\Psi')$ *by*

$$A'(s,t)=A(\varepsilon s,\varepsilon t),\quad \Phi'(s,t)=\varepsilon\Phi(\varepsilon s,\varepsilon t),\quad \Psi'(s,t)=\varepsilon\Psi(\varepsilon s,\varepsilon t),$$

and similarly for $\xi'=(\alpha',\phi',\psi')$ *and* $\tilde\xi'=(\tilde\alpha',\tilde\phi',\tilde\psi')$. *Denote* $\lambda'(s,t)=\lambda(\varepsilon s,\varepsilon t)$ *and* ${\nabla_s}'=\partial_s+\Psi'$ *and* ${\nabla_t}'=\partial_t+\Psi'$. *Then* $\tilde\xi=\mathcal{D}_\varepsilon\xi$ *if and only if*

$$
\begin{aligned}
\tilde\alpha' &= {\nabla_s}'\alpha'-\mathrm{d}_{A'}\phi'+*_{\varepsilon s,\varepsilon t}({\nabla_t}'\alpha'-\mathrm{d}_{A'}\psi'),\\
\tilde\phi' &= {\nabla_s}'\phi'+{\nabla_t}'\psi'+(\lambda')^2*\mathrm{d}_{A'}*_{\varepsilon s,\varepsilon t}\alpha',\\
\tilde\psi' &= {\nabla_s}'\psi'-{\nabla_t}'\phi'+(\lambda')^2*\mathrm{d}_{A'}\alpha'.
\end{aligned}\tag{61}
$$

Step 2. *Let* Ξ', ξ', *and* $\tilde\xi'$ *be as in Step 1, and denote* $J'(s,t)=J(\varepsilon s,\varepsilon t)$. *Then (60) is equivalent to*

$$
\begin{aligned}
&\int_0^{1/\varepsilon}\int_{-\infty}^{\infty}\lambda^2\left(\|\mathrm{d}_{A'}\alpha'\|^p_{L^p(\Sigma)}+\|\mathrm{d}_{A'}(\alpha'\circ J')\|^p_{L^p(\Sigma)}\right)\\
&\qquad+\int_0^{1/\varepsilon}\int_{-\infty}^{\infty}\lambda^{2-p}\left(\left\|{\nabla_s}'\alpha'\right\|^p_{L^p(\Sigma,J')}+\left\|{\nabla_t}'\alpha'\right\|^p_{L^p(\Sigma,J')}\right)\\
&\qquad+\int_0^{1/\varepsilon}\int_{-\infty}^{\infty}\lambda^{2-p}\left(\|\mathrm{d}_{A'}\phi'\|^p_{L^p(\Sigma,J')}+\|\mathrm{d}_{A'}\psi'\|^p_{L^p(\Sigma,J')}\right)\\
&\qquad+\int_0^{1/\varepsilon}\int_{-\infty}^{\infty}\lambda^{2-2p}\left(\left\|{\nabla_s}'\phi'\right\|^p_{L^p(\Sigma)}+\left\|{\nabla_t}'\phi'\right\|^p_{L^p(\Sigma)}\right)\\
&\qquad+\int_0^{1/\varepsilon}\int_{-\infty}^{\infty}\lambda^{2-2p}\left(\left\|{\nabla_s}'\psi'\right\|^p_{L^p(\Sigma)}+\left\|{\nabla_t}'\psi'\right\|^p_{L^p(\Sigma)}\right)\le
\end{aligned}\tag{62}
$$

$$
\begin{aligned}
&\leq c \int_0^{1/\varepsilon} \int_{-\infty}^{\infty} \Big(\lambda^2 \, \|\pi_{A'}(\alpha')\|^p_{L^p(\Sigma,J')} \\
&\quad + \lambda^{2-p} \, \|\tilde{\alpha}'\|^p_{L^p(\Sigma,J')} + \lambda^{2-2p} \|\tilde{\phi}'\|^p_{L^p(\Sigma)} + \lambda^{2-2p} \|\tilde{\psi}'\|^p_{L^p(\Sigma)} \Big).
\end{aligned}
$$

Step 3. *Assume $v = w = 0$ and $\Phi = \Psi = 0$. Fix an integer $j \in \{0, \dots, n\}$ and denote by*

$$Q_j = \iota_j((-\infty, -T) \times Q_{\tilde{f}_j}) \subset Q$$

the jth cylindrical end. Then there exist constants $c > 0$ and $\varepsilon_0 > 0$ such that every $\xi \in W^{1,p}$ that vanishes oustside Q_j satisfies (60) for $0 < \varepsilon < \varepsilon_0$.

Suppose, without loss of generality, that $j = 0$. Hence assume that $\xi(s,t) = 0$ for $s \geq -T$, and recall that $\lambda(s,t) = 1$ for $s \leq -T$. Choose a cutoff function $\beta : \mathbb{R} \to [0,1]$ such that

$$\beta(t) = \begin{cases} 1, & \text{for } |t| \leq \delta, \\ 0, & \text{for } |t| \geq 1/2 - \delta, \end{cases} \qquad \beta(-t) = \beta(t),$$

and

$$0 \leq t \leq 1/2 \qquad \Longrightarrow \qquad \beta(t) + \beta(t - 1/2) = 1.$$

Consider the functions

$$\xi_0(s,t) = \beta(t)\xi(s,t), \qquad \xi_0(s,t) = \beta(t - 1/2)\xi(s,t).$$

Both functions have compact support. By Step 1, the rescaled function ξ_0' satisfies

$$
\begin{aligned}
\beta(\varepsilon t)\tilde{\alpha}' + \varepsilon\dot{\beta}(\varepsilon t) *_{\varepsilon s, \varepsilon t} \alpha' &= \partial_s' \alpha_0' - d_{A'}\phi_0' + *_{\varepsilon s,\varepsilon t}(\partial_t \alpha_0' - d_{A'}\psi_0'), \\
\beta(\varepsilon t)\tilde{\phi}' + \varepsilon\dot{\beta}(\varepsilon t)\psi' &= \partial_s'\phi_0' + \partial_t\psi_0' + *d_{A'} *_{\varepsilon s,\varepsilon t} \alpha_0', \\
\beta(\varepsilon t)\tilde{\psi}' - \varepsilon\dot{\beta}(\varepsilon t)\phi' &= \partial_s\psi_0' - \partial_t\phi_0' + *d_{A'}\alpha_0',
\end{aligned}
$$

and similarly for ξ_1'. Now apply Proposition C.1 to ξ_0' and ξ_1' to obtain that ξ' satisfies (62) for $\varepsilon > 0$ sufficiently small. This proves Step 3 for $j = 0$. For general j, apply the same argument to the pullback 1-form $\xi_j = \tilde{\iota}_j{}^*\xi$ on Q_j. This shows that ξ_j satisfies (62) with $\lambda = 1$. Step 3 then follows by expressing this estimate in terms of ξ.

Step 4. *Assume $v = w = 0$ and $\Phi = \Psi = 0$. Let $B = \{s + it \in \mathbb{C} \,|\, s^2 + t^2 \leq 1\}$, and let $\iota : B \hookrightarrow S$ be a holomorphic embedding. Then there exist constants $c > 0$ and $\varepsilon_0 > 0$ such that every $\xi \in W^{1,p}$ with support in $\iota(B)$ satisfies (60) for $0 < \varepsilon < \varepsilon_0$.*

Trivialize the bundle X over B to obtain an embedding

$$\iota : B \times \Sigma \to X.$$

Then trivialize the bundle Q over the image of ι to obtain an embedding

$$\tilde{\iota} : B \times P \to Q.$$

By Step 1, the rescaled pullback 1-form $(\tilde{\iota}^*\xi)'$ satisfies (61). Hence, by Proposition C.1, it satisfies (62). Hence, by Step 2, it satisfies (60).

Step 5. *Assume $v = w = 0$ and $\Phi = \Psi = 0$. Then there exist constants $c > 0$ and $\varepsilon_0 > 0$ such that every $\xi \in W^{1,p}$ with compact support in X satisfies (60) for $0 < \varepsilon < \varepsilon_0$.*

Cover S by finitely many open sets U_ν, each of which either is a cylindrical end or is contained in the holomorphic image of a ball B. Choose a partition of unity

$\rho_\nu : S \to [0,1]$ subordinate to the cover, and apply Steps 3 and 4 to the functions $(\rho_\nu \circ \pi)\xi$. Then the error terms in the estimate (60) arising from the cutoff functions $\rho_\nu = \rho_\nu(s,t)$ are arbitrarily small as ε tends to zero. They can be dominated by positive terms on the left hand side of the inequality, and hence ξ satisfies (60).

Step 6. *We prove the lemma.*

By Lemma B.3, there exists a constant $c_0 > 0$ such that for every $(s,t) \in \mathbb{R}^2$, every $\phi \in \Omega^0(\Sigma, \mathfrak{g}_P)$, and every $\alpha \in \Omega^1(\Sigma, \mathfrak{g}_P)$,

$$\|\phi\|_{W^{1,p}(\Sigma)} \le c_0 \, \|\mathrm{d}_A \phi\|_{L^p(\Sigma,J)} \, ,$$

$$\|\alpha\|_{L^p(\Sigma,J)} \le c_0 \left(\|\mathrm{d}_A \alpha\|_{L^p(\Sigma)} + \|\mathrm{d}_A(\alpha \circ J)\|_{L^p(\Sigma)} + \|\pi_A(\alpha)\|_{L^p(\Sigma,J)} \right).$$

Moreover, the limit condition (6) implies that there exists a constant $c_1 > 0$ such that, for every $(s,t) \in \mathbb{R}^2$ and every $\alpha \in \Omega^1(\Sigma, \mathfrak{g}_P)$,

$$\begin{aligned} &\|\Phi\|_{L^\infty(\Sigma)} + \|\Psi\|_{L^\infty(\Sigma)} \le c_1 \lambda, \\ &\|\mathrm{d}v_{s,t}(A)\alpha\|_{L^p(\Sigma,J)} + \|\mathrm{d}w_{s,t}(A)\alpha\|_{L^p(\Sigma,J)} \le c_1 \lambda \, \|\alpha\|_{L^p(\Sigma,J)} \, . \end{aligned} \tag{63}$$

Hence the error terms in (60) arising from nonzero terms v, w, Φ, or Ψ can be controlled by larger positive terms on the left hand side, provided that $\varepsilon > 0$ is sufficiently small. This proves the lemma. □

PROOF OF PROPOSITION 9.1. The inequality (16) follows directly from Lemma D.1 and Lemma B.3. To prove (17), write

$$\alpha = \pi_A(\alpha) + \mathrm{d}_A \zeta - (\mathrm{d}_A \eta) \circ J$$

for $\eta, \zeta \in \Omega^0(\Sigma, \mathfrak{g}_P)$, and abbreviate

$$B_s = \partial_s A - \mathrm{d}_A \Phi, \qquad B_t = \partial_t A - \mathrm{d}_A \Psi, \qquad B = B_s - B_t \circ J.$$

By (6), there exists a constant $c_1 > 0$ such that, for every $(s,t) \in \mathbb{R}^2$,

$$\|B_s\|_{L^\infty(\Sigma,J)} + \|B_t\|_{L^\infty(\Sigma,J)} \le c_1 \lambda. \tag{64}$$

Moreover,

$$\mathrm{d}_A *_{s,t} \mathrm{d}_A \eta = \mathrm{d}_A(\alpha - \pi_A(\alpha)), \qquad \mathrm{d}_A *_{s,t} \mathrm{d}_A \zeta = \mathrm{d}_A *_{s,t} (\alpha - \pi_A(\alpha)).$$

Hence, by Lemma B.4, there exists a $c_2 > 0$ such that, for all $(s,t) \in \mathbb{R}^2$,

$$\|\mathrm{d}_A \eta\|_{L^p(\Sigma,J)} + \|\mathrm{d}_A \zeta\|_{L^p(\Sigma,J)} \le c_2 \, \|(\alpha - \pi_A(\alpha))\|_{L^p(\Sigma,J)} \, . \tag{65}$$

Moreover,

$$\begin{aligned} \nabla_s(\alpha - \pi_A(\alpha)) &= \nabla_s(\mathrm{d}_A \zeta - (\mathrm{d}_A \eta) \circ J) \\ &= \mathrm{d}_A \nabla_s \zeta - (\mathrm{d}_A \nabla_s \eta) \circ J + [B_s, \zeta] - [B_s \circ J, \eta] - (\mathrm{d}_A \eta) \circ \partial_s J, \end{aligned} \tag{66}$$

and similarly for $\nabla_t(\alpha - \pi_A(\alpha))$. Hence

$$\begin{aligned} \mathrm{d}_A *_{s,t} \mathrm{d}_A \nabla_s \eta &= \mathrm{d}_A \nabla_s \alpha + [B_s \wedge \pi_A(\alpha)] \\ &\quad - \mathrm{d}_A \left([B_s, \zeta] + [*_{s,t} B_s, \eta] - (\mathrm{d}_A \eta) \circ \partial_s J \right), \\ \mathrm{d}_A *_{s,t} \mathrm{d}_A \nabla_s \zeta &= \mathrm{d}_A *_{s,t} \nabla_s \alpha + [B_s \wedge *_{s,t} \pi_A(\alpha)] \\ &\quad - \mathrm{d}_A \left([*_{s,t} B_s, \zeta] - [B_s, \eta] - (\mathrm{d}_A \eta) \circ J \partial_s J - \pi_A(\alpha) \circ \partial_s J \right). \end{aligned}$$

Hence, by Lemma B.4, Lemma B.3, (64), and (65), there exists a constant $c_3 > 0$ such that, for all $(s,t) \in \mathbb{R}^2$,

$$\|\mathrm{d}_A \nabla_s \eta\|_{L^p(\Sigma,J)} + \|\mathrm{d}_A \nabla_s \zeta\|_{L^p(\Sigma,J)} \le c_3 \left(\|\nabla_s \alpha\|_{L^p(\Sigma,J)} - \lambda \, \|\alpha\|_{L^p(\Sigma,J)} \right).$$

By (65) and (66), this implies

$$\|\nabla_s(\alpha - \pi_A(\alpha))\|_{L^p(\Sigma,J)} \le c_4 \left(\|\nabla_s \alpha\|_{L^p(\Sigma,J)} + \lambda \|\alpha\|_{L^p(\Sigma,J)} \right).$$

A similar inequality holds for $\nabla_t(\alpha - \pi_A(\alpha))$. Hence (17) follows from Lemma D.1 and Lemma B.3.

To prove (18), note that

$$\pi_A(\mathcal{D}_\varepsilon \xi) - \mathcal{D}_0 \pi_A(\xi) = \pi_A(\theta_0 - \theta_1), \tag{67}$$

where

$$\begin{aligned} \theta_0 &= \nabla_s(\alpha - \pi_A(\alpha)) + *_{s,t} \nabla_t(\alpha - \pi_A(\alpha)), \\ \theta_1 &= \mathrm{d}v_{s,t}(A)(\alpha - \pi_A(\alpha)) + *_{s,t} \mathrm{d}w_{s,t}(\alpha - \pi_A(\alpha)). \end{aligned}$$

By (66),

$$\begin{aligned} \pi_A(\theta_0) &= \pi_A(\nabla_s(\alpha - \pi_A(\alpha)) - (\nabla_t(\alpha - \pi_A(\alpha))) \circ J) \\ &= \pi_A\left([B,\zeta] - [B \circ J, \eta] - (\mathrm{d}_A \eta) \circ (\partial_s J + J \partial_t J)\right). \end{aligned}$$

Hence, by (63), (64), (65), (67), and Lemma B.3,

$$\begin{aligned} &\|\pi_A(\mathcal{D}_\varepsilon \xi) - \mathcal{D}_0 \pi_A(\xi)\|_{\tilde{L}^p}^p \\ &\quad \le \|\pi_A(\theta_0)\|_{\tilde{L}^p}^p + \|\pi_A(\theta_1)\|_{\tilde{L}^p}^p \\ &\quad = \int_0^1 \int_{-\infty}^{\infty} \lambda^{2-p} \left(\|\pi_A(\theta_0)\|_{L^p(\Sigma,J)}^p + \|\pi_A(\theta_1)\|_{L^p(\Sigma,J)}^p \right) \\ &\quad \le c_5 \int_0^1 \int_{-\infty}^{\infty} \lambda^2 \left(\|\mathrm{d}_A \zeta\|_{L^p(\Sigma,J)}^p + \|\mathrm{d}_A \eta\|_{L^p(\Sigma,J)}^p + \|\alpha - \pi_A(\alpha)\|_{L^p(\Sigma,J)}^p \right) \\ &\quad \le c_6 \int_0^1 \int_{-\infty}^{\infty} \lambda^2 \|\alpha - \pi_A(\alpha)\|_{L^p(\Sigma,J)}^p \\ &\quad \le c_6 \|\xi - \pi_A(\xi)\|_{0,p,\varepsilon}^p. \end{aligned}$$

This proves the proposition. □

Acknowledgement. This paper was completed while I visited the Research Institute of Mathematical Sciences at Kyoto University. I would like to thank the JSPS for their generous support and my hosts for their wonderful hospitality.

References

[1] M. F. Atiyah, *New invariants of three and four dimensional manifolds*, The Mathematical Heritage of Hermann Weyl, Proc. Sympos. Pure Math., vol. 48, Amer. Math. Soc., Providence, RI, 1988, pp. 285–299.

[2] S. K. Donaldson, *Symplectic submanifolds and almost complex geometry*, J. Diff. Geom. **44** (1996), 666–705.

[3] ———, *Symplectic submanifolds and topological Lefschetz fibrations*, In preparation.

[4] S. Dostoglou and D. A. Salamon, *Instanton homology and symplectic fixed points*, Symplectic Geometry, D. Salamon (ed.), LMS Lecture Notes Series, vol. 192, Cambridge Univ. Press, Cambridge, 1993, pp. 57–93.

[5] ———, *Cauchy–Riemann operators, self-duality, and the spectral flow*, First European Congress of Mathematics, Volume I, Invited Lectures (Part 1), A. Joseph, F. Mignot, F. Murat, B. Prum, R. Rentschler (eds.), Birkhäuser, Basel, 1994, pp. 511–545.

[6] ———, *Self-dual instantons and holomorphic curves*, Ann. of Math. (2) **139** (1994), 581–640.

[7] A. Floer, *An instanton invariant for 3-manifolds*, Comm. Math. Phys. **118** (1988), 215–240.

[8] ———, *Symplectic fixed points and holomorphic spheres*, Comm. Math. Phys. **120** (1989), 575–611.

[9] A. Floer and H. Hofer, *Coherent orientations for periodic orbit problems in symplectic geometry*, Math. Z. **212** (1993), 13–38.

[10] A. Floer, H. Hofer, and D. Salamon, *Transversality in elliptic Morse theory for the symplectic action*, Duke Math. J. **80** (1996), 251–292.

[11] F. G. Handfield, *Adiabatic limits of the anti-seld-dual equation*, PhD thesis, Austin, Texas, 1998.

[12] D. McDuff and D. Salamon, *J-holomorphic Curves and Quantum Cohomology*, University Lecture Series, vol. 6, Amer. Math. Soc., Providence, RI, 1994.

[13] ———, *Introduction To Symplectic Topology*, 2nd edition, Oxford University Press, Oxford, 1998.

[14] V. Munoz, *Quantum cohomology of the moduli space of stable bundles over a Riemann surface*, Duke Math. J. 98 (1999), 525–540.

[15] ———, *Fukaya-Floer homology of* $\Sigma \times S^1$ *and applications*, Preprint, 1998.

[16] S. Piunikhin, D. Salamon, and M. Schwarz, *Symplectic Floer-Donaldson theory and quantum cohomology*, Contact and Symplectic Geometry, C. B. Thomas (ed.), Publ. Newton Institute, Cambridge University Press, Cambridge, 1996, pp. 171–200.

[17] D. A. Salamon, *Lagrangian intersections, 3-manifolds with boundary, and the Atiyah–Floer conjecture*, Proc. Internat. Congr. Math. (Zürich, 1994), Vol. 1, Birkhäuser, Basel, 1995, pp. 526–536.

[18] ———, *Lectures on Floer homology*, Symplectic Geometry and Topology (Ya. Eliashberg and L. Traynor, eds.). ISA/Park City Math. Series, vol. 7, Amer. Math. Soc., Providence, RI, 1999, pp. 143–225.

[19] ———, *Spin Geometry and Seiberg-Witten Invariants*, Birkhäuser, Basel (to appear).

[20] D. A. Salamon and E. Zehnder, *Morse theory for periodic solutions of Hamiltonian systems and the Maslov index*, Comm. Pure Appl. Math. **45** (1992), 1303–1360.

DEPARTEMENT MATHEMATIK, ETH-ZENTRUM, RÄMISTRASSE 101, CH-8092 ZÜRICH, SWITZERLAND

E-mail address: `salamon@math.ethz.ch`

Amer. Math. Soc. Transl.
(2) Vol. **196**, 1999

On the Group of Symplectic Automorphisms of $\mathbb{C}\mathrm{P}^m \times \mathbb{C}\mathrm{P}^n$

Paul Seidel

1. Introduction

Let (M, ω) be a compact symplectic manifold and $\mathrm{Aut}(M, \omega)$ the group of symplectic automorphisms, with the C^∞-topology. The inclusion $\mathrm{Aut}(M, \omega) \subset \mathrm{Diff}(M)$ induces homomorphisms

$$\beta_k : \pi_k(\mathrm{Aut}(M, \omega)) \longrightarrow \pi_k(\mathrm{Diff}(M, \omega)).$$

A non-injectivity result for β_k means that by the inclusion some topology becomes lost.

Let ω_n be the standard symplectic form on $\mathbb{C}\mathrm{P}^n$, normalized in such a way that $[\omega_n]$ is Poincaré dual to a hyperplane. We denote the product $(\mathbb{C}\mathrm{P}^m, \omega_m) \times (\mathbb{C}\mathrm{P}^n, \omega_n)$, for $m, n \geq 1$, by (P_{mn}, η_{mn}). Let $\mathrm{Diff}(P_{mn})$ be the group of diffeomorphisms with the C^∞-topology, $\mathrm{Aut}(P_{mn}, \eta_{mn})$ the subgroup of symplectic automorphisms, and

$$\beta_k : \pi_k(\mathrm{Aut}(P_{mn}, \eta_{mn})) \longrightarrow \pi_k(\mathrm{Diff}(P_{mn}))$$

the homomorphisms induced by inclusion.

THEOREM 1.1. *Let k be odd and $\leq \max\{2m-1, 2n-1\}$. Then β_k is not surjective. In fact*

$$\begin{aligned}\mathrm{rank}(\mathrm{coker}\, \beta_k) \geq\ & b_{2m+1-k}(P_{mn}) - b_{2m+1-k}(\mathbb{C}\mathrm{P}^m) \\ & + b_{2n+1-k}(P_{mn}) - b_{2n+1-k}(\mathbb{C}\mathrm{P}^n) > 0,\end{aligned}$$

where b_* are the Betti numbers.

For $m = n = 1$ the theorem says that

$$\mathrm{rank}(\mathrm{coker}\, \beta_1) \geq 2. \tag{1.1}$$

This can be derived from a result of Gromov [**5**, 0.3.C] which says that the group of Kähler isometries of P_{11} (with respect to the standard metric) is a deformation retract of $\mathrm{Aut}(P_{11}, \eta_{11})$. Since the isometry group is an extension of $\mathbb{Z}/2$ by $SO(3) \times SO(3)$, it follows that $\pi_1(\mathrm{Aut}(P_{11}, \eta_{11})) \cong \mathbb{Z}/2 \oplus \mathbb{Z}/2$. The topology of $\mathrm{Diff}(P_{11})$ is unknown, but by looking at its image in the space of continuous self-maps of P_{11} one can show that $\pi_1(\mathrm{Diff}(P_{11}))$ is a group of rank ≥ 2, which implies (1.1).

1991 *Mathematics Subject Classification.* Primary 58D05.
Supported by NSF grant DMS 9304580.

©1999 American Mathematical Society

Gromov's theorem on $\mathrm{Aut}(P_{11}, \eta_1)$ and its cousins for $(\mathbb{C}\mathrm{P}^2, \omega_2)$ were the first non-trivial results about the topology of symplectic automorphism groups. More recently Abreu [**1**] and Abreu–McDuff [**2**] have studied the group of automorphisms of the symplectic structures $\eta_{11}^{(\lambda)} = \lambda(\omega_1 \times 1) + 1 \times \omega_1$ on P_{11} for $\lambda \neq 1$, using an approach suggested by Gromov [**5**, 2.4.C_2]. The automorphism groups of blowups of $\mathbb{C}\mathrm{P}^2$ are closely related to the spaces of symplectic embeddings of balls, which have been studied by Biran [**3**] and McDuff [**10**]. These results rely on specific properties of rational or ruled symplectic four-manifolds, and little is known outside this class (although see [**14**] for some information about π_0 of the automorphism groups of symplectic four-manifolds). It seems that Theorem 1.1 provides the first examples of symplectic manifolds of dimension > 4 for which the map $\pi_1(\mathrm{Aut}) \longrightarrow \pi_1(\mathrm{Diff})$ is known not to be surjective. The fact that no such examples were known was pointed out to the author by F. Lalonde. Lalonde also suggested that it may be possible to construct such examples by exploiting the "rigidity theorem" of [**7**] instead of the methods used here (any such examples would have positive first Betti number).

By definition, the symplectic automorphism group is the structure group of symplectic fibre bundles. The study of pseudo-holomorphic curves in each fibre of such a fibre bundle E yields "parametrized Gromov–Witten invariants" which are multilinear maps on $H^*(E)$. These invariants have been considered by Lê [**8**] and others. They satisfy (at least in principle) axioms similar to those of Kontsevich and Manin [**6**]. In this paper we argue that in certain cases the existence of such a system of invariants restricts the possibilities for what $H^*(E)$ can be. To simplify the technical issues, we do not consider all Gromov–Witten invariants, but just a single particularly simple one.

Although our argument relies on Gromov–Witten invariants, it does not suppose any knowledge of what the actual value of the invariants is. In this respect it resembles the recent Lalonde–McDuff–Polterovich "rigidity theorem" [**7**]. Note that the proof of the "rigidity theorem" involves pseudo-holomorphic curves of a different kind, namely, pseudo-holomorphic sections of a fibre bundle whose base is a Riemann surface (in that case, S^2). It is possible that a combination of the two approaches would yield more information about symplectic automorphism groups.

This paper is structured as follows. The next section reviews some basic facts about the cohomology rings of fibrations. In Section 3 we apply these considerations to the case where the fibre is P_{mn}. Up to this point the argument is purely topological. The relevant Gromov–Witten invariant is introduced in Sections 4 and 5. Section 6 contains the main computation. In the final section we derive Theorem 1.1 and discuss a related result and some possible further developments.

Acknowledgements. The author is indebted to François Lalonde, Dusa McDuff, and Richard Thomas for helpful discussions.

2. The cohomology rings of fibrations over spheres

Fix a field $\mathbb{F}$. By a graded $\mathbb{F}$-algebra $R = \bigoplus_{i \in \mathbb{Z}} R^i$ we mean one which is finite-dimensional, commutative (in the graded sense), and has a unit. All homomorphisms between graded algebras preserve units.

DEFINITION 2.1. Let R be a graded $\mathbb{F}$-algebra. A *deformation* of R of dimension $d > 0$ consists of

(1) a graded $\mathbb{F}$-algebra $\widetilde{R}$,

(2) an element $t \in \widetilde{R}^d$ with $t^2 = 0$, such that the sequence

$$0 \longrightarrow \widetilde{R}/t\widetilde{R} \xrightarrow{\cdot t} \widetilde{R} \longrightarrow \widetilde{R}/t\widetilde{R} \longrightarrow 0$$

is exact, and

(3) a homomorphism of graded algebras $j : \widetilde{R} \longrightarrow R$ which is surjective with kernel $t\widetilde{R}$.

To be precise, these objects should be called "first order infinitesimal deformations"; we use the shorter name for brevity's sake. The exactness of the sequence $\widetilde{R}/t\widetilde{R} \longrightarrow \widetilde{R} \longrightarrow \widetilde{R}/t\widetilde{R}$ is equivalent to the flatness of $\widetilde{R}$ as a module over $\mathbb{F}[\epsilon]/\epsilon^2$, where ϵ acts by multiplication with t.

Given two graded algebras R_1, R_2 and a homomorphism $f : R_1 \longrightarrow R_2$, one defines a *morphism over* f from a d-dimensional deformation $(\widetilde{R}_1, t_1, j_1)$ of R_1 to a d-dimensional deformation $(\widetilde{R}_2, t_2, j_2)$ of R_2 to be a homomorphism of graded algebras $\tilde{f} : \widetilde{R}_1 \longrightarrow \widetilde{R}_2$ such that $\tilde{f}(t_1) = t_2$ and $j_2\tilde{f} = fj_1$. In the special case when $R_1 = R_2 = R$ and $f = \mathrm{id}$, $\tilde{f}$ is called a morphism of deformations of R. All morphisms of deformations of R are isomorphisms.

The set $\mathrm{Def}_d(R)$ of isomorphism classes of d-dimensional deformations of R carries a natural structure of an $\mathbb{F}$-vector space. The sum of two d-dimensional deformations $(\widetilde{R}_i, t_i, j_i)$, $i = 1, 2$, is the deformation $(\widetilde{R}, t, j)$ defined as follows. Let $\Delta \subset \widetilde{R}_1 \oplus \widetilde{R}_2$ be the subalgebra of those (x_1, x_2) such that $j_1(x_1) = j_2(x_2)$. Then $\widetilde{R} = \Delta/(t_1, -t_2)\Delta$, $t = [t_1, 0] = [0, t_2] \in \widetilde{R}$, and $j : \widetilde{R} \longrightarrow R$ is given by $j([x_1, x_2]) = j_1(x_1) = j_2(x_2)$.

REMARK 2.2. There is a second and more explicit definition of $\mathrm{Def}_d(R)$. Let $Z_d(R)$ be the space of $\mathbb{F}$-bilinear maps $\psi : R \times R \longrightarrow R$ of degree $-d$ which are graded symmetric and satisfy

$$(-1)^{d\deg(x)} x\psi(y, z) - \psi(xy, z) + \psi(x, yz) - \psi(x, y)z = 0,$$

and $B_d(R) \subset Z_d(R)$ the subset of maps of the form

$$\psi(x, y) = (-1)^{d\deg(x)} x\xi(y) - \xi(xy) + \xi(x)y,$$

where $\xi : R \longrightarrow R$ is some linear map of degree $-d$. Then $\mathrm{Def}_d(R) = Z_d(R)/B_d(R)$. The equivalence of this definition with the previous one is proved by choosing, for a deformation $(\widetilde{R}, t, j)$ of R, an isomorphism of $\widetilde{R}$ with $R[\epsilon]/\epsilon^2$ as an $\mathbb{F}[\epsilon]/\epsilon^2$-module. Then the product on $\widetilde{R}$ has the form

$$(x_0 + \epsilon x_1)(y_0 + \epsilon y_1) = x_0y_0 + \epsilon((-1)^{d\deg(x_0)} x_0y_1 + x_1y_0 + \psi(x_0, y_0))$$

for some $\psi \in Z_d(R)$. The equivalence class of ψ in $Z_d(R)/B_d(R)$ is independent of the choice of isomorphism. This alternative description shows that if R is concentrated in even dimensions, then $\mathrm{Def}_d(R) = 0$ for all odd d.

In the deformation theory of algebras (see **[4]** for a survey) $\mathrm{Def}_*(R)$ is called the second Harrison cohomology group of the graded algebra R.

Let M be a compact manifold. A smooth fibre bundle over a sphere S^d with fibre M consists of a manifold E, a submersion $\pi : E \longrightarrow S^d$, and a diffeomorphism $i : M \longrightarrow E_{b_0}$ for some fixed $b_0 \in S^d$. The cohomology of E (unless otherwise stated, all cohomology groups have $\mathbb{F}$-coefficients) sits in a Wang sequence

$$\cdots \xrightarrow{\delta} H^*(M) \xrightarrow{i_!} H^{*+d}(E) \xrightarrow{i^*} H^{*+d}(M) \xrightarrow{\delta} H^{*+1}(M) \longrightarrow \ldots,$$

where $i_!$ is the cohomology transfer or pushforward. Let ϵ be the standard generator of $H^d(S^d)$, and $t = \pi^*(\epsilon) \in H^d(E)$.

LEMMA 2.3. *If i^* is onto, then $(H^*(E), t, i^*)$ is a d-dimensional deformation of $H^*(M)$.*

PROOF. Because t is Poincaré dual to the fibre of E, $i_!(i^*(x)) = xt$ for all $x \in H^*(E)$. Hence $\ker(i^*) = \mathrm{im}(i_!) = \mathrm{im}(i_! i^*) = tH^*(E)$. It remains to show that any x with $tx = 0$ lies in $tH^*(E)$. Because $i_!$ is injective, $tx = i_! i^*(x) = 0$ implies that $i^*(x) = 0$, and we have already seen that $\ker(i^*) = tH^*(E)$. □

The isomorphism classes of smooth fibre bundles over S^d with fibre M form a group under the operation of fibre connected sum. This group is isomorphic to $\pi_{d-1}(\mathrm{Diff}(M))$; the isomorphism is given by a clutching construction which associates to a map $\phi : (S^{d-1}, b_0) \longrightarrow (\mathrm{Diff}(M), \mathrm{id})$ a fibre bundle $(E_\phi, \pi_\phi, i_\phi)$. On the level of cohomology ϕ and E_ϕ are related in the following way: since ϕ can be written as a map $M \times S^{d-1} \longrightarrow M$, it induces a homomorphism $\delta_\phi : H^{*+d-1}(M) \longrightarrow H^*(M)$, and this homomorphism is the connecting map in the Wang sequence for E_ϕ. In particular, the subgroup $G_{d-1} \subset \pi_{d-1}(\mathrm{Diff}(M))$ of classes $[\phi]$ such that $\delta_\phi = 0$ corresponds to the isomorphism classes of fibre bundles (E, π, i) for which i^* is onto. Lemma 2.3 associates to any such fibre bundle an element of $\mathrm{Def}_d(H^*(M))$. Since the sum of deformations imitates the behaviour of cohomology under fibre connected sum, this defines a homomorphism

$$\alpha_{d-1} : G_{d-1} \longrightarrow \mathrm{Def}_d(H^*(M)).$$

Although we have used smooth fibre bundles throughout, the construction does not really depend on the smooth structure of M. In fact the group of diffeomorphisms can be replaced by the bigger semigroup of homotopy equivalences $M \longrightarrow M$. To take into account the smooth structure of M would mean to consider deformations with certain distinguished elements (the Pontryagin classes). For $\mathbb{F} = \mathbb{Q}$ these questions can be treated in a much more satisfactory way in the framework of rational homotopy theory; see [**16**, pp. 313–314, 322–326].

EXAMPLE 2.4. Let $R = \mathbb{F}[u]/u^{n+1}$, where u has dimension 2. Since R is concentrated in even dimensions, $\mathrm{Def}_d(R) = 0$ for all odd d. Now fix an even d. To any element $a \in R^{2n+2-d}$ one can associate a d-dimensional deformation $(\widetilde{R}_a, t, j)$ of R. This deformation is defined in the following way: $a = \alpha u^{n+1-d/2}$ for some $\alpha \in \mathbb{F}$. Then

$$\widetilde{R}_a = \mathbb{F}[\tilde{u}, t]/(\tilde{u}^{n+1} + \alpha t \tilde{u}^{n+1-d/2}, t^2),$$

and $j : \widetilde{R}_a \longrightarrow R$ is the algebra homomorphism with $j(\tilde{u}) = u$ and $j(t) = 0$. The sum of $\widetilde{R}_a$ and $\widetilde{R}_b$ is isomorphic to $\widetilde{R}_{a+b}$.

Every d-dimensional deformation $(\widetilde{R}, t, j)$ of R is isomorphic to $\widetilde{R}_a$ for some a. This is proved as follows: choose a $\tilde{u} \in \widetilde{R}^2$ with $j(\tilde{u}) = u$. Since $u^{n+1} = 0$, $\tilde{u}^{n+1} = -t\tilde{a}$ for some $\tilde{a}$. At this point it is easy to construct a morphism of deformations of R from $\widetilde{R}_a$, where $a = j(\tilde{a})$, to $\widetilde{R}$; recall that any such morphism is an isomorphism.

If $d \neq 2$, then $\widetilde{R}_a$ is isomorphic to $\widetilde{R}_b$ iff $a = b$. This can be proved either directly or by observing that, for an arbitrary deformation, the element $\tilde{u}$ is unique and hence $a = j(\tilde{a})$ is an isomorphism invariant of the deformation. For $d = 2$, $\widetilde{R}_a$

and $\widetilde{R}_b$ are isomorphic iff $b - a$ is a multiple of $(n+1)u^n$. Therefore

$$\mathrm{Def}_d(R) \cong \begin{cases} H^{2n+2-d}(R), & d \neq 2, \\ H^{2n}(R)/(n+1)H^{2n}(R), & d = 2. \end{cases}$$

We will now interpret this result geometrically. $R = H^*(\mathbb{C}\mathrm{P}^n)$. Because the cohomology is concentrated in even dimensions, any smooth fibre bundle (E, π, i) with fibre $\mathbb{C}\mathrm{P}^n$ over S^d satisfies the condition that i^* is onto if d is even. In other words, $G_{d-1} = \pi_{d-1}(\mathrm{Diff}(\mathbb{C}\mathrm{P}^n))$ for all even d. Hence one has homomorphisms

$$\alpha_{d-1} : \pi_{d-1}(\mathrm{Diff}(\mathbb{C}\mathrm{P}^n)) \longrightarrow \mathrm{Def}_d(R) \cong \begin{cases} \mathbb{F}, & d = 4, \dots, 2n+2, \\ \mathbb{F}/(n+1)\mathbb{F}, & d = 2. \end{cases}$$

If we restrict them to the subgroup $PU(n+1) \subset \mathrm{Diff}(\mathbb{C}\mathrm{P}^n)$, these homomorphisms are the Chern classes $c_2, \dots, c_{n+1}$ and the first Chern class mod $n+1$. This follows from the standard formula for the cohomology ring of the projective bundle associated to a vector bundle.

The argument above is an instance of a general way of computing the deformation spaces of an algebra given by generators and relations. We will use this method again in the next section.

3. Deformations of $H^*(P_{mn})$

Let $R = H^*(P_{mn}) = \mathbb{F}[u, v]/(u^{m+1}, v^{n+1})$. Fix an even d. One can associate to any pair $(a, b) \in R^{2m+2-d} \oplus R^{2n+2-d}$ a d-dimensional deformation of R in the following way: write $a = \sum_i a_i u^{i-d/2} v^{m+1-i}$ and $b = \sum_i b_i u^{i-d/2} v^{n+1-i}$ with $a_i, b_i \in \mathbb{F}$. Let $\widetilde{R}_{a,b}$ be the algebra with generators $\tilde{u}, \tilde{v}, t$ of degrees $2, 2, d$ and relations $t^2 = 0$,

$$\tilde{u}^{m+1} + \sum_i a_i t \tilde{u}^{i-d/2} \tilde{v}^{m+1-i} = 0, \quad \tilde{v}^{n+1} + \sum_i b_i t \tilde{u}^{i-d/2} \tilde{v}^{n+1-i} = 0.$$

Then $\widetilde{R}_{a,b}$, t, and the homomorphism $j : \widetilde{R}_{a,b} \longrightarrow R$ with $j(\tilde{u}) = u$, $j(\tilde{v}) = v$, and $j(t) = 0$, define a deformation of R.

Using the same idea as in Example 2.4, one can check that for $d \neq 2$ these examples form a complete list, with no repetitions, of the d-dimensional deformations of R. For $d = 2$ there are again some isomorphisms between the $\widetilde{R}_{a,b}$ for different (a, b). In this way one obtains isomorphisms

$$\mathrm{Def}_d(R) \cong \begin{cases} R^{2m+2-d} \oplus R^{2n+2-d}, & d \neq 2, \\ R^{2m}/\mathbb{F}(m+1)u^m \oplus R^{2n}/\mathbb{F}(n+1)v^n, & d = 2. \end{cases} \tag{3.1}$$

Similarly to the case of $\mathbb{C}\mathrm{P}^n$, the homomorphisms α_{d-1} are defined on all of $\pi_{d-1}(\mathrm{Diff}(P_{mn}))$ if d is even. For the rest of this section let $\mathbb{F} = \mathbb{Q}$.

PROPOSITION 3.1. *For any even d the homomorphism*

$$\alpha_{d-1} \otimes \mathrm{id}_{\mathbb{Q}} : \pi_{d-1}(\mathrm{Diff}(P_{mn})) \otimes \mathbb{Q} \longrightarrow \mathrm{Def}_d(R)$$

is surjective.

Clearly the proof consists in finding sufficiently many examples of smooth fibre bundles. We will use the following construction. Let $\xi \longrightarrow \mathbb{C}\mathrm{P}^m \times S^d$ be a complex vector bundle of rank $n+1$, such that $\xi|\mathbb{C}\mathrm{P}^m \times \{b_0\}$ is trivial, and $\mathbb{P}(\xi)$ the associated bundle of projective spaces. The map $\mathbb{P}(\xi) \longrightarrow \mathbb{C}\mathrm{P}^m \times S^d \longrightarrow S^d$ makes $\mathbb{P}(\xi)$

into a smooth fibre bundle with fibre $\mathbb{C}\mathrm{P}^m \times \mathbb{C}\mathrm{P}^n$. $H^*(\mathbb{P}(\xi))$ is a module over $H^*(\mathbb{C}\mathrm{P}^m \times S^d)$ with one two-dimensional generator v and one relation

$$v^{n+1} + \sum_{i=0}^{n} c_i(\xi) v^{n+1-i} = 0.$$

Since $\xi|\mathbb{C}\mathrm{P}^m \times \{b_0\}$ is trivial, the Chern classes can be written as $c_i(\xi) = \gamma_i(u^{i-d/2} \otimes \epsilon)$, where $\gamma_i \in \mathbb{Z}$, u is the generator of $H^2(\mathbb{C}\mathrm{P}^m)$, and ϵ is the generator of $H^2(S^d)$. It follows that the deformation of R determined by $\mathbb{P}(\xi)$ is isomorphic to $\widetilde{R}_{0,b}$ with $b = \sum_i \gamma_i u^{i-d/2} v^{n+1-i}$. The Chern classes which can be nonzero are those with $d/2 \le i \le \nu = \min\{n+1, m+d/2\}$. Now assume that for every such i there is a vector bundle ξ_i such that $c_i(\xi_i) \neq 0$ and $c_j(\xi_i) = 0$ for all $j > i$. Since $R^{2n+2-d} = \bigoplus_{i=d/2}^{\nu} \mathbb{Q} u^{i-d/2} v^{n+1-i}$, it would follow that $0 \oplus R^{2n+2-d} \subset \mathrm{Def}_d(R)$ lies in the image of $\alpha_{d-1} \otimes \mathrm{id}_{\mathbb{Q}}$. By exchanging m and n, one would obtain the same result for the complementary subspace $R^{2m+2-d} \oplus 0$, and this would prove Proposition 3.1. Hence it remains to construct the ξ_i. In order to fulfil $c_j(\xi_i) = 0$ for $j > i$, we will take ξ_i to be the sum of a vector bundle η_i of rank i and a trivial bundle. All we need to prove is

LEMMA 3.2. *For any integer i with $d/2 \le i \le m + d/2$ there is a complex vector bundle $\eta_i \longrightarrow \mathbb{C}\mathrm{P}^m \times S^d$ of rank i such that $\eta_i|\mathbb{C}\mathrm{P}^m \times \{b_0\}$ is trivial and $c_i(\eta_i) \neq 0$.*

PROOF. Vector bundles $\eta \longrightarrow \mathbb{C}\mathrm{P}^k \times S^d$ of rank i equipped with a trivialization of $\eta|\mathbb{C}\mathrm{P}^k \times \{b_0\}$ are classified by the homotopy classes of (based) maps

$$\mathbb{C}\mathrm{P}^k \longrightarrow \Omega^{d-1} U(i) \simeq \Omega^d BU(i). \tag{3.2}$$

Since $\Omega^{d-1}U(i)$ is a topological group, two such maps can be multiplied. This multiplication associates to two vector bundles a third one, whose i-th Chern class is the sum of the i-th Chern classes of the original ones.

The obstruction to extending a map (3.2) from $\mathbb{C}\mathrm{P}^k$ to $\mathbb{C}\mathrm{P}^{k+1}$ is an element of the finite group $\pi_{2k+1}(\Omega^{d-1}U(i)) = \pi_{2k+d}(U(i))$. Hence the extension can always be carried out after replacing the original map by a positive multiple. Start with a map $\mathbb{C}\mathrm{P}^{i-d/2} \longrightarrow \Omega^{d-1}U(i)$ which collapses everything except the top-dimensional cell to a point and represents a nontrivial element of $\pi_{2i-d}(\Omega^{d-1}U(i)) \otimes \mathbb{Q} \cong \pi_{2i-1}(U(i)) \otimes \mathbb{Q} \cong \mathbb{Q}$. This map corresponds to a vector bundle over $\mathbb{C}\mathrm{P}^{i-d/2} \times S^d$ with nonzero i-th Chern class. After passing to a suitable multiple if necessary, one can extend the map to $\mathbb{C}\mathrm{P}^m$. □

4. Symplectic fibre bundles

In this section and the next one, (M, ω) is a compact symplectic manifold, $R = H^*(M)$, and $A \in H_2^s(M;\mathbb{Z}) = \mathrm{im}(\pi_2(M) \longrightarrow H_2(M;\mathbb{Z}))$ is a spherical homology class such that $\omega(A)$ is positive and generates the period group $\omega(H_2^s(M;\mathbb{Z})) \subset \mathbb{R}$. In other words, we assume that there is no $A' \in H_2^s(M;\mathbb{Z})$ with $0 < \omega(A') < \omega(A)$. Of course, such a class can only exist if the period group is discrete.

The Gromov–Witten invariant which counts rational pseudo-holomorphic A-curves with three marked points can be written as a bilinear map $\psi_A : R \times R \longrightarrow R$. This map is (graded) symmetric, of degree $-2\langle c_1(M,\omega), A\rangle$, and has the following properties:

(1) $x\psi_A(y,z) - \psi_A(xy,z) + \psi_A(x,yz) - \psi_A(x,y)z = 0$,
(2) $\psi_A(1,x) = 0$ for all x, and

(3) if $u \in R^2$ satisfies $\langle u, A\rangle = 0$, then $\psi_A(u,x) = 0$ for all x.

Define a product $*_A$ on $R \otimes \mathbb{F}[q]/q^2$ by

$$(x_0 + x_1 q) *_A (y_0 + y_1 q) = (x_0 y_0) + (x_0 y_1 + x_1 y_0 + \psi_A(x_0, y_0))q.$$

Property (1) is equivalent to the associativity of $*_A$. Property (2) says that $1 \in R$ is a unit for $*_A$, and property (3) can be written as $u *_A (x_0 + x_1 q) = u x_0 + (u x_1) q$.

DEFINITION 4.1. Let $(\widetilde{R}, t, j)$ be a deformation of R. An *extension* of ψ_A to $\widetilde{R}$ is a bilinear map $\widetilde{\psi}_A : \widetilde{R} \times \widetilde{R} \longrightarrow \widetilde{R}$ of degree $-2\langle c_1(M,\omega), A\rangle$ which is (graded) symmetric and has the following properties:

(1′) $x\widetilde{\psi}_A(y,z) - \widetilde{\psi}_A(xy,z) + \widetilde{\psi}_A(x,yz) - \widetilde{\psi}_A(x,y)z = 0$;

(2′) $\widetilde{\psi}_A(1,x) = \widetilde{\psi}_A(t,x) = 0$ for all x;

(3′) if $u \in \widetilde{R}^2$ satisfies $\langle j(u), A\rangle = 0$, then $\widetilde{\psi}_A(u,x) = 0$ for all x;

(4′) $j(\widetilde{\psi}_A(x,y)) = \psi_A(j(x), j(y))$.

These properties can again be interpreted in terms of a suitably defined product $\tilde{*}_A$ on $\widetilde{R} \otimes \mathbb{F}[q]/q^2$. (4′) says that

$$j \otimes \mathrm{id}_{\mathbb{F}[q]/q^2} : (\widetilde{R} \otimes \mathbb{F}[q]/q^2, \tilde{*}_A) \longrightarrow (R \otimes \mathbb{F}[q]/q^2, *_A)$$

is a ring homomorphism. The deformations of R which admit an extension of ψ_A form a subset $\mathrm{Def}_d(R, \psi_A) \subset \mathrm{Def}_d(R)$. We leave it to the reader to verify that this is actually a linear subspace.

Let (E, π, i) be a smooth fibre bundle over S^d with fibre M. A fibrewise symplectic structure on E is a smooth family $\Omega = (\Omega_b)_{b \in S^d}$ of symplectic forms on the fibres E_b such that the cohomology class $[\Omega_b] \in H^2(E_b; \mathbb{R})$ is locally constant in b. If in addition $i^*(\Omega_{b_0}) = \omega$, we call (E, π, i, Ω) a symplectic fibre bundle with fibre (M, ω). The isomorphism classes of such bundles form a group under fibre connected sum, and this group is isomorphic to $\pi_{d-1}(\mathrm{Aut}(M,\omega))$. Forgetting Ω corresponds to passing from $\mathrm{Aut}(M,\omega)$ to $\mathrm{Diff}(M)$.

PROPOSITION 4.2. *Let (E, π, i, Ω) be a symplectic fibre bundle over S^d $(d \geq 2)$ with fibre (M, ω), such that i^* is onto. Then the deformation of R determined by $H^*(E)$ admits an extension of ψ_A.*

In the case when $G_{d-1} = \pi_{d-1}(\mathrm{Diff}(M))$, the situation can be summarized in the commutative diagram

$$\begin{array}{ccc} \pi_{d-1}(\mathrm{Aut}(M,\omega)) & \longrightarrow & \pi_{d-1}(\mathrm{Diff}(M)) \\ \big\downarrow & & \big\downarrow{\scriptstyle \alpha_{d-1}} \\ \mathrm{Def}_d(R, \psi_A) & \hookrightarrow & \mathrm{Def}_d(R). \end{array}$$

5. Proof of Proposition 4.2

We begin by recalling the definition of the Gromov–Witten invariant ψ_A. Let J be an ω-compatible almost complex structure on M, $\mathcal{M}(A,J)$ the moduli space of J-holomorphic maps $u : \mathbb{C}\mathrm{P}^1 \longrightarrow M$ which represent A, and

$$\mathcal{M}_r(A,J) = \mathcal{M}(A,J) \times_{PSL_2(\mathbb{C})} (\mathbb{C}\mathrm{P}^1)^r$$

$(r = 0, 1, 2, \dots)$ the associated moduli spaces of marked J-holomorphic curves. Our assumptions on A imply that any $u \in \mathcal{M}(A,J)$ is simple (not multiply covered) and that A cannot be represented by a J-holomorphic cusp curve. It follows from the standard theory of pseudo-holomorphic curves (see **[12]** for an exposition) that for generic J the spaces $\mathcal{M}_r(A, J)$ are compact oriented smooth manifolds. Let $e_r : \mathcal{M}_r(A, J) \longrightarrow M^r$ be the r-fold evaluation map. ψ_A is defined by

$$\langle \psi_A(x, y)z, [M] \rangle = \langle x \times y \times z, (e_3)_*[\mathcal{M}_3(A, J)] \rangle$$

for $x, y, z \in H^*(M)$ and generic J. A cobordism argument [**12**, Theorem 3.1.3] shows that ψ_A is independent of the choice of J.

Now let (E, π, i, Ω) be a symplectic fibre bundle over S^d ($d \geq 2$) with fibre (M, ω). The assumption that $d \geq 2$ implies that the homology of any fibre E_b can be canonically identified with the homology of E_{b_0} and hence (using i) with that of M. Therefore it makes sense to say that a map $u : \mathbb{CP}^1 \longrightarrow E_b$ represents $A \in H_2^s(M; \mathbb{Z})$. Let $\mathbf{J} = (J_b)_{b \in S^d}$ be a family of almost complex structures on the fibres E_b such that J_b is Ω_b-compatible for all b. Equivalently, one can think of $\mathbf{J}$ as an almost complex structure on the subbundle $TE^v = \ker(T\pi) \subset TE$, with the property that $\Omega(\cdot, \mathbf{J}\cdot)$ is a metric on TE^v. Let $\mathcal{M}^p(A, \mathbf{J})$ be the space of pairs (b, u) where $b \in S^d$ and $u : \mathbb{CP}^1 \longrightarrow E_b$ is a J_b-holomorphic map which represents A in the sense explained above. $\mathcal{M}^p(A, \mathbf{J})$ is called the *parametrized moduli space* of rational $\mathbf{J}$-holomorphic A-curves on E. As before, we define

$$\mathcal{M}_r^p(A, \mathbf{J}) = \mathcal{M}^p(A, \mathbf{J}) \times_{PSL_2(\mathbb{C})} (\mathbb{CP}^1)^r.$$

The basic theory of pseudo-holomorphic curves carries over to the parametrized situation (this is familiar from the cobordism argument mentioned above, in which one considers one-parameter families of almost complex structures on M). In particular, for generic $\mathbf{J}$ the spaces $\mathcal{M}_r^p(A, \mathbf{J})$ are again compact smooth oriented manifolds; their dimension is

$$\dim \mathcal{M}_r^p(A, \mathbf{J}) = \dim E + 2\langle c_1(M, \omega), A \rangle + 2(r - 3).$$

The space $\mathcal{M}_r^p(A, \mathbf{J})$ comes with two canonical maps: the projection $P_r : \mathcal{M}_r^p(A, \mathbf{J}) \longrightarrow S^d$ and the r-fold evaluation map $e_r^p : \mathcal{M}_r^p(A, \mathbf{J}) \longrightarrow E^r$. Note that in general P_r is not a smooth fibration. We define the *parametrized Gromov–Witten invariant* $\widetilde{\psi}_A : H^*(E) \times H^*(E) \longrightarrow H^*(E)$ by

$$\langle \widetilde{\psi}_A(x, y)z, [E] \rangle = \langle x \times y \times z, (e_3^p)_*[\mathcal{M}_3^p(A, \mathbf{J})] \rangle$$

for generic $\mathbf{J}$. Here $[E]$ is the orientation induced by the symplectic orientation of M and the standard orientation of S^d. An argument similar to that for ψ_A proves that $\widetilde{\psi}_A$ is independent of $\mathbf{J}$.

The three properties of ψ_A listed in Section 4 are special cases of properties which are known to hold for far more general Gromov–Witten invariants. Nevertheless, we will outline proofs of them, making use of the special features of our case to simplify the argument.

(1) Consider the cycles $Z, Z' \subset \mathcal{M}_4(A, J)$ consisting of those $[u, z_1, z_2, z_3, z_4]$ such that $z_1 = z_2$ or $z_3 = z_4$ (for Z) (resp. $z_2 = z_3$ or $z_1 = z_4$ (for Z')). One can prove easily, using the cross-ratio

$$(z_1, z_2, z_3, z_4) \longmapsto \frac{(z_1 - z_2)(z_3 - z_4)}{(z_2 - z_3)(z_1 - z_4)},$$

that Z and Z' are homologous. On the other hand,

$$\begin{aligned}\langle x\times y\times z\times w,(e_4)_*[Z]\rangle &= \langle xy\times z\times w + x\times y\times zw,(e_3)_*[\mathcal{M}_3(A,J)]\rangle\\ &= \langle \psi_A(xy,z)w+\psi_A(x,y)zw,[M]\rangle,\end{aligned}$$

and similarly

$$\langle x\times y\times z\times w,(e_4)_*[Z']\rangle = \langle \psi_A(x,yz)w + x\psi_A(y,z)w,[M]\rangle.$$

(2) Let $\phi:\mathcal{M}_3(A,J)\longrightarrow\mathcal{M}_2(A,J)$ be the map which forgets the first marked point. There is a commutative diagram

$$\begin{array}{ccc}\mathcal{M}_3(A,J) & \xrightarrow{\;e_3\;} & M^3\\ {\scriptstyle\phi}\downarrow & & \downarrow{\scriptstyle p}\\ \mathcal{M}_2(A,J) & \xrightarrow{\;e_2\;} & M^2\end{array}$$

where p is the projection. Because $\dim\mathcal{M}_2(A,J)=\dim\mathcal{M}_3(A,J)-2$, we have $\phi_*[\mathcal{M}_3(A,J)]=0$. This implies that

$$\langle 1\times y\times z,(e_3)_*[\mathcal{M}_3(A,J)]\rangle = \langle y\times z,p_*(e_3)_*[\mathcal{M}_3(A,J)]\rangle = 0.$$

(3) The forgetful map ϕ is a fibration with fibre $\mathbb{C}P^1$. Let $\phi_!$ be the cohomology transfer. Then

$$\begin{aligned}\langle\psi_A(u,y)z,[M]\rangle &= \langle e_3^*(u\times y\times z),[\mathcal{M}_3(A,J)]\rangle\\ &= \langle e_3^*(u\times 1\times 1)\phi^*(e_2^*(y\times z)),[\mathcal{M}_3(A,J)]\rangle\\ &= \langle \phi_!(e_3^*(u\times 1\times 1))e_2^*(y\times z),[\mathcal{M}_2(A,J)]\rangle\end{aligned}$$

for all $u,y,z\in H^*(M)$. Now assume that u is two-dimensional. Since e_3 maps the homology class of the fibre of ϕ to $A\times[\mathit{point}]\times[\mathit{point}]\in H_2(M^3)$, one has

$$\phi_!e_3^*(u\times 1\times 1)=\langle u,A\rangle 1\in H^0(\mathcal{M}_2(A,J)).$$

Hence $\psi_A(u,y)$ vanishes if $\langle u,A\rangle=0$.

To prove Proposition 4.2 we have to show that $\widetilde{\psi}_A$ is an extension of ψ_A in the sense of Definition 4.1. The first three properties listed there are analogues of the properties of ψ_A, and the proofs given above can be easily adapted to the parametrized case. The remaining property is

$$i^*(\widetilde{\psi}_A(x,y))=\psi_A(i^*(x),i^*(y)). \tag{5.1}$$

As usual, let $t\in H^d(E)$ be the pullback of the fundamental class of S^d; then $1\times 1\times t\in H^d(E^3)$ is Poincaré dual to the submanifold $C=E\times E\times E_{b_0}\subset E^3$. Therefore

$$\begin{aligned}\langle i^*(\widetilde{\psi}_A(x,y))i^*(z),[M]\rangle &= \langle\widetilde{\psi}_A(x,y)z,[E_{b_0}]\rangle\\ &= \langle\widetilde{\psi}_A(x,y)zt,[E]\rangle\\ &= \langle x\times y\times zt,(e_3^p)_*[\mathcal{M}_3^p(A,\mathbf{J})]\rangle\\ &= \langle x\times y\times z,(e_3^p)_*[\mathcal{M}_3^p(A,\mathbf{J})]\circ[C]\rangle\end{aligned}$$

for all $x,y,z\in H^*(E)$. Here $\circ$ denotes the intersection product on $H_*(E^3)$. Now, if we assume that

$$(e_3^p)_*[\mathcal{M}_3^p(A,\mathbf{J})]\circ[C]=(i\times i\times i)_*(e_3)_*[\mathcal{M}_3(A,J)], \tag{5.2}$$

we obtain

$$\begin{aligned}\langle i^*(\widetilde{\psi}_A(x,y))i^*(z),[M]\rangle &= \langle i^*(x)\times i^*(y)\times i^*(z),(e_3)_*[\mathcal{M}_3(A,\mathbf{J})]\rangle\\ &= \langle \psi_A(i^*(x),i^*(y))i^*(z),[M]\rangle.\end{aligned}$$

Since $i^*(z)$ can be any element of $H^*(M)$ (we have assumed in Proposition 4.2 that i^* is onto), this implies (5.1).

Equation (5.2) expresses the (rather obvious) fact that for suitable choices of J and $\mathbf{J}$, the unparametrized moduli space $\mathcal{M}_3(A,J)$ can be identified with a fibre of the projection $P_3 : \mathcal{M}_3^p(A,\mathbf{J}) \longrightarrow S^d$. A more precise formulation of the argument goes as follows: choose J and $\mathbf{J}$ in such a way that the moduli spaces $\mathcal{M}_3(A,J)$ and $\mathcal{M}_3^p(A,\mathbf{J})$ are smooth and $i_*(J) = J_{b_0}$; this is possible. Then e_3^p is transverse to C, and $(e_3^p)^{-1}(C) \subset \mathcal{M}_3^p(A,\mathbf{J})$, which consists of those elements $[b,u,z_1,z_2,z_3]$ with $b=b_0$, can be identified with $\mathcal{M}_3(A,J)$ in such a way that the diagram

$$\begin{array}{ccc}\mathcal{M}_3(A,J) & \xrightarrow{e_3} & M^3\\ \Big\downarrow{=} & & \Big\downarrow{i\times i\times i}\\ (e_3^p)^{-1}(C) & \xrightarrow{e_3^p} & E^3\end{array}$$

commutes. This proves (5.2) and hence completes the proof of Proposition 4.2.

6. The main computation

Let R_1, R_2 be a pair of graded $\mathbb{F}$-algebras, and $R = R_1\otimes R_2$ their graded tensor product. Given a deformation $(\widetilde{R}_1, t_1, j_1)$ of R_1 and a deformation $(\widetilde{R}_2, t_2, j_2)$ of R_2 of the same dimension, one can define a deformation $(\widetilde{R}, t, j)$ of R by $\widetilde{R} = \widetilde{R}_1 \otimes_{\mathbb{F}[\epsilon]/\epsilon^2} \widetilde{R}_2$ (where ϵ acts on $\widetilde{R}_i$ by multiplication with t_i), $t = t_1\otimes 1 = 1\otimes t_2$, and $j = j_1\otimes j_2$. This operation, which we call the exterior product of deformations, defines homomorphisms $\times_d : \mathrm{Def}_d(R_1)\oplus \mathrm{Def}_d(R_2) \longrightarrow \mathrm{Def}_d(R)$. A deformation of R is said to be *split* if it is isomorphic to the exterior product of deformations of R_i, that is, if its isomorphism class lies in the image of $\times_d$.

REMARK 6.1. In the alternative picture of Remark 2.2 the exterior product is defined by assigning to a pair of bilinear maps $\psi_i : R_i\otimes R_i \longrightarrow R_i$ the product

$$\begin{aligned}\psi(x_1\otimes x_2, y_1\otimes y_2) = (-1)^{\deg(x_2)\deg(y_1)}\big(&\psi_1(x_1,y_1)\otimes x_2y_2\\ &+ (-1)^{\deg(x_1y_1)d}x_1y_1\otimes\psi_2(x_2,y_2)\big).\end{aligned}$$

The main result of this section is

PROPOSITION 6.2. *Let (E,π,i,Ω) be a symplectic fibre bundle over S^d, for some even d, with fibre (P_{mn},η_{mn}). Then the deformation of $H^*(P_{mn}) = H^*(\mathbb{CP}^m)\otimes H^*(\mathbb{CP}^n)$ determined by $H^*(E)$ is split.*

We will obtain this as a consequence of a more general result. To state that result we need to introduce one more algebraic notion. Let $f_i : R_i \longrightarrow R$ be the obvious inclusions.

DEFINITION 6.3. Let $(\widetilde{R},t,j)$ be a deformation of R. We say that $\widetilde{R}$ is *semi-split* with respect to R_i ($i = 1$ or 2) if there exist a deformation $(\widetilde{R}_i, t_i, j_i)$ of R_i and a morphism $\tilde{f}_i : \widetilde{R}_i \longrightarrow \widetilde{R}$ which lies over f_i.

A split deformation is clearly semi-split with respect to both R_1 and R_2. The converse it also true: a deformation of R which is semi-split with respect to both R_1 and R_2 is split. Later, we will use the following elementary criterion:

LEMMA 6.4. *Let $(\widetilde{R}, t, j)$ be a deformation of R. If $\widetilde{R}$ has a subalgebra $\widetilde{R}_1$ with $t \in \widetilde{R}_1$, $j(\widetilde{R}_1) = \mathrm{im}(f_1)$ and $\dim(\widetilde{R}_1 \cap t\widetilde{R}) \leq \dim R_1$ then it is semi-split with respect to R_1.* □

Let M' be a compact manifold whose cohomology ring is generated by $H^2(M')$, and ω' a symplectic form on M' such that $\omega'(H_2^s(M';\mathbb{Z})) \subset \mathbb{Z}$. We consider the product $(M, \omega) = (M', \omega') \times (\mathbb{C}P^n, \omega_n)$ for some n. Let $A \in H_2^s(M;\mathbb{Z})$ be the homology class of a line in $\mathbb{C}P^n$. A satisfies the conditions of section 3, and the corresponding Gromov–Witten invariant is well known:

$$\psi_A(x \otimes u^i, y \otimes u^j) = \begin{cases} xy \otimes u^{i+j-n-1} & \text{if } i+j \geq n+1, \\ 0 & \text{otherwise,} \end{cases} \tag{6.1}$$

for all $x, y \in H^*(M')$ and $0 \leq i, j \leq n$ (as usual, u denotes the standard generator of $H^2(\mathbb{C}P^n)$).

PROPOSITION 6.5. *Let (E, π, i, Ω) be a symplectic fibre bundle with fibre (M, ω) over S^d for some even d. Then the deformation of $R = H^*(M)$ given by $H^*(E)$ is semi-split with respect to $H^*(M')$.*

Setting $(M', \omega') = (\mathbb{C}P^m, \omega_m)$, it follows that a symplectic fibre bundle over S^d with fibre (P_{mn}, η_{mn}) determines a deformation of $H^*(P_{mn})$ which is semi-split with respect to $H^*(\mathbb{C}P^m)$. Since the situation is symmetric with respect to m and n, it follows that the deformation is also semi-split with respect to $H^*(\mathbb{C}P^n)$, hence split. Therefore Proposition 6.5 implies Proposition 6.2.

PROOF OF PROPOSITION 6.5. Note that i^* is onto because d is even and $H^*(M)$ is concentrated in even dimensions. Let $(\widetilde{R}, t, j)$ be the deformation of R determined by $H^*(E)$. Proposition 4.2 ensures the existence of an extension $\widetilde{\psi}_A$ of ψ_A to $\widetilde{R}$. Choose a basis $\xi_1, \ldots, \xi_g$ of $H^2(M')$ and elements $\tilde{\xi}_1, \ldots, \tilde{\xi}_g, \tilde{u} \in \widetilde{R}^2$ such that $j(\tilde{\xi}_i) = \xi_i \otimes 1$ and $j(\tilde{u}) = 1 \otimes u$. Let $\widetilde{R}_1 \subset \widetilde{R}$ be the subalgebra generated by $t, \tilde{\xi}_1, \ldots, \tilde{\xi}_g$. Clearly $j(\widetilde{R}_1) = H^*(M') \otimes 1$. We want to apply Lemma 6.4 to this subalgebra. What remains to be shown is that the dimension of $\widetilde{R}_1 \cap t\widetilde{R}$ is not greater than the dimension of $H^*(M')$. This is a consequence of the following fact.

ASSERTION. *Every $x \in \widetilde{R}_1 \cap t\widetilde{R}$ is of the form $x = ty$ for some y such that $j(y) \in H^*(M') \otimes 1$.*

The assertion is proved in two steps:

Step 1: *Every $x \in \widetilde{R}_1$ satisfies $\widetilde{\psi}_A(x, \tilde{u}^n) = 0$.* Since $\langle j(\tilde{\xi}_i), A \rangle = \langle \xi_i, A \rangle = 0$, we have $\widetilde{\psi}_A(\tilde{\xi}_i, x) = 0$ for all x by property (3′) of $\widetilde{\psi}_A$. Therefore the equation

$$\tilde{\xi}_i \widetilde{\psi}_A(y, \tilde{u}^n) - \widetilde{\psi}_A(\tilde{\xi}_i y, \tilde{u}^n) + \widetilde{\psi}_A(\tilde{\xi}_i, y\tilde{u}^n) - \widetilde{\psi}_A(\tilde{\xi}_i, y)\tilde{u}^n = 0,$$

which is a special case of property (1′) of $\widetilde{\psi}_A$, simplifies to

$$\widetilde{\psi}_A(\tilde{\xi}_i y, \tilde{u}^n) = \tilde{\xi}_i \widetilde{\psi}_A(y, \tilde{u}^n). \tag{6.2}$$

Note that by property (2′) $\widetilde{\psi}_A(1, \tilde{u}^n) = \widetilde{\psi}_A(t, \tilde{u}^n) = 0$. Arguing inductively using (6.2), we find that $\widetilde{\psi}_A(x, \tilde{u}^n) = 0$ for all $x \in \widetilde{R}_1$.

Step 2. *If $x = ty \in t\widetilde{R}$ satisfies $\widetilde{\psi}_A(x, \tilde{u}^n) = 0$, then $j(y) \in H^*(M') \otimes 1$.* Since $\widetilde{\psi}_A(t, w) = 0$ for all w, the equation $t\widetilde{\psi}_A(y, \tilde{u}^n) - \widetilde{\psi}_A(x, \tilde{u}^n) + \widetilde{\psi}_A(t, y\tilde{u}^n) - \widetilde{\psi}_A(t, y)\tilde{u}^n = 0$ reduces to

$$\widetilde{\psi}_A(x, \tilde{u}^n) = t\widetilde{\psi}_A(y, \tilde{u}^n). \tag{6.3}$$

Now write $j(y) = \sum_{i=0}^n c_i \otimes u^i$ with $c_i \in H^*(M')$, and choose $\tilde{c}_0, \dots, \tilde{c}_n \in \widetilde{R}$ with $j(\tilde{c}_i) = c_i$. Using property (4′) of $\widetilde{\psi}_A$ and the formula (6.1) for ψ_A, one sees that

$$j(\widetilde{\psi}_A(y, \tilde{u}^n)) = \psi_A(j(y), u^n) = \sum_{i=1}^n c_i \otimes u^{i-1}.$$

Therefore $\widetilde{\psi}_A(y, \tilde{u}^n) = \sum_{i=1}^n \tilde{c}_i \tilde{u}^{i-1} + tz$ for some $z \in \widetilde{R}$. Using (6.3) it follows that

$$\widetilde{\psi}_A(x, \tilde{u}^n) = t\Big(\sum_{i=1}^n \tilde{c}_i \tilde{u}^{i-1}\Big).$$

If this vanishes, then $t\tilde{c}_i = 0$ and hence $c_i = 0$ for all $i > 0$, which means that $j(y) \in H^*(M') \otimes 1$. □

7. Conclusion

Now we can prove Theorem 1.1. Set $\mathbb{F} = \mathbb{Q}$, and let $R = H^*(P_{mn})$. Fix an even d, and let $\mathrm{Def}_d^s(R) \subset \mathrm{Def}_d(R)$ be the subspace of split deformations. According to Proposition 6.2 there is a commutative diagram

$$\begin{array}{ccc} \pi_{d-1}(\mathrm{Aut}(P_{mn}, \eta_{mn})) \otimes \mathbb{Q} & \xrightarrow{\beta_{d-1} \otimes \mathrm{id}_{\mathbb{Q}}} & \pi_{d-1}(\mathrm{Diff}(M)) \otimes \mathbb{Q} \\ \downarrow & & \downarrow {\scriptstyle \alpha_{d-1} \otimes \mathrm{id}_{\mathbb{Q}}} \\ \mathrm{Def}_d^s(R) & \hookrightarrow & \mathrm{Def}_d(R) \end{array}$$

Proposition 3.1 shows that $\alpha_{d-1} \otimes \mathrm{id}_{\mathbb{Q}}$ is onto. Therefore

$$\mathrm{rank}(\mathrm{coker}\, \beta_{d-1}) \geq \dim_{\mathbb{Q}} \mathrm{Def}_d(R) - \dim_{\mathbb{Q}} \mathrm{Def}_d^s(R).$$

Assume first that $d > 2$. Using Example 2.4 and the results of section 3, one can identify the subspace $\mathrm{Def}_d^s(R) \subset \mathrm{Def}_d(R)$ with $H^{2m+2-d}(\mathbb{C}\mathrm{P}^m) \oplus H^{2n+2-d}(\mathbb{C}\mathrm{P}^n) \subset R^{2m+2-d} \oplus R^{2n+2-d}$. Therefore

$$\begin{aligned} \mathrm{rank}(\mathrm{coker}\, \beta_{d-1}) \geq\; & b_{2m+2-d}(P_{mn}) - b_{2m+2-d}(\mathbb{C}\mathrm{P}^m) \\ & + b_{2n+2-d}(P_{mn}) - b_{2n+2-d}(\mathbb{C}\mathrm{P}^n). \end{aligned}$$

This is positive as long as $d \leq \max\{2m, 2n\}$. For $d = 2$ one has $\mathrm{Def}_2^s(R) = 0$ (by Example 2.4), and hence

$$\begin{aligned} \mathrm{rank}(\mathrm{coker}\, \beta_1) &\geq \dim_{\mathbb{Q}} \mathrm{Def}_2(R) = \dim_{\mathbb{Q}} R^{2m} + \dim_{\mathbb{Q}} R^{2n} - 2 \\ &= b_{2m}(P_{mn}) - b_{2m}(\mathbb{C}\mathrm{P}^m) + b_{2n}(P_{mn}) - b_{2n}(\mathbb{C}\mathrm{P}^n). \end{aligned}$$

□

One can ask what happens if one considers P_{mn} with a weighted symplectic form $\eta_{mn}^{(\lambda)} = \lambda(\omega_m \times 1) + 1 \times \omega_n$, where $\lambda > 1$. Proposition 6.2 is no longer true for symplectic fibre bundles with fibre $(P_{mn}, \eta_{mn}^{(\lambda)})$, at least if λ is sufficiently large. Indeed, all the examples constructed in Proposition 3.1, which were of the form $E = \mathbb{P}(\xi)$ for some $\xi \longrightarrow \mathbb{C}\mathrm{P}^m \times S^d$, admit a fibrewise symplectic structure modelled on $\eta_{mn}^{(\lambda)}$ for some large λ, while the corresponding deformations are not always split.

If λ is integral, we can still use Proposition 6.5 to obtain some information about symplectic fibre bundles with fibre $(P_{mn}, \eta_{mn}^{(\lambda)})$. This leads to

THEOREM 7.1. *Let $\lambda \geq 2$ be an integer. Then the homomorphism*

$$\beta_k^{(\lambda)} : \pi_k(\mathrm{Aut}(P_{mn}, \eta_{mn}^{(\lambda)})) \longrightarrow \pi_k(\mathrm{Diff}(P_{mn}))$$

induced by inclusion is not surjective for any odd k with $1 \leq k \leq 2m-1$. In fact,

$$\mathrm{rank}(\mathrm{coker}\, \beta_k^{(\lambda)}) \geq b_{2m+1-k}(P_{mn}) - b_{2m+1-k}(\mathbb{C}\mathrm{P}^m) > 0.$$

We omit the proof, which is similar to that of Theorem 1.1. For $\lambda \notin \mathbb{N}$ it becomes possible that the class $A \in H_2^s(P_{mn}; \mathbb{Z})$ coming from the generator of $H_2(\mathbb{C}\mathrm{P}^n; \mathbb{Z})$ can be represented by a rational pseudo-holomorphic cusp-curve. However, our argument would still work if one could prove that these cusp-curves can be removed by perturbing the almost complex structure. More precisely, one needs to show that no such cusp-curves occur in a generic family of compatible almost complex structures depending on a certain number of parameters.

EXAMPLE 7.2. Let (E, π, i, Ω) be a symplectic fibre bundle over S^2 with fibre $(P_{21}, \eta_{21}^{(\lambda)})$. Choose a family $\mathbf{J} = (J_b)_{b \in S^2}$ of compatible almost complex structures on its fibres. If A is represented by a rational J_b-holomorphic cusp-curve for some b, there must be $A_1, A_2 \in H_2^s(P_{21}; \mathbb{Z})$ with $A_1 + A_2 = A$, such that $\omega(A_1), \omega(A_2) > 0$ and $\mathcal{M}^f(A_1, \mathbf{J}), \mathcal{M}^f(A_2, \mathbf{J}) \neq \emptyset$. The virtual dimensions of these moduli spaces are $\mathrm{v.dim}.\mathcal{M}^f(A_1, \mathbf{J}) = 8 + 2\langle c_1, A_1 \rangle$, $\mathrm{v.dim}.\mathcal{M}^f(A_2, \mathbf{J}) = 8 + 2\langle c_1, A_2 \rangle$. A pair (A_1, A_2) such that both dimensions are nonnegative exists iff there are integers k, l with $1 > \lambda k + l > 0$ and $6 \geq 3k + 2l \geq -4$. An elementary argument shows that this is possible only if $\lambda < 3$. Therefore it is plausible that if $\lambda \geq 3$ then the deformation of $H^*(P_{21})$ determined by $H^*(E)$ is semi-split with respect to $H^*(\mathbb{C}\mathrm{P}^2)$. To make this argument rigorous, one has to deal with the problem of multiply-covered pseudo-holomorphic maps of negative Chern number, in the manner of **[9]**, **[13]**, **[15]**.

It is interesting to compare our approach with the geometric methods used by Gromov and others in the four-dimensional case. For instance, consider the following result, which is a version of the main step in the proof of Gromov's theorem on $\mathrm{Aut}(P_{11}, \eta_{11})$.

PROPOSITION 7.3 (Gromov). *Any symplectic fibre bundle over S^d $(d \geq 2)$ with fibre (P_{11}, η_{11}) is the fibre product of two symplectic fibre bundles with fibre $(\mathbb{C}\mathrm{P}^1, \omega_1)$.*

Proposition 6.2 is a higher-dimensional analogue of this in which the splitting takes place only on the level of cohomology rings. The reason for the stronger nature of the four-dimensional result is that the geometric behaviour of pseudo-holomorphic curves on a symplectic four-manifold is tightly controlled by the positive intersection theorem and the adjunction inequality **[11]**, both of which are used in the proof of Proposition 7.3. We end this discussion with a general conjecture suggested by Proposition 6.5.

CONJECTURE 7.4. *Let (M, ω) and (N, η) be compact symplectic manifolds. Let (E, π, i) be a smooth fibre bundle over S^d with fibre $M \times N$, such that $i^* : H^*(E) \longrightarrow H^*(M \times N)$ is surjective. Assume that for all sufficiently large λ there is a fibre-wise symplectic structure $\Omega^{(\lambda)}$ on E such that $i^*(\Omega^{(\lambda)}) = \lambda(\omega \times 1) + 1 \times \eta$. Then* *the deformation of* $H^*(M \times N)$ *determined by* $H^*(E)$ *is semi-split with respect to* $H^*(M)$.

References

[1] M. Abreu, *Topology of symplectomorphism groups of* $S^2 \times S^2$, Invent. Math. **131** (1998), 1–24.

[2] M. Abreu and D. McDuff, *Topology of symplectomorphism groups of rational ruled surfaces*, In preparation.

[3] P. Biran, *Connectedness of spaces of symplectic embeddings*, Internat. Math. Res. Notices **1996**, 487–491.

[4] M. Gerstenhaber and S. Schack, *Algebraic cohomology and deformation theory*, Deformation theory of algebras and structures and applications, NATO Adv. Sci. Inst. Ser. C, vol. 247, Kluwer, Dorderecht, 1988, pp. 11–264.

[5] M. Gromov, *Pseudoholomorphic curves in symplectic manifolds*, Invent. Math. **82** (1985), 307–347.

[6] M. Kontsevich and Yu. I. Manin, *Gromov-Witten classes, quantum cohomology, and enumerative geometry*, Commun. Math. Phys. **164** (1994), 525–562.

[7] F. Lalonde, D. McDuff, and L. Polterovich, *Topological rigidity of Hamiltonian loops and quantum homology*, Invent. Math. **135** (1999), 369–385.

[8] Hong-Van Le, *Topology of symplectomorphism groups and pseudoholomorphic curves*, Preprint, 1997.

[9] J. Li and G. Tian, *Virtual moduli cycles and Gromov-Witten invariants of general symplectic manifolds*, Topics in Symplectic 4-manifolds (Irvine, CA, 1996), Internat. Press, Cambridge, MA, 1998, pp. 47–83.

[10] D. McDuff, *From symplectic deformation to isotopy*, Topics in Symplectic 4-manifolds (Irvine, CA, 1996), Internat. Press, Cambridge, MA, 1998, pp. 85–99.

[11] ———, *Singularities and positivity of intersections of J-holomorphic curves*, Holomorphic Curves in Symplectic Geometry (M. Audin and J. Lafontaine, eds.), Progress in Mathematics, vol. 117, Birkhäuser, 1994, pp. 191–215.

[12] D. McDuff and D. Salamon, *J-holomorphic curves and quantum cohomology*, University Lecture Notes, vol. 6, Amer. Math. Soc., Providence, RI, 1994.

[13] Y. Ruan, *Virtual neighborhoods and pseudo-holomorphic curves*, Preprint alg-geom/9611021.

[14] P. Seidel, *Floer homology and the symplectic isotopy problem*, Ph.D. thesis, Oxford University, 1997.

[15] B. Siebert, *Gromov-Witten invariants of general symplectic manifolds*, Preprint dg-ga/9608005.

[16] D. Sullivan, *Infinitesimal computations in topology*, Publ. Math. IHES **47** (1977), 269–332.

INSTITUTE FOR ADVANCED STUDY, PRINCETON
E-mail address: pseidel@math.ias.edu

Amer. Math. Soc. Transl.
(2) Vol. **196**, 1999

On the Cohomology Rings of Hamiltonian T-spaces

Susan Tolman and Jonathan Weitsman

ABSTRACT. Let M be a symplectic manifold equipped with a Hamiltonian action of a torus T. Let F denote the fixed point set of the T-action and let $i : F \hookrightarrow M$ denote the inclusion. By a theorem of Kirwan, the induced map $i^* : H_T^*(M) \longrightarrow H_T^*(F)$ in equivariant cohomology is an injection. We give a simple proof of a formula of Goresky, Kottwitz, and MacPherson for the image of the map i^*.

1. Introduction

The classification of manifolds equipped with group actions presents difficulties beyond those inherent in the study of manifolds *per se*. Even the basic questions—What is the equivariant cohomology ring of the manifold? What can be said about the fixed manifolds of the group action? What is the cohomology ring of the quotient?—turn out to be delicate and involved.

Much more is known in the case of a symplectic manifold (M^{2m}, ω) equipped with a Hamiltonian action of a torus T. Let $H_T^*(M)$ denote the rational equivariant cohomology ring of M.

The following theorem of F. Kirwan relates the equivariant cohomology of M with the equivariant cohomology of its fixed point set.

THEOREM 1.1 (Kirwan, [**K**]). *Let a torus T act on a compact symplectic manifold (M, ω) in a Hamiltonian fashion. Denote the fixed point set of the action by F. The natural inclusion $i : F \longrightarrow M$ of the fixed point set in the manifold induces an injection $i^* : H_T^*(M) \hookrightarrow H_T^*(F)$.*

A related result, also due to Kirwan, relates the equivariant cohomology of M and the cohomology of the symplectic quotient of M.

THEOREM 1.2 (Kirwan, [**K**]). *Let a torus T act on a compact symplectic manifold (M, ω) with moment map $\mu : M \longrightarrow \mathfrak{t}^* = \mathrm{Lie}(T)^*$. Suppose that $\xi \in \mathfrak{t}^*$ is a regular value of μ, and let $M_\xi := \mu^{-1}(\xi)/T$ be the symplectic quotient of M.*

1991 *Mathematics Subject Classification.* Primary 55N91.

S. Tolman was partially supported by an NSF Mathematical Sciences Postdoctoral Research Fellowship, by an NSF grant, and by an Alfred P. Sloan Foundation Fellowship. J. Weitsman was supported in part by NSF grant DMS 94/03567, by NSF Young Investigator grant DMS 94/57821, and by an Alfred P. Sloan Foundation Fellowship.

©1999 American Mathematical Society

The inclusion map $K : \mu^{-1}(\xi) \longrightarrow M$ *induces a surjection* $K^* : H_T^*(M) \longrightarrow H_T^*(\mu^{-1}(\xi)) = H^*(M_\xi)$.

These two results give rise to two natural questions. What is the image of i^*? And what is the kernel of K^*? The purpose of this paper is to address the first question. In a companion paper [**TW**], we answer the second.

The description we give of the image of the injective map i^* is due to Goresky, Kottwitz, and MacPherson [**GKM**], using the work of Chang and Skjelbred ([**CS**]; see also [**AP, BV, H**]).

DEFINITION 1.3. Let $N \subset M$ denote the subset of M consisting of points whose T-orbits are one-dimensional, i.e.,

$$N := \{p \in M \mid T \cdot p \sim S^1\}.$$

By the local normal form theorem, each connected component N_α of N is an open symplectic manifold whose closure $\overline{N_\alpha}$ is a compact, symplectic submanifold of M. The restriction of μ to $\overline{N_\alpha}$ is a moment map for the restricted torus action. Furthermore the closure of N is given by $\overline{N} = N \cup F$; this is referred to as the **one-skeleton** of M.

EXAMPLE 1.4. Consider the case where M^{2m} is a $2m$ dimensional toric variety, equipped with the appropriate Hamiltonian action of a torus T of rank m. The image of the moment map is the moment polytope $\Delta = \mu(M)$. Let $v(\Delta)$ denote the union of of the vertices of Δ, and $e(\Delta)$ the union of the interior of the edges. Then $F = \mu^{-1}(v(\Delta))$, while $N = \mu^{-1}(e(\Delta))$.

The main result of this paper is the following

THEOREM 1. *Let* (M, ω) *be a compact symplectic manifold equipped with a Hamiltonian action of a torus* T. *Let* F *be the fixed point set and let* $\overline{N}$ *be the one-skeleton. Let* $i : F \longrightarrow M$ *and* $j : F \longrightarrow \overline{N}$ *be the natural inclusions, and let* $H_T^*(M) \xrightarrow{i^*} H_T^*(F)$ *and* $H_T^*(\overline{N}) \xrightarrow{j^*} H_T^*(F)$ *be the pull-back maps in equivariant cohomology. Then the images of* i^* *and* j^* *are the same.*

This theorem is proved in considerable generality in [**GKM**]. The purpose of this paper is to give a simple proof of this result in the symplectic setting, which will enable us to obtain a description of the cohomology ring of M in a form that will make the structure of the map K^* to the cohomology ring of the symplectic quotient transparent. Our methods should yield similar statements in integral cohomology. Additionally, our methods give an algorithm for turning this description of the cohomology ring into an explicit set of generators and relations.

Before outlining the proof of Theorem 1, let us consider a few special examples.

EXAMPLE 1.5. If the torus T is one-dimensional, then $\overline{N} = M$, so the theorem is obviously true, but trivial.

EXAMPLE 1.6. Suppose that the closure $\overline{N_i}$ of each of the components of N is a copy of the two-sphere P^1. For each such component there exists a corank-one subgroup $K_i \subset T$ which acts trivially on $\overline{N_i}$. The quotient T/K_i is isomorphic to S^1, and the corresponding action on P^1 must be the usual action, so that $H_T^*(\overline{N_i})$ is given as follows. Let $\gamma_i : K_i \longrightarrow T$ denote the inclusion, and let $\gamma_i^* : H_T^*(\text{pt}) \longrightarrow H_{K_i}^*(\text{pt})$ denote the induced map in equivariant cohomology. For each i, the set $\overline{N_i} \cap F$ consists of two points n_i, s_i, and the image of $H_T^*(\overline{N_i})$ in $H_T^*(\overline{N_i} \cap F) =$

$H_T^*(n_i) \oplus H_T^*(s_i)$ consists of those elements $(a,b) \in H_T^*(n_i) \oplus H_T^*(s_i)$ such that $\gamma_i^* a = \gamma_i^* b$. Let the fixed points of the T-action be given by F_i, $i = 1, \dots, N$. Then the image of $H_T^*(M)$ in $H_T^*(F) = \bigoplus H_T^*(F_i)$ consists of

$$(1.7) \qquad (a_1, \dots, a_N) \in H_T^*(F_1) \oplus \dots H_T^*(F_N)$$

such that for each $\overline{N_i}$ we have $\gamma_i^* a_{n_i} = \gamma_i^* a_{s_i}$ (see [**GKM**], [**GS**]). This gives a completely combinatorial algorithm for computing $H_T^*(M)$.

The main tool needed to prove Theorem 1, as well as the injectivity and surjectivity Theorems 1.1 and 1.2, is a repeated use of equivariant Morse theory. The key fact in all these cases is that components of the moment map μ give equivariantly self-perfecting Morse functions whose critical set is precisely F. Since the same is true for each of the $\overline{N_i}$'s, similar statements can be made for the one-skeleton $\overline{N}$.

These self-perfecting Morse functions give us a very useful way of constructing the equivariant cohomology ring of M from the cohomology rings of the fixed manifolds F_i: roughly speaking, the contribution of each fixed manifold to the cohomology ring of M consists of those classes in $H_T^*(M)$ that vanish on all fixed points "below" F_i, and whose value on F_i is a multiple of the downward Euler class of the Morse flow. Since a similar statement can be made about the cohomology ring of $\overline{N}$, we may compare the images of $H_T^*(M)$ and $H_T^*(\overline{N})$ to prove our result.

2. Morse Theory and the Moment Map

In this section we state several results which will be the key steps in the proof of Theorem 1. Among them is Kirwan's injectivity theorem (for which we supply a proof). All of these results follow directly from the equivariant Morse complex associated to the choice of a Hamiltonian as a Morse function. We note that several of the results of this section have analogs in integral cohomology; however we are only concerned with rational cohomology in this paper.

Let us recall our set-up. Let (M, ω) be a compact symplectic manifold with a moment map μ for the action of a torus T^n. Let $F \subset M$ denote the fixed point set of the torus action. Given a generic element $\xi \in \mathfrak{t}$, the function $f = \langle \mu, \xi \rangle : M \longrightarrow \mathbf{R}$ is a Morse function on M whose critical set coincides with the fixed point set F.

Let us consider the fundamental exact sequence corresponding to the Morse function f. Let us denote by C the critical set of f, and choose $c \in C$. We may assume that an interval $[c-\epsilon, c+\epsilon]$ contains no critical values of f other than c. Let $M_c^+ = f^{-1}(-\infty, c+\epsilon)$, $M_c^- = f^{-1}(-\infty, c-\epsilon)$. Then we have the following lemma, which is the main technical fact behind our results.

Proposition 2.1. *Let a torus T act on a compact manifold M with moment map $\mu : M \longrightarrow \mathfrak{t}^*$. Given $\xi \in \mathfrak{t}$, choose a critical value c of the projection $f := \mu^\xi$. Let F be the set of fixed points, and let F_c be the component of F with value c.*

The long exact sequence in equivariant cohomology for the pair (M_c^+, M_c^-) splits into short exact sequences:

$$(2.2) \qquad 0 \longrightarrow H_T^*(M_c^+, M_c^-) \longrightarrow H_T^*(M_c^+) \xrightarrow{k^*} H_T^*(M_c^-) \longrightarrow 0.$$

Moreover, the restriction from $H_T^*(M_c^+)$ to $H_T^*(F_c)$ induces an isomorphism from the kernel of k^* to those classes in $H_T^*(F_c)$ that are multiples of e_c, the equivariant Euler class of the negative normal bundle of F_c.

PROOF. By our assumptions, f is a Morse function, and there is a unique critical value of f contained in the interval $[c-\epsilon, c+\epsilon]$. Denote the corresponding critical manifold by F_c, and the negative disc and sphere bundles of F_c by D_c and S_c, respectively. The pair (M_c^+, M_c^-) can be retracted onto the pair (D_c, S_c), so there is an isomorphism

$$H_T^*(M_c^+, M_c^-) \cong H_T^*(D_c, S_c). \tag{2.3}$$

By the Thom isomorphism theorem, we have

$$H_T^*(D_c, S_c) \cong H_T^{*-\lambda_c}(D_c) \ = H_T^{*-\lambda_c}(F_c), \tag{2.4}$$

where λ_c is the Morse index of the critical manifold F_c; so we obtain a commutative diagram

$$\begin{array}{ccccccc}
\longrightarrow & H_T^*(M_c^+, M_c^-) & \xrightarrow{\gamma_c} & H_T^*(M_c^+) & \xrightarrow{\beta_c} & H_T^*(M_c^-) & \longrightarrow \\
& \downarrow & & \downarrow & & & \\
& H_T^*(D_c, S_c) & \xrightarrow{\delta_c} & H_T^*(D_c) & & & \\
& \downarrow & \nearrow_{\cup e_c} & & & & \\
& H_T^{*-\lambda_c}(D_c) & & & & &
\end{array} \tag{2.5}$$

where $e_c = e(D_c)$ is the equivariant Euler class of the bundle $D_c \longrightarrow F_c$. The cup product map $\cup e_c$ is *injective*; the same therefore is true of the maps δ_c and γ_c, proving the lemma. □

The following corollary is then immediate.

COROLLARY 1 (see [**AB1, AB2, K**]). *The function f is an equivariantly perfect Morse function on M.*

Another application of this proposition is the proof of the following theorem of Kirwan.

THEOREM 2. *Let a torus T act on a symplectic manifold (M, ω) with proper bounded-below moment map $\mu : M \longrightarrow \mathfrak{t}^*$. Let $i : F \longrightarrow M$ denote the natural inclusion of the set F of fixed points. The pullback map $i^* : H_T^*(M) \longrightarrow H_T^*(F)$ is injective.*

PROOF. Order the critical values of f as $c_1 < c_2 < \cdots < c_n$. The theorem obviously holds for $f^{-1}(-\infty, c_1) = \emptyset$. Assume the proposition holds for the manifold $M^- := f^{-1}(-\infty, c_i)$. We will show that it will hold for $M^+ := f^{-1}(-\infty, c_{i+1})$; the result then follows by induction.

By Proposition 2.1, we have a map of short exact sequences

$$\begin{array}{ccccccccc}
0 & \longrightarrow & H_T^*(M^+, M^-) & \longrightarrow & H_T^*(M^+) & \longrightarrow & H_T^*(M^-) & \longrightarrow & 0 \\
& & \downarrow \simeq & & \downarrow i_+^* & & \downarrow i_-^* & & \\
0 & \longrightarrow & H_T^*(F_i) & \longrightarrow & H_T^*(F \cap M^+) & \longrightarrow & H_T^*(F \cap M^-) & \longrightarrow & 0,
\end{array} \tag{2.6}$$

where F_i denotes the critical set with value c_i. By induction, the inclusion i_- of $F \cap M^-$ into M^- induces an injection in cohomology. By Proposition 2.1 the image of $H_T^*(M^+, M^-)$ in $H_T^*(M^+)$ is embedded injectively in $H_T^*(F_i)$. The theorem then follows by diagram chasing. □

3. Proof of the Main Theorem

We are now ready to prove our main theorem: the restriction map to the fixed point set induces an isomorphism from the equivariant cohomology of the original Hamiltonian manifold to the image of the equivariant cohomology of the one-skeleton, under its restriction map to the fixed point set. The key idea is to use the tools developed in the last section to compare the graded rings associated to these images using the filtration given by the Morse function obtained by choosing a projection of the moment map. The result will then follow from the naturality of these objects, induction on the critical points, and the injectivity of the restriction map.

Recall that the one-skeleton is given by

$$\overline{N} := \{p \in M \mid T \cdot p \text{ is one-dimensional or zero-dimensional}\}.$$

Clearly, the image of $i^* : H_T^*(M) \longrightarrow H_T^*(F)$ is a subset of the image of $j^* : H_T^*(\overline{N}) \longrightarrow H_T^*(F)$. Therefore, i^* induces a map, which we will also call i^*, from $H_T^*(M)$ to $\operatorname{im} j^* \subset H_T^*(F)$. By Theorem 1.1, this map is injective. Therefore, to prove the theorem, it suffices to show that this map is surjective.

On the level of the graded rings associated to the Morse filtration, surjectivity will follow from comparing the proposition below with Proposition 2.1.

PROPOSITION 3.1. *Let a torus T act on a compact symplectic manifold M with moment map $\mu : M \longrightarrow \mathfrak{t}^*$. Given $\xi \in \mathfrak{t}$, choose a critical point c of the projection $f := \mu^\xi$. Let F denote the fixed point set and let F_c denote the component of F with value c. Define $F^- := F \cap f^{-1}(-\infty, c-\epsilon)$ and $\overline{N^+} := \overline{N} \cap f^{-1}(-\infty, c+\epsilon)$.*

Let η be a cohomology class in $H_T^*(\overline{N^+})$ which vanishes when restricted to $H_T^*(F^-)$. Its restriction to $H_T^*(F_c)$ is a multiple of $e_c = e(D_c)$, the equivariant Euler class of the downward normal bundle D_c of F_c (in M).

PROOF. Consider any component N_α of the set N of one-dimensional orbits such that the closure $\overline{N_\alpha}$ contains F_c. The closure $\overline{N_\alpha}$ is a smooth T-invariant symplectic manifold with moment map μ. The class η induces a cohomology class on $\overline{N_\alpha^+} := \overline{N_\alpha} \cap f^{-1}(-\infty, c+\epsilon)$ which vanishes when restricted to $\overline{N_\alpha} \cap F^-$, and hence (by injectivity, which we need to state in this version), when restricted to $\overline{N_\alpha^-} := \overline{N_\alpha} \cap f^{-1}(-\infty, c-\epsilon)$. Thus, by Proposition 2.1, any element of the kernel of the natural map $H_T^*(N_\alpha^+) \longrightarrow H_T^*(N_\alpha^-)$ is, when restricted to $H_T^*(F_c)$, a multiple of the equivariant Euler class e_α (here $e_\alpha = e(D_c \cap \overline{N_\alpha})$ is the equivariant Euler class e_c of the downward normal bundle $D_c \cap \overline{N_\alpha}$ of F_c in $\overline{N_\alpha}$).

So the restriction of η to F_c is a multiple of the equivariant Euler class of the downward normal bundle of F_c in N_α. Since this holds for each component N_α, and each of these components must have a different stabilizer, we may apply Lemma 3.2 below. Therefore, the class η must be a multiple of the product of the equivariant Euler classes of the negative normal bundles to F_c in all the components of N whose closure contains F_c. But this is precisely the equivariant Euler class of the negative normal bundle to F_c in M. □

LEMMA 3.2. *Let a torus T act on a complex vector bundle E over a manifold F, so that the fixed set is precisely F. Decompose E into the direct sum of bundles E_α, where each E_α is acted on with a different weight $\alpha \in \mathfrak{t}^*$. Let e_α be the Euler class of the subbundle E_α.*

Then if $y \in H_T^*(F)$ is a multiple of e_α for each α, then y is a multiple of the product of the e_α.

PROOF. Assume first that F is a single point.

Let $\alpha \in \mathfrak{t}^*$ be the weight with which T acts on the subbundle E_α. Since α is a linear function on $\mathfrak{t}$, it lies naturally in $H^*(BT) = H_T^*(F) = \mathrm{Sym}(\mathfrak{t}^*)$, the algebra of symmetric polynomials on $\mathfrak{t}$. The equivariant Euler class of E_α is given by $e_\alpha = \alpha^{n_\alpha}$, where n_α is the complex dimension of E_α. The α are distinct by assumption, and nonzero since no point not in the zero section is fixed by T. Therefore, the e_α are pairwise relatively prime. (Recall that every polynomial ring over $\mathbf{Q}$ is a unique factorization domain.)

More generally, since F is fixed by T, $H_T^*(F) = H^*(F) \otimes H^*(BT)$. Thus, $H_T^*(F)$ is bigraded. In particular, given any integer i, any cohomology class $a \in H_T^*(F)$ has a well-defined component $a_i \in H^i(F) \otimes H^*(BT)$, and the sum of all such components is a itself; we will call a_i **the component of a with F-degree i.**

Note that the component of e_α with F-degree 0 is precisely α^{n_α}. By the previous discussion, these are nonzero and pairwise relatively prime. Therefore it is enough to prove that if e and f are two cohomology classes whose components e_0 and f_0 with F-degree zero are relatively prime, and if e and f both divide α, then so does $e \cdot f$. We will prove this by induction.

We claim that if $e(f \cdot w + x) = f(e \cdot w + y)$, where the components of x and y with F-degree i vanish for all $i < k$, then there exist x', y' such that $e(f \cdot w + x') = f(e \cdot w + y')$, and such that the components of x' and y' with F-degree i vanish for all $i < k + 1$. To see this, compare the component of F-degree k on the two sides of the equation $e(f \cdot w + x) = f(e \cdot w + y)$. Cancelling out terms which appear on both sides, we get $e_0 x_k = f_0 y_k$. Since e_0 and f_0 are relatively prime polynomials, this shows that there exists z_k such that $x_k = f_0 z_k$ and $y_k = e_0 z_k$. □

We now proceed to prove that the map

$$i^* : H_T^*(M) \longrightarrow \operatorname{im} j^* \subset H_T^*(F)$$

is surjective. We proceed, as usual, by induction.

Consider any critical point c of the projection $f := \mu^\xi$. Define

$$M^+ := f^{-1}(-\infty, c + \epsilon) \quad \text{and} \quad M^- := f^{-1}(-\infty, c - \epsilon)$$

for any sufficiently small ϵ. Let

$$\overline{N^+} := \overline{N} \cap M^+, \qquad \overline{N^-} := \overline{N} \cap M^-,$$
$$F^+ := F \cap M^+, \qquad F^- := F \cap M^-.$$

Let

$$i^+ : F^+ \longrightarrow M^+, \qquad i^- : F^- \longrightarrow M^-,$$
$$j^+ : F^+ \longrightarrow \overline{N^+}, \qquad j^- : F^- \longrightarrow \overline{N^-}$$

denote the corresponding inclusion maps. It is enough to assume that the induced map

$$i^{-*} : H_T^*(M^-) \longrightarrow \operatorname{im} j^{-*} \subset H_T^*(F^-)$$

is surjective, and to prove that the induced map

$$i^{+*} : H_T^*(M^+) \longrightarrow \operatorname{im} j^{+*} \subset H_T^*(F^+)$$

is also surjective.

Since the images of $H_T^*(M^-)$ and $H_T^*(N^-)$ inside $H_T^*(F^-)$ are the same, it follows that the natural restriction map r from $\operatorname{im} j^{+*} \subset H_T^*(F^+)$ to $\operatorname{im} j^{-*} \subset H_T^*(F^-)$ is surjective. Thus, taking the exact sequence (2.2) of Proposition 2.1, we have a map of short exact sequences

(3.3)

$$\begin{array}{ccccccccc} 0 & \longrightarrow & H_T^*(M^+, M^-) & \longrightarrow & H_T^*(M^+) & \longrightarrow & H_T^*(M^-) & \longrightarrow & 0 \\ & & \downarrow & & \downarrow {\scriptstyle i_+^*} & & \downarrow {\scriptstyle i_-^*} & & \\ 0 & \longrightarrow & \ker r & \longrightarrow & \operatorname{im} j^{+*} \subset H_T^*(F^+) & \xrightarrow{r} & \operatorname{im} j^{-*} \subset H_T^*(F^-) & \longrightarrow & 0. \end{array}$$

By our inductive assumption, i^{-*} is surjective.

By Proposition 3.1, every element in $\ker r$ is a multiple of e_c, the equivariant Euler class of the negative normal bundle of F_c, the component of the fixed point set with value c. On the other hand, by Proposition 2.1, every multiple of e_c is in the image of the restriction of $H_T^*(M^+, M^-)$ to $H_T^*(F_c)$ Thus, the arrow from $H_T^*(M^+, M^-)$ to $\ker r$ is surjective too.

The result follows by a diagram chase.

4. Some Comments about Integral Cohomology

We close with some comments about integral cohomology. Unfortunately, the integer version of Theorem 1 is not true in general. In fact even the injectivity Theorem 1.1 will not hold in integral cohomology without some assumptions. However, both injectivity and a version of Theorem 1 can be proved over the integers when certain restrictions are placed on the allowable stabilizer subgroups.

Perhaps the easiest example where Theorem 1 is not true for integer cohomology is that of $S^1 \times S^1$ acting on $S^2 \times S^2$, with speed two on each sphere.[1] Using the moment map for the diagonal action as our Morse function, we see that every cohomology class which vanishes outside a neighborhood of the north pole $\times$ the north pole must be a multiple of $4x^2$ when restricted to that point. In contrast, there exists a cohomology class on the one-skeleton which vanishes outside this neighborhood but is only a multiple of $2x^2$. Essentially, the problem is that the weights are not relatively prime. It is easy to place a condition on the stabilizer groups at each fixed point in a way that negates this possibility. This is essentially all that can go wrong, and a version of Theorem 1 can be expected to hold if such an assumption of relative primality is made.

References

[AP] C. Allday and V. Puppe, *Cohomological methods in transformation groups,* Cambridge Univ. Press, Cambridge, 1993

[AB1] M. Atiyah and R. Bott, *The moment map and equivariant cohomology,* Topology **23** (1984), 1–28.

[AB2] ———, *The Yang-Mills equations over Riemann surfaces,* Phil. Trans. Roy. Soc. Lond. **A308** (1982), 523–615.

[BV] M. Brion and M. Vergne, *On the localization theorem in equivariant cohomology,* dg-ga/9711003.

[CS] T. Chang and T. Skjelbred, *The topological Schur Lemma and related results.* Ann. Math. **100** (1974), 307–321

[1] Any reader who is disturbed by the fact that this action is not effective may tack on another couple of S^2's spinning at speed one.

[GKM] M. Goresky, R. Kottwitz, and R. MacPherson. *Equivariant cohomology, Koszul duality, and the localization theorem*, Invent. Math., **131** (1998), 25–83.

[GS] V. Guillemin and S. Sternberg, *Supersymmetry and equivariant DeRham theory*, Springer-Verlag, Berlin, 1999.

[H] W.-Y. Hsiang, *Cohomology theory of topological transformation groups*, Springer-Verlag, Berlin, 1975.

[K] F. C. Kirwan, *The cohomology of quotients in symplectic and algebraic geometry*, Princeton Univ. Press, Princeton, NJ, 1984.

[TW] S. Tolman and J. Weitsman, *The cohomology rings of abelian symplectic quotients,* preprint.

Department of Mathematics, University of Illinois at Urbana-Champaign, Urbana, IL 61801

E-mail address: stolman@math.uiuc.edu

Department of Mathematics, University of California, Santa Cruz, CA 95064

E-mail address: weitsman@cats.ucsc.edu

Selected Titles in This Subseries

(Continued from the front of this publication)

19 **V. A. Marchenko, Editor,** Spectral operator theory and related topics, 1994
18 **Oleg Viro, Editor,** Topology of manifolds and varieties, 1994
17 **Dmitry Fuchs, Editor,** Unconventional Lie algebras, 1993
16 **Sergei Gelfand and Simon Gindikin, Editors,** I. M. Gelfand seminar, Parts 1 and 2, 1993
15 **A. T. Fomenko, Editor,** Minimal surfaces, 1993
14 **Yu. S. Il′yashenko, Editor,** Nonlinear Stokes phenomena, 1992
13 **V. P. Maslov and S. N. Samborskiĭ, Editors,** Idempotent analysis, 1992
12 **R. Z. Khasminskiĭ, Editor,** Topics in nonparametric estimation, 1992
11 **B. Ya. Levin, Editor,** Entire and subharmonic functions, 1992
10 **A. V. Babin and M. I. Vishik, Editors,** Properties of global attractors of partial differential equations, 1992
9 **A. M. Vershik, Editor,** Representation theory and dynamical systems, 1992
8 **E. B. Vinberg, Editor,** Lie groups, their discrete subgroups, and invariant theory, 1992
7 **M. Sh. Birman, Editor,** Estimates and asymptotics for discrete spectra of integral and differential equations, 1991
6 **A. T. Fomenko, Editor,** Topological classification of integrable systems, 1991
5 **R. A. Minlos, Editor,** Many-particle Hamiltonians: spectra and scattering, 1991
4 **A. A. Suslin, Editor,** Algebraic K-theory, 1991
3 **Ya. G. Sinaĭ, Editor,** Dynamical systems and statistical mechanics, 1991
2 **A. A. Kirillov, Editor,** Topics in representation theory, 1991
1 **V. I. Arnold, Editor,** Theory of singularities and its applications, 1990